DYE CURVES: THE THEORY AND PRACTICE OF INDICATOR DILUTION

Edited by

Dennis A. Bloomfield, M.B., M.R.C.P., M.R.C.P.(E)

Director, Cardiac Catheterization Laboratory, Maimonides Medical Center, Brooklyn, New York; Assistant Professor of Medicine, Downstate Medical Center, State University of New York

UNIVERSITY PARK PRESS
Baltimore · London · Tokyo

UNIVERSITY PARK PRESS
International Publishers in Science and Medicine
Chamber of Commerce Building
Baltimore, Maryland 21202

Library of Congress Cataloging in Publication Data

Bloomfield, Dennis A.
  Dye curves: the theory and practice of indicator
dilution.

  1. Indicator dilution.   I. Title.
[DNLM: 1. Dye dilution technic.   WG106 B655d 1973]
RC683.B55   616.1'2'0754   73-18178
ISBN 0-8391-0638-6

CONTENTS

General and Practical Considerations

Special Applications

Equipment and Materials

CONTRIBUTORS

Stuart Antman, Ph.D. Professor of Mathematics, University of Maryland, College Park, Maryland

Dennis A. Bloomfield, M.B., M.R.C.P., M.R.C.P.(E.) Assistant Professor of Medicine, Downstate Medical Center, State University of New York, and Director, Cardiac Catheterization Laboratory, Maimonides Medical Center, Brooklyn, New York

Joachim H. Bürsch, M.D. Research Associate, Department of Pediatric Cardiology, University Childrens Hospital, Kiel, West German Federal Republic; and Visiting Research Fellow, Mayo Graduate School of Medicine, University of Minnesota, Rochester, Minnesota

Irwin J. Fox, M.D., Ph.D. Professor of Physiology, University of Minnesota Medical School, Minneapolis, Minnesota

William Ganz, M.D., C.Sc. Senior Research Scientist, Department of Cardiology, Cedars-Sinai Medical Center, and Professor of Medicine, University of California School of Medicine, Los Angeles, California

Ephraim Glassman, M.D. Director, Cardiac Catheterization Laboratory, Associate Professor of Medicine, New York University Medical Center, New York, New York

Paul G. Hugenholtz, M.D. Professor of Cardiology, Medical Faculty of Rotterdam, University of Rotterdam, The Netherlands

A. M. Safiqul Khan, M.B., B.S., M.R.C.P. Fellow in Cardiology, Maimonides Medical Center, Brooklyn, New York; Presently at King George Highway, Surrey, B.C., Canada

L. Jerome Krovetz, M.D., Ph.D. Associate Professor of Pediatrics, (Cardiology) and Biomedical Engineering, and Director, Pediatric Cardiovascular Research Laboratory, The Johns Hopkins University School of Medicine, Baltimore, Maryland

G. T. Meester, M.D. Associate Professor of Clinical Patho-Physiology of the Circulatory System, University of Rotterdam, The Netherlands

Norman F. Paradise, Ph.D. Postdoctoral Fellow in Clinical Cardiology, University of Minnesota Medical School, Minneapolis, Minnesota

Michael L. Polanyi, Ph.D. Senior Research Physicist, American Optical Corporation, Research Division, Framingham, Massachusetts

Elliot Rapaport, M.D. Professor of Medicine, Member of the Senior Staff, Cardiovascular Research Institute, University of California, San Francisco, California

Erik L. Ritman, M.D., B.S.C. Research Associate, Department of Physiology and Biophysics, Mayo Graduate School of Medicine, University of Minnesota, Rochester, Minnesota

Loring B. Rowell, Ph.D. Professor of Physiology and Biophysics, Adjunct Professor of Medicine, School of Medicine, University of Washington, Seattle, Washington

Bruce C. Sinclair-Smith, M.B., F.R.C.P.(E.), F.R.A.C.P. Associate Professor of Medicine, Vanderbilt University School of Medicine, Nashville, Tennessee

Ralph E. Sturm, Associate Professor of Physiology, Mayo Graduate School of Medicine, University of Minnesota, and Consultant in Biophysics, Mayo Clinic, Rochester, Minnesota

William F. Sutterer, B.S. Project Engineer, Waters Instruments, Inc.; Presently with the Department of Physiology, Mayo Clinic, Rochester, Minnesota

H. J. C. Swan, M.D., Ph.D. Director, Department of Cardiology, Cedars-Sinai Medical Center, and Professor of Medicine, University of California School of Medicine, Los Angeles, California

Claude R. Swayze, M.D. Diplomate, American Board of Anesthesiology, and Research Assistant, Department of Physiology, University of Minnesota Medical School, Minneapolis, Minnesota

Michael R. Tripp, M.D. Minnesota Heart Association Postdoctoral Fellow, Department of Physiology, University of Minnesota Medical School, Minneapolis, Minnesota

P. D. Verdouw, Ph.D. Scientific Civil Servant, University of Rotterdam, The Netherlands

Earl H. Wood, M.D., Ph.D. Professor of Physiology, Mayo Graduate School of Medicine, University of Minnesota, Rochester, Minnesota

Tada Yipintsoi, M.D. Associate Consultant, Department of Physiology and Biophysics, Mayo Clinic, and Mayo Foundation, Rochester, Minnesota; Presently, Associate Professor of Medicine, Montefiore Hospital and Medical Center, Bronx, New York

Paul N. Yu, M.D. Sarah McCort Ward Professor of Medicine, and Head, Cardiology Unit, University of Rochester Medical Center, Rochester, New York

PREFACE

In 1960, the American Heart Association recognized a need for a series of critical reviews of methods widely used and important at that time for the evaluation of cardiovascular functions. It is significant that the indicator-dilution technique was chosen as the first subject in the series. There were a number of reasons for this choice—that many new advances had been made, that practice of the method was both widespread and nonstandardized, and that no thoughtful or penetrating evaluation of the general topic had been attempted for some years.

Now the circumstances have been recreated and the need for a summing-up has returned. Twelve years ago, the challenge was answered in the shape of a published symposium, the American Heart Association Monograph Number Four. Today, it is my intent that this text should serve a similar purpose.

The need for such a contemporary appraisal of indicator-dilution techniques is attested to by the readiness and willingness with which the many contributors not only accepted their task but made valuable suggestions which enhanced the content and scope of this work beyond my original intent or capability. The significance of their effort can be gauged immediately by scanning the Table of Contents. My gratitude and pride in their participation cannot be expressed adequately.

The book is constructed in three parts. The first section provides a practical manual for performing dye curves for the measurement of cardiac output. Set in an historic perspective and on a mathematical foundation, this part stresses the limitations and the conditions which must be met for validity of the technique.

The second section concerns the special applications of nondiffusable indicators beyond that of cardiac output measurement. The newer techniques of thermal dilution, fiberoptics, and video densitometry are detailed. A single chapter summarizes the applications of diffusable indicators.

The final section discusses the chemistry, the physics, and the mathematics behind the materials and the equipment used in the practice of indicator dilution, with some comparative appraisal of the instruments commercially available.

Many people have aided me in the preparation of this book.

Anneta Duveen, T.O.F., F.R.A.S., provided the drawings, and Harold Friedman and his staff of the visual aids department furnished the photography for the illustrations in many of the chapters. Ms. Elaine Enge typed the final draft. Mrs. Natalie Kochan and the library staff were ever helpful. To all of them I am greatly indebted. It is a pleasure to acknowledge the help of Marcos Rivera, chief technician, and the staff of the cardiac catheterization laboratory, for their able assistance in the studies, and my secretaries, colleagues, and fellows for their encouragement and their shouldering of many of my responsibilities in order that I should have time for this work.

I especially acknowledge the assistance of my wife, who shared responsibilities of the proofreading and indexing, and Mr. Braxton Mitchell of University Park Press. To both of them I am grateful for the advice, encouragement, and great patience that they have shown during the collection and the writing of this book.

Dennis Bloomfield
Maimonides Medical Center
Brooklyn, New York

part 1/ GENERAL AND PRACTICAL CONSIDERATIONS

chapter 1/ The History of Circulatory Indicator Dilution

Tada Yipintsoi/Earl H. Wood
Department of Physiology and Biophysics/
Mayo Clinic and Mayo Foundation/Rochester, Minnesota

If a recognizable substance is added to a flowing stream, the time course of its concentration downstream gives a quantitative index of the flow which has diluted it. (The product of the flow and average traversal time of all the particles of the substance gives a quantitative assessment of the volume of that stream between the points of addition and sampling of the substance.) The realization of this concept and the development of the methodology for its practical application to circulatory physiology provide the background for this book and the foundations for the further elaboration of the "indicator-dilution principle" contained in the pages to follow.

Indicators were first used in the vascular system for the purpose of determining circulation times (Haller, 1761, quoted by Stewart, G. N., 1894; Hering, 1827, quoted by Fox, I. J., 1962a). A great variety of these substances was later employed in the estimation of circulating blood volume; Keith et al. (1915) reported that Valentin (1847) used water, Mallassez (1874) and Quincke (1877) used red cells, Gréhant and Quinquand (1882) used carbon monoxide inhalation, and Behring (1911) used tetanus antitoxin. The only prerequisite for the indicator was that it should remain in the circulation long enough to be practical for the measurement of concentration.

THE STEWART EXPERIMENTS

In the 1890s, a series of papers on circulation time were published by George Stewart (1894, 1897*a,b*). Stewart (Fig. 1.1) had trained under the English physiologist, William Stirling, and later occupied the Chair of Physiology at Western Reserve University.

Stewart injected hypertonic saline solution (1.5 to 4.0%) intravenously into anesthetized animals and detected the arrival of this indicator at arterial or venous sampling sites by the change in the electrical conductance of the flowing blood. This change could be seen in the deflection of a galvanometer or heard as a changing tone in a telephone circuit attached to a Wheatstone bridge. He also injected colored dyes (aniline and methylene blue) and detected their first appearance visually with an ophthalmoscope on the retinal vessels or chemically by blood sampling. The circulation time reported was mistakenly assumed to be closely related to the mean transit time, inasmuch as Stewart (1897*a*) stated that he could not understand the

Figure 1.1
George Neil Stewart (1860–1930).

reason why, in a system with known capacity, flow, and measured circulation time, the flow was always less than the capacity divided by time. With his simple apparatus, Stewart proceeded to calculate cardiac output by using a step input of indicator ("continuous infusion") and sampling the arterial concentration plateau when the telephone tone informed him that the electrical conductance had risen to a uniform level (Stewart, 1921a,b,c).

He recognized the potential faults of the method (Stewart, 1897a), enumerating incomplete mixing of the indicator at the injection site (the initial sample may be from the faster flowing central axial stream), problems with intravascular dispersion, perturbation of the system by the injection process (pressure effect being less with injection into the arterial tree), the osmotic effect of a high concentration of sodium chloride solution, the duration of injection, and the relationship of the duration of injection to that of the cardiac cycle.

Sampling time was determined subjectively by visual or audio-visual means. Nevertheless, Stewart (1894) was able to postulate that, if two concentration peaks were detected (injecting portal vein and sampling carotid artery), the first was probably due to a short-circuited path taken by the indicator. Subsequently, with the advent of cardiac catheterization in the mid-20th century, this observation was quantified and used in the detection of shunts and regurgitation.

Stewart's work met with apparent disinterest for three decades. This is a surprising fact, in the light of the monumental advances in circulatory physiology which had been made by the turn of the century, and is reminiscent of the neglect which enveloped Fick's principle for many years after its enunciation.

However, some reports extending Stewart's concepts appeared in the German literature, and the significance of this work (Henriques, 1913; Bock and Buchholtz, 1920; Romm, 1924) has been discussed by Fox (1962a). Henriques made a most significant contribution in developing Stewart's "sudden single injection technique." Employing thiocyanate, he recognized the necessity for confining the indicator to the intravascular compartment when measuring cardiac output and even devised a technique for estimating coronary blood flow.

THE HAMILTON EXPERIMENTS

In the 1920s and 1930s, however, cardiac output measurements were generally performed by modified or "indirect" Fick methods and eventually became stylized about the procedures and results of

Grollman (1929), who used inhaled acetylene gas as the indicator. However, a group led by William Hamilton (Fig. 1.2), in Louisville, Kentucky, recognizing the short recirculation time of this gas and convinced that accurate output measurements would be impractical with respiratory procedures, turned again to the injection method of Stewart. In a series of experiments in 1928 and 1929, using the bolus injection modification, Hamilton, Moore, and Kinsman restated and defined the requirements and assumptions for validity of the indicator-dilution principle. They described the exponential character of the descending limb of the concentration-time curve and devised a method of compensating for recirculation artifact in the output calculations (Hamilton *et al.,* 1928*a,b,* 1932, 1948; Kinsman, Moore, and Hamilton, 1929; Moore *et al.,* 1929). Their grasp of the concept can be succinctly illustrated by Fig. 1.3 from their publication of 1929, in which the dependence on a mixing chamber, the inconsequential anatomic site of arterial sampling, the form of the curve with and without recirculation, and the method of reproducing the latter from the former are all clearly shown. Their technique for measurement of the time-concentration curve has rightly become known as the Stewart-Hamilton method.

Hamilton also demonstrated the necessity to use mean transit times (the first moment of the curve) in calculating circulatory blood vol-

Figure 1.2 William Ferguson Hamilton (1893–1964).

umes, as distinct from the circulation times of Stewart. His group also showed that, if a diffusible dye, phenoltetraiodophthalein sodium, was used, some of which was lost from the circulation between its entrance and exit, the decay of the dye concentration-time curve after bolus injection would deviate from a single exponential. The importance of this observation became more evident when the indicator-dilution principle was extended to include diffusible indicators.

More than a decade was to pass before the validity of the indicator-dilution method was generally accepted. This was due to the tediousness of the sample analysis, to the lack of accord with outputs mea-

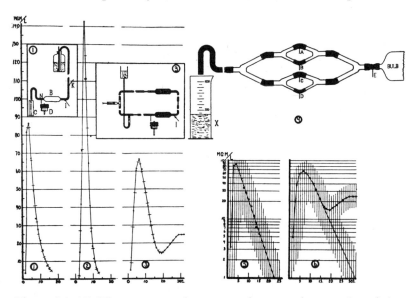

Figure 1.3 (*1*) Time-concentration curve of consecutive samples of dye injected at *I* (*inset 1*) into water flowing through bulb *B*. Samples are taken through needle *N* into tubes on rotating kymograph drum *D*. (*2*) Same as (*1*), except that bulb B was filled with beads. (*3*) Time-concentration curve of consecutive samples taken from apparatus (*inset 3*) designed to simulate animal circulation. (*4*) Apparatus used in experiment designed to show that dye concentration in different branches of a stream does not vary significantly. (*5*) Same curve as (*1*) plotted on semilogarithmic scale. Downstroke is straight line. (*6*) Same curve as (*3*) plotted on a semilogarithmic scale. When straight downstroke is prolonged, it serves to differentiate once-circulated from twice-circulated dye. (From Kinsman, J. M., J. W. Moore, and W. F. Hamilton. 1929. Studies on the circulation. I. Injection method: physical and mathematical considerations. Amer. J. Physiol. 89:322–330. By permission.)

sured by the then popular Grollman acetylene method, and to the necessity to await two further developments. These were the capability of sampling from and introducing indicators into different sites in the circulation, and the synthesis of a truly intravascular indicator.

Cournand and Ranges (1941) met the first requirement by the introduction of clinical cardiac catheterization, and Dawson and co-workers (1920) and Gregersen and Gibson (1937) satisfied the other requirement by introducing the dye, T1824, which was rapidly and firmly bound to plasma proteins.

In 1948, Hamilton's group moved to New York to join with Cournand's team in a comparative study of cardiac output measured simultaneously by the direct Fick and indicator-dilution methods. This productive collaboration between what were, in fact, two competing schools established indicator-dilution as a reliable and legitimate technique for measuring cardiac output.

The next era can be divided into two chronologically parallel parts. The first concerned the improvement of practical techniques so that routine physiologic and diagnostic applications of circulatory dilution methods were possible. The second was the mathematical counterpart of this development, which included proof of the methods, the restrictions imposed, allowable violations of the principle, and the potential for future development.

PRACTICAL APPLICATIONS

A vital improvement in the practical capabilities of indicator-dilution centered on the development of improved dyes, such as Indocyanine green (Fox *et al.*, 1957*a*) and Coomassie blue (Taylor and Thorp, 1959). The specific properties of the indicators are dealt with in Chapter 18.

Continuous sampling devices evolved to eliminate the time-consuming and cumbersome intermittent sampling techniques. They included ear oximetry (Matthes and Malikiosis, 1936; Knutson *et al.*, 1950), external isotopic or conductivity methods (Romm, 1924; White, 1947; Heller, Lochner, and Schoedel, 1951; MacIntyre *et al.*, 1951), and monochromatic and compensated dichromatic cuvette densitometers (Friedlich, Heimbecker, and Bing, 1950; Nicholson, Burchell, and Wood, 1951; Gilford *et al.*, 1953; Sutterer, 1960; Sinclair *et al.*, 1961). The principles of operation of the densitometers are treated in Chapter 19.

These developments made the use of bolus injections of indicators possible for determinations of blood flow, total and regional volumes of circulating blood in both normal and diseased states (Cournand and Ranges, 1941; McMichael and Sharpey-Schafer, 1944; Cournand *et al.,* 1945; Stead *et al.,* 1945; Bing, Vandam, and Gray, 1947; Dexter *et al.,* 1947; White, 1947; Borden *et al.,* 1949; Lagerlöf *et al.,* 1949; Werkö *et al.,* 1949; Friedlich *et al.,* 1950; Nylin and Celander, 1950; Kopelman and Lee, 1951; Taylor and Shillingford, 1959), and the diagnosis and quantification of congenital and acquired cardiac abnormalities such as intracavitary and intravascular shunts and valvular regurgitation (Benenson and Hitzig, 1938; Prinzmetal, 1941; Bing *et al.,* 1947; Dexter *et al.,* 1947; Friedlich *et al.,* 1950; Knutson *et al.,* 1950; Symposium, 1957, 1958; Taylor and Shillingford, 1959; Wood, 1962). These applications required the accurate recording of all portions of indicator-dilution curves so that the necessity of high fidelity sampling and recording of dilution curves became evident. These advances were reported in the "Symposium on Use of Indicator-Dilution Technics in the Study of the Circulation" (1962), wherein detailed properties of the different dyes and the oximeters and densitometers used to measure their concentration were also described. Discussions of other indicators such as radioactive tracers, gases, heat, hemodilution, and the early development of miniature detectors were also included.

During this period, the applications of the indicator-dilution principle by using diffusible indicators were still in their infancy. Fegler (1954) and later Hosie (1962) used thermodilution to measure flow, and they indicated that, in poorly insulated pathways, heat may act as a diffusible indicator. Credit must be given to Kety and Schmidt (1945) for popularizing the use of highly diffusible inert gases to measure blood flow in organs with multiple entrances and exits, such as the brain and the heart. The measurement of blood flow by diffusible-indicator techniques (see Chapter 17) requires that the partition coefficient (the ratio of tissue to blood solubility of the indicator) be known and that the values of flow obtained are expressed per unit weight or unit volume of the organ, as contrasted with the intravascular indicator techniques, which measure total blood flows to or through the regions under study. The initial theories and techniques for measuring flows through organs such as heart, brain, liver, and kidney and other peripheral sites by using diffusible tracers were summarized by Bruner (1960).

Chinard and his colleagues (Chinard, Vosburgh, and Flexner, 1950; Chinard et al., 1959; Chinard, Enns, and Nolan, 1960, 1962) demonstrated additional practical and analytic applications of the dilution principle after bolus injections of multiple indicators, one of which had to be truly intravascular. Comparison of the shapes of the dilution curves of the diffusible tracers in relation to the concentration-time curve of the intravascular indicator allowed assessment of the relative permeability of the various components of the tissue to these tracers. In addition to the relative permeability at the exchanging vessels, the mean transit time volumes and, hence, the extravascular volumes of distribution of particular substances could be determined.

Fiberoptic cardiac catheters (Hugenholtz et al., 1965; Singh et al., 1970), used initially for intracavitary oximetry, were extended to the measurement of Indocyanine green concentration in the blood. This technique eliminates the loss of blood due to sampling and the distortion of dilution curves associated with conventional catheter-densitometer techniques. Theoretically, with zero sampling volume and a rapid response time, high fidelity records of indicator concentrations should be obtained. However, nonspecific changes in the optical density of blood (Sinclair et al., 1961) produced by variations in blood flow and eddy formation at the catheter tip (Mook et al., 1968), together with the possibility of formation of blood clots at the tip of the catheter, have limited attainment of the potential value initially envisaged for this technique. The theoretical basis and present practice with these instruments are summarized in Chapter 15.

The use of thermal dilution (Fegler, 1954) not only obviates blood withdrawal and provides rapid response times, but is devoid of the complication of recirculating indicator. The application of this technique to cardiac output and regional blood flow measurement is contained in Chapter 14.

Active research and development continues in the practical application of newer techniques in indicator dilution. Specific examples of this activity are the introduction of video-densitometry, described in Chapter 16, and the use of light-emitting diodes in both standard and fiberoptic densitometry. These diodes produce light of a predictable wave length with a narrow band width and are replacing the cumbersome filtering devices previously necessary in these instruments.

Investigations related to isotopic dilution methods have been concerned with estimations of circulatory blood, plasma, or red cell volume (Gillett and Halmagyi, 1970); measurement of extracellular tracer spaces (Furneaux and Tracy, 1970); and radio-immuno assays (Felber and Aubert, 1969); and the diagnosis, localization, and therapy of diseases associated with space-occupying lesions using specially tagged particles. The last application is a stimulus for the continued search for better isotopic detector systems (Freedman *et al.,* 1969; Ueda *et al.,* 1969; Burke, Halko, and Peskin, 1971; Cykowski *et al.,* 1971). A further development in isotopic dilution is the rapidly developing tagged microembolization technique. Radioactively tagged particles, such as carbonized microspheres and macroaggregated albumin-microspheres, are injected into the blood stream, and their distribution in the microcirculation is determined (Wagner *et al.,* 1969; Burdine *et al.,* 1971). The technique is based on the assumption that, after their injection, these particles of less than 60-μm diameter are dispersed uniformly in the blood stream (Phibbs and Dong, 1970) and are distributed to different organs and regions of the same organ in proportion to the distribution of blood flow (Rudolf and Heymann, 1967; Wood, 1970). Recent data suggest, however, that distribution of these particles is influenced by the geometry of the vasculature, such as vessel branchings, so that the density of the vascular deposits does not mimic accurately the distribution of blood flow when small tissue regions are studied (Domenech *et al.,* 1969; Phibbs and Dong, 1970). This technique is further considered in Chapter 17.

MATHEMATICAL DEVELOPMENT

It is evident that this practical development could not and would not have been worthwhile if the mathematical implications were not simultaneously developed. The devising of specific mathematical models, based on experimental data, results in further knowledge being extracted from existing data. Some form of approximation frequently can be used to arrive at specific measurements, and new techniques can be evolved or developed on the basis of these formulations. This type of approach encouraged the application of multidisciplinary techniques such as the systems analysis methods of electrical engineering, fluid dynamicist methods of defining blood flow, and statistical methods for describing indicator dilution in terms of stochastic models.

Using a purely theoretic approach, Visscher and Johnson (1953) and Stow (1954) showed the possible magnitude of errors that can occur in calculations of cardiac output by using conventional applications of the Fick method. Such errors can occur whenever variations in both flow and concentration of the indicator are present at the sampling site, such as frequently result from the effects of the cardiac and respiratory cycles. In this circumstance, the concentration of the tracer in a pooled blood sample withdrawn at a constant rate of speed from a sampling site (time-averaged sample) may not be equal to the concentration in a volume-averaged sample, which is required for application of the Fick principle.

As the diagnostic implications of the contours of different portions of dye-dilution curves were demonstrated, the degree of distortion of dilution curves caused by the volume of the sampling system and the rate of sampling through this system were recognized. Fox, Sutterer, and Wood (1957b), Lacy, Emanuel, and Newman (1957), Sheppard, Jones, and Couch (1959), and Sherman et al. (1959) demonstrated these effects and proved, by theory, the limiting values of these parameters at which the degree of distortion of the recorded dilution curve was acceptable.

The observations of Coulter and Pappenheimer (1949), Rossi, Powers, and Dwork (1953), Taylor (1953, 1954), and Aris (1960) tested and formulated factors that affect the dispersion of substances in a fluid under various flow conditions (laminar and turbulent flow), varied tube diameters, and different solvents.

On the basis of these theoretical developments for idealized systems, various mathematical representations of intravascular indicator-dilution curves were proposed, all of which provided the capability of reproducing experimental curves but made use of various assumptions, not all of which have directly demonstrable physiologic bases (Bassingthwaighte, 1970; Harris and Newman, 1970). The various mathematical expressions that have been described are the random walk of Sheppard (1954, 1962), the log normal of Stow and Hetzel (1954), the equivalent system response of Stephenson (1948), Meier and Zierler (1954), and González-Fernández (1962), the perfect mixer in series of Newman et al. (1951) and Schlossmacher et al. (1967), the gamma variate of Thompson et al. (1964), and the lagged normal density model (Nicholes and Warner, 1964; Bassingthwaighte, Ackerman, and Wood, 1966).

Proofs based on these models have been derived for the practical applications of dilution curves to estimate blood flows and volumes, for the relationship between intravascular and extravascular processes

(Zierler, 1958, 1962, 1965), for the frequency functions of particles residing in and leaving the system, for the transport function of segments of circulatory pathways (Nicholes and Warner, 1964; Bassingthwaighte, 1966; Coulam *et al.,* 1966; Gómez *et al.,* 1965; Knopp and Bassingthwaighte, 1969), and finally the restrictions and assumptions used so that these theories will not be grossly violated when their practical application is attempted (Bassingthwaighte, Knopp, and Anderson, 1970).

The recording and storage of high fidelity tracer-dilution data, their analysis by high-speed digital computers, and the different mathematical techniques used have made possible further description of circulatory segments, together with the probable physiologic implications of these findings and the potential faults in the practical applications (Bassingthwaighte, 1966; Coulam *et al.,* 1966; Knopp and Bassingthwaighte, 1969; Bassingthwaighte *et al.,* 1970; Tancredi and Zierler, 1971). Further historic information on aspects of indicator dilution is contained in the reviews written by Keith, Rowntree, and Geraghty (1915), Hamilton (1953), Dow (1956) and Fox 1962*a*).

ACKNOWLEDGMENTS

The photographs of Drs. Hamilton and Stewart were kindly provided by Mr. A. H. Vaughn, Supervisor of Publications of the Medical College of Georgia, and Patsy Gerstner, Curator of the Howard Dittrick Museum of Historical Medicine, Cleveland, respectively.

The authors acknowledge the assistance of Norbett T. Ott, M.D., in the translation of the German literature.

This review was supported in part by Research Grants HL-3532 and HL-43128 from the National Institutes of Health, Public Health Service, and CI-10 from the American Heart Association.

REFERENCES

Aris, R. 1960. On the dispersion of a solute in pulsating flow through a tube. Roy. Soc. (London) Proc. A. 259:370–376.

Bassingthwaighte, J. B. 1966. Plasma indicator dispersion in arteries of the human leg. Circ. Res. 19:332–346.

Bassingthwaighte, J. B. 1970. Blood flow and diffusion through mammalian organs. Science 167:1347–1353.

Bassingthwaighte, J. B., F. H. Ackerman, and E. H. Wood. 1966. Applications of the lagged normal density curve as a model for arterial dilution curves. Circ. Res. 18:398–415.

Bassingthwaighte, J. B., T. J. Knopp, and D. U. Anderson. 1970. Flow estimation by indicator dilution (bolus injection): reduction of errors due to time-averaged sampling during unsteady flow. Circ. Res. 27:277–291.

Benenson, W., and W. M. Hitzig. 1938. The diagnosis of venous-arterial shunt by ether circulation time method. Proc. Soc. Exp. Biol. Med. 38:256–258.

Bing, R. J., L. D. Vandam, and F. D. Gray, Jr. 1947. Physiological studies in congenital heart disease. I. Procedures. Johns Hopkins Med. J. 80: 107–120.

Bock, J., and J. Buchholtz. 1920. Über das Minutenvolum des Herzens beim Hunde und über den Einfluss des Coffeins auf die Grösse des Minutenvolums. Arch. Exp. Pathol. Pharmakol. 88:192–215.

Borden, C. W., R. V. Ebert, R. H. Wilson, and H. S. Wells. 1949. Studies of the pulmonary circulation. II. The circulation time from the pulmonary artery to the femoral artery and the quantity of blood in the lungs in patients with mitral stenosis and in patients with left ventricular failure. J. Clin. Invest. 28:1138–1143.

Bruner, H. D. (Ed.). 1960. Peripheral blood flow measurement. Methods Med. Res. 8:222.

Burdine, J. A., L. A. Ryder, R. E. Sonnemaker, G. DePuey, and M. Calderon. 1971. ^{99m}Tc-human albumin microspheres (HAM) for lung imaging. J. Nucl. Med. 12:127–130.

Burke, G., A. Halko, and G. Peskin. 1971. Determination of cardiac output by radioisotope angiography and image-intensifier scintillation camera. J. Nucl. Med. 12:112–116.

Chinard, F. P., T. Enns, and M. F. Nolan. 1960. Contributions of bicarbonate ion and of dissolved CO_2 to expired CO_2 in dogs. Amer. J. Physiol. 198:78–88.

Chinard, F. P., T. Enns, and M. F. Nolan. 1962. Indicator-dilution studies with "diffusible" indicators. Circ. Res. 10:473–490.

Chinard, F. P., W. R. Taylor, M. F. Nolan, and T. Enns. 1959. Renal handling of glucose in dogs. Amer. J. Physiol. 196:535–544.

Chinard, F. P., G. H. Vosburgh, and L. B. Flexner. 1950. The mechanism of passage of various substances from blood to interstitial spaces in legs, liver and head of the dog. Eighteenth International Physiology Congress, Copenhagen, p. 155–156.

Coulam, C. M., H. R. Warner, E. H. Wood, and J. B. Bassingthwaighte. 1966. A transfer function analysis of coronary and renal circulation calculated from upstream and downstream indicator-dilution curves. Circ. Res. 19:879–890.

Coulter, N. A., Jr., and J. R. Pappenheimer. 1949. Development of turbulence in flowing blood. Amer. J. Physiol. 159:401–408.

Cournand, A., and H. A. Ranges. 1941. Catheterization of the right auricle in man. Proc. Soc. Exp. Biol. Med. 46:462–466.

Cournand, A., R. L. Riley, E. S. Breed, E. de F. Baldwin, D. W. Richards, Jr., M. S. Lester, and M. Jones. 1945. Measurement of cardiac output in man using the technique of catheterization of the right auricle or ventricle. J. Clin. Invest. 24:106–116.

Cykowski, C. B., D. L. Kirch, C. E. Polhemus, and D. W. Brown. 1971. Image enhancement of radionuclide scans by optical spatial filtering. J. Nucl. Med. 12:85–87.

Dawson, A. B., H. M. Evans, and G. H. Whipple. 1920. Blood volume studies. III. Behavior of large series of dyes introduced into the circulating blood. Amer. J. Physiol. 51:232–256.

Dexter, L., F. W. Haynes, C. S. Burwell, E. C. Eppinger, R. E. Seibel, and J. M. Evans. 1947. Studies of congenital heart disease. I. Technique of venous catheterization as a diagnostic procedure. J. Clin. Invest. 26: 547–553.

Domenech, R. J., J. I. E. Hoffman, M. I. M. Noble, K. B. Saunders, J. R. Henson, and S. Subijanto. 1969. Total and regional coronary blood flow measured by radioactive microspheres in conscious and anesthetized dogs. Circ. Res. 25:581–596.

Dow, P. 1956. Estimations of cardiac output and central blood volume by dye dilution. Physiol. Rev. 36:77–102.

Fegler, G. 1954. Measurement of cardiac output in anesthetized animals by a thermo-dilution method. Quart. J. Exp. Physiol. 39:153–164.

Felber, J. P., and M. L. Aubert. 1969. Radioimmunoassays: general principles. J. Nucl. Biol. Med. 13:1–9.

Fox, I. J. 1962a. History and developmental aspects of the indicator-dilution technic. Circ. Res. 10:381–392.

Fox, I. J., L. G. S. Brooker, D. W. Heseltine, H. E. Essex, and E. H. Wood. 1957a. A tricarbocyanine dye for continuous recording of dilution curves in whole blood independent of variation in blood oxygen saturation. Proc. Staff Meet. Mayo Clin. 32:478–484.

Fox, I. J., W. F. Sutterer, and E. H. Wood. 1957b. Dynamic response characteristics of systems for continuous recording of concentration changes in a flowing liquid (for example, indicator-dilution curves). J. Appl. Physiol. 11:390–404.

Freedman, G. S., P. N. Goodwin, P. M. Johnson, and R. N. Pierson, Jr. 1969. The evaluation of the image-intensifier scintillation camera with some comparisons to the single crystal camera. Radiology 92:21–29.

Friedlich, A., R. Heimbecker, and R. J. Bing. 1950. A device for continuous recording of concentration of Evans blue dye in whole blood and its application to determination of cardiac output. J. Appl. Physiol. 3:12–20.

Furneaux, R. W., and G. D. Tracy. 1970. The validity of the isotope dilution method of measuring extracellular fluid volume after acute hemorrhage. Aust. J. Exp. Biol. Med. Sci. 48:407–415.

Gilford, S. R., D. E. Gregg, O. W. Shadle, T. B. Ferguson, and L. A. Marzetta. 1953. An improved cuvette densitometer for cardiac output determination by the dye-dilution method. Rev. Sci. Instrum. 24:696–702.

Gillett, D. J., and D. F. J. Halmagyi. 1970. Accuracy of single-label blood volume measurement before and after corrected blood loss in sheep and dogs. J. Appl. Physiol. 28:213–215.

Gómez, D. M., M. Demeester, P. R. Steinmetz, J. Lowenstein, B. P. Sammons, D. S. Baldwin, and H. Chasis. 1965. Functional blood volume and distribution of specific blood flow in the kidney of man. J. Appl. Physiol. 20:703–708.

González-Fernández, J. M. 1962. Theory of the measurement of the dispersion of an indicator in indicator-dilution studies. Circ. Res. 10:409–428.

Gregersen, M. I., and J. G. Gibson, 2nd. 1937. Conditions affecting the absorption spectra of vital dyes in plasma. Amer. J. Physiol. 120:494–513.

Grollman, A. 1929. The determination of the cardiac output of man by the use of acetylene. Amer. J. Physiol. 88:432–445.

Hamilton, W. F. 1953. The physiology of the cardiac output. Circulation 8:527–543.

Hamilton, W. F., J. W. Moore, J. M. Kinsman, and R. G. Spurling. 1928a. Simultaneous determination of the pulmonary and systemic circulation times in man and of a figure related to the cardiac ouput. Amer. J. Physiol. 84:338–344.

Hamilton, W. F., J. W. Moore, J. M. Kinsman, and R. G. Spurling. 1928b. Simultaneous determination of the greater and lesser circulation times, of the mean velocity of blood flow through the heart and lungs, of the cardiac output and an approximation of the amount of blood actively circulating in the heart and lungs. Amer. J. Physiol. 85:377–378.

Hamilton, W. F., J. W. Moore, J. M. Kinsman, and R. G. Spurling. 1932. Studies on the circulation. IV. Further analysis of the injection method, and of changes in hemodynamics under physiological and pathological conditions. Amer. J. Physiol. 99:534–551.

Hamilton, W. F., R. L. Riley, A. M. Attyah, A. Cournand, D. M. Fowell, A. Himmelstein, R. P. Noble, J. W. Remington, D. W. Richards, Jr., N. C. Wheeler, and A. C. Witham. 1948. Comparison of the Fick and dye injection methods of measuring the cardiac output in man. Amer. J. Physiol. 153:309–321.

Harris, T. R., and E. V. Newman. 1970. An analysis of mathematical models of circulatory indicator-dilution curves. J. Appl. Physiol. 28: 840–850.

Heller, S., W. Lochner, and W. Schoedel. 1951. Die Bestimmung des Herzzeitvolumens mittels Injektions-methode bei fortlaufender photometrischer Registrierung der Zeit-Konzentrationskurven. Pfluegers Arch. 253:181–193.

Henriques, V. 1913. Über die Verteilung des blutes vom linken Herzen zwischen dem Herzen und dem übrigen Organismus. Biochem. Z. 56: 230–248.

Hosie, K. F. 1962. Thermal-dilution technics. Circ. Res. 10:491–504.

Hugenholtz, P. G., W. J. Gamble, R. G. Monroe, and M. Polanyi. 1965. The use of fiberoptics in clinical cardiac catherization. H. In vivo dye-dilution curves. Circulation 31:344–355.

Keith, N. M., L. G. Rowntree, and J. T. Geraghty. 1915. A method for the determination of plasma and blood volume. Arch. Intern. Med. 16: 547–576.

Kety, S. S., and C. F. Schmidt. 1945. Determination of cerebral blood flow in man by use of nitrous oxide in low concentrations. Amer. J. Physiol. 143:53–66.

Kinsman, J. M., J. W. Moore, and W. F. Hamilton. 1929. Studies on the circulation. I. Injection method: physical and mathematical considerations. Amer. J. Physiol. 89:322–330.

Knopp, T. J., and J. B. Bassingthwaighte. 1969. Effect of flow on transpulmonary circulatory transport functions. J. Appl. Physiol. 27:36–43.

Knutson, J. R. B., B. E. Taylor, E. J. Ellis, and E. H. Wood. 1950. Studies on circulation time with the aid of the oximeter. Proc. Staff Meet. Mayo Clin. 25:405–412.

Kopelman, H., and G. de J. Lee. 1951. The intrathoracic blood volume in mitral stenosis and left ventricular failure. Clin. Sci. 10:383–403.

Lacy, W. W., R. W. Emanuel, and E. V. Newman. 1957. Effect of the sampling system on the shape of indicator dilution curves. Circ. Res. 5:568–572.

Lagerlöf, H., L. Werkö, H. Bucht, and A. Holmgren. 1949. Separate determination of the blood volume of the right and left heart and the lungs in man with the aid of the dye injection method. Scand. J. Clin. Lab. Invest. 1:114–125.

MacIntyre, W. J., W. H. Pritchard, R. W. Eckstein, and H. L. Friedell. 1951. The determination of cardiac output by a continuous recording system utilizing iodinated (I^{131}) human serum albumin. I. Animal studies. Circulation 4:552–556.

Matthes, K., and X. Malikiosis. 1936. Untersuchungen über die Strömungsgeschwindigkeit des Blutes in menschlichen Arterien. Deut. Arch. Klin. Med. 179:500–517.

McMichael, J., and E. P. Sharpey-Schafer. 1944. Cardiac output in man by a direct Fick method: effects of posture, venous pressure change, atropine, and adrenaline. Brit. Heart J. 6:33–40.

Meier, P., and K. L. Zierler. 1954. On the theory of the indicator-dilution method for measurement of blood flow and volume. J. Appl. Physiol. 6:731–744.

Mook, G. A., P. Osypka, R. E. Sturm, and E. H. Wood. 1968. Fibre optic reflection photometry on blood. Cardiovasc. Res. 2:199–209.

Moore, J. W., J. M. Kinsman, W. F. Hamilton, and R. G. Spurling. 1929. Studies on the circulation. II. Cardiac output determinations: comparison of injection method with direct Fick procedure. Amer. J. Physiol. 89:331–339.

Newman, E. V., M. Merrell, A. Genecin, C. Monge, W. R. Milnor, and W. P. McKeever. 1951. The dye dilution method for describing the central circulation: an analysis of factors shaping the time-concentration curves. Circulation 4:735–746.

Nicholes, K. K., and H. R. Warner. 1964. A study of dispersion of an indicator in the circulation. Ann. N. Y. Acad. Sci. 115:721–737.

Nicholson, J. W., III, H. B. Burchell, and E. H. Wood. 1951. A method for the continuous recording of Evans blue dye curves in arterial blood, and its application to the diagnosis of cardiovascular abnormalities. J. Lab. Clin. Med. 37:353–364.

Nylin, G., and H. Celander. 1950. Determination of blood volume in the heart and lungs and the cardiac output through the injection of radiophosphorus. Circulation 1:76–83.

Phibbs, R. H., and L. Dong. 1970. Nonuniform distribution of microspheres in blood flowing through a medium-size artery. Can. J. Physiol. Pharmacol. 48:415–421.

Prinzmetal, M. 1941. Calculation of the venous-arterial shunt in congenital heart disease. J. Clin. Invest. 20:705–708.

Romm, S. O. 1924. Zur Bestimmungsmethode der Umlaufszeit des Blutes im Kreislauf. Arch. Gesamte Physiol. 202:14–24.

Rossi, H. H., S. H. Powers, and B. Dwork. 1953. Measurement of flow in straight tubes by means of the dilution technique. Amer. J. Physiol. 173:103–108.

Rudolf, A. M., and M. A. Heymann. 1967. The circulation of the fetus in utero: methods for studying distribution of blood flow, cardiac output and organ blood flow. Circ. Res. 21:163–184.

Schlossmacher, E. J., H. Weinstein, S. Lochaya, and A. B. Schaffer. 1967. Perfect mixers in series model for fitting venoarterial indicator-dilution curves. J. Appl. Physiol. 22:327–332.

Sheppard, C. W. 1954. Mathematical considerations of indicator dilution techniques. Minn. Med. 37:93–104.

Sheppard, C. W. 1962. Stochastic models for the tracer experiments in the circulation: parallel random walks. J. Theor. Biol. 2:33–47.

Sheppard, C. W., M. P. Jones, and B. L. Couch. 1959. Effect of catheter sampling on the shape of indicator-dilution curves: mean concentration versus mean flux of outflowing dye. Circ. Res. 7:895–906.

Sherman, H., R. C. Schlant, W. L. Kraus, and C. B. Moore. 1959. A figure of merit for catheter sampling systems. Circ. Res. 7:303–313.

Sinclair, J. D., W. F. Sutterer, I. J. Fox, and E. H. Wood. 1961. Apparent dye-dilution curves produced by injection of transparent solutions. J. Appl. Physiol. 16:669–673.

Singh, R., A. J. Ranieri, Jr., H. R. Vest, Jr., D. L. Bowers, and J. F. Dammann, Jr. 1970. Simultaneous determinations of cardiac output by thermal dilution, fiberoptic and dye-dilution methods. Amer. J. Cardiol. 25:579–587.

Stead, E. A., Jr., J. V. Warren, A. J. Merrill, and E. S. Brannon. 1945. The cardiac output in male subjects as measured by the technique of right atrial catheterization: normal values with observations on the effect of anxiety and tilting. J. Clin. Invest. 24:326–331.

Stephenson, J. L. 1948. Theory of the measurement of blood flow by the dilution of an indicator. Bull. Math. Biophys. 10:117–121.

Stewart, G. N. 1894. Researches on the circulation time in organs and on the influences which affect it. J. Physiol. 15:1–89.

Stewart, G. N. 1897a. Researches on the circulation time and on the influences which affect it. IV. The output of the heart. J. Physiol. 22:159–183.

Stewart, G. N. 1897b. The measurement of the output of the heart. Science 5:137.

Stewart, G. N. 1921a. The output of the heart in dogs. Amer. J. Physiol. 57:27–50.

Stewart, G. N. 1921b. The pulmonary circulation time, the quantity of blood in the lungs and the output of the heart. Amer. J. Physiol. 58:20–44.

Stewart, G. N. 1921c. Researches on the circulation time and on the influences which affect it. V. The circulation time of the spleen, kidney, intestine, heart (coronary circulation) and retina, with some further observations on the time of the lesser circulation. Amer. J. Physiol. 58: 278–295.

Stow, R. W. 1954. Systematic errors in flow determinations by the Fick method. Minn. Med. 37:30–35.

Stow, R. W., and P. S. Hetzel. 1954. An empirical formula for indicator-dilution curves as obtained in human beings. J. Appl. Physiol. 7:161–167.

Sutterer, W. F. 1960. A compensated dichromic densitometer for Indocyanine green. Physiologist 3:159.

Symposium on diagnostic applications of indicator-dilution technics. 1957. Proc. Staff Meet. Mayo Clin. 32:463–553.

Symposium on diagnostic applications of indicator-dilution curves recorded from the right and left sides of the heart. 1958. Proc. Staff Meet. Mayo Clin. 33:535–577; 581–610.

Symposium on use of indicator-dilution technics in the study of the circulation. 1962. In E. H. Wood (ed.). American Heart Association, Inc., New York, p. 377–581.

Tancredi, R. G., and K. L. Zierler. 1971. Indicator-dilution, flow-pressure and volume-pressure curves in excised dog lung. Fed. Proc. 30:380.

Taylor, G. 1953. Dispersion of soluble matter in solvent flowing slowly through a tube. Roy. Soc. (London) Proc. A 219:186–203.

Taylor, G. 1954. The dispersion of matter in turbulent flow through a pipe. Roy. Soc. (London) Proc. A 223:446–468.

Taylor, S. H., and J. P. Shillingford. Clinical applications of Coomassie blue. Brit. Heart J. 21:497–504.

Taylor, S. H., and J. M. Thorp. 1959. Properties and biological behavior of Coomassie blue. Brit. Heart J. 21:492–496.

Thompson, H. K., Jr., C. F. Starmer, R. E. Whallen, and H. D. McIntosh. 1964. Indicator transit time considered as a gamma variate. Circ. Res. 14:502–515.

Ueda, H., Y. Sasaki, M. Iio, S. Kaihara, K. Machida, I. Ito, S. Takayanagi, T. Kabayashi, and T. Sugita. 1969. Catheter semiconductor radiation detector for continuous measurement of cardiac output. J. Nucl. Med. 10: 502–515.

Visscher, M. B., and J. A. Johnson. 1953. The Fick principle: analysis of potential errors in its conventional application. J. Appl. Physiol. 5: 635–638.

Wagner, H. N., Jr., B. A. Rhodes, Y. Sasaki, and J. P. Ryan. 1969. Studies of the circulation with radioactive microspheres. Invest. Radiol. 4:374–386.

Werkö, L., H. Lagerlöf, H. Bucht, B. Wehle, and A. Holmgren. 1949. Comparison of the Fick and Hamilton methods for the determination of cardiac output in man. Scand. J. Clin. Lab. Invest. 1:109–113.

White, H. L. 1947. Measurement of cardiac output by a continuously recording conductivity method. Amer. J. Physiol. 151:45–57.

Wood, E. H. 1962. Diagnostic applications of indicator-dilution technics in congenital heart disease. Circ. Res. 10:531–568.

Wood, E. H. 1970. Scintiscanning system for study of regional distribution of blood flow. Report No. SAM-TR-70-6. United States Air Force School of Aerospace Medicine, Brooks Air Force Base, Texas.

Zierler, K. L. 1958. A simplified explanation of the theory of indicator-dilution for measurement of fluid flow and volume and other distributive phenomena. Johns Hopkins Med. J. 103:199–217.

Zierler, K. L. 1962. Theoretical basis of indicator-dilution methods for measuring flow and volume. Circ. Res. 10:393–407.

Zierler, K. L. 1965. Tracer-dilution techniques in the study of microvascular behavior. Fed. Proc. 24:1085–1091.

chapter 2/ Foundations of Indicator-Dilution Theory

Stuart Antman
Department of Mathematics/University of Maryland/
College Park, Maryland

Indicator-dilution theories are models for the one-dimensional flow of mixtures. This chapter gives an elementary exposition of those aspects of the theory leading to practical techniques for the study of circulatory processes. The purpose is not to construct the most sophisticated and general models, but to develop the underlying concepts.

A dye injected into a body of fluid at some point identifies thereafter those fluid particles that have passed through that point during the interval of the dye injection. When the observed motion of the dyed particles is interpreted by the theory of fluid dynamics (and possibly by the theory of diffusion), a precise picture of the fluid motion can result. This program can be modified to yield techniques for the analysis of the flow of blood. Difficulties such as those due to the complicated constitution of the blood, the deformability of blood vessels, and the unsteady character of the flow are more than compensated for by simplifications arising from the need for only gross properties of the flow and from the thinness of many of the vessels and orifices (i.e. the one-dimensionality of the flow).

A number of simplifying assumptions are introduced in the formulation of tractable models for the flow of blood; the conditions under which such assumptions are to hold are explicitly defined. It is the responsibility of the investigator who wishes to use the results based upon such a set of assumptions to verify that the assumptions adequately describe the physiological situation under study.

INTRODUCTORY CONCEPTS

Consider a mixture of two substances filling a region of space with a volume V. If one of the substances, the "indicator," has a mass m contained within the space, then the *average* concentration C of the indicator is defined as

$$C = \frac{m}{V}. \tag{2.1}$$

The *concentration* of the indicator at a *point* in space is defined as the limit of the concentrations of the indicator in a sequence of regions of decreasing volume that contain that point. It is approximately the concentration in a small volume containing that point.

CONDITIONS AND ASSUMPTIONS

If no material can enter or leave the region and if the contents of the region are thoroughly mixed so that the particles of indicator are homogeneously distributed, the concentration of indicator is constant throughout that region. Thus, if the mass of indicator and its concentration at any point are known, the volume of the region can be determined from (2.1):

$$V = \frac{m}{C}.$$

The entire indicator-dilution theory is but a generalization of this simple fact. The important physiological information required is the volume or rate of change of volume in parts of the circulatory system and these are inaccessible to direct measurement. Generalizations of (2.1) allow such data to be determined indirectly through measurements of concentration, which are easy to obtain. In practice, the volume contribution of the indicator is negligible compared to that of the other component of the mixture, the native fluid. This initial assumption holds for the entire discussion and the native fluid is henceforth referred to simply as the fluid.

There is no change in the theory if m is regarded as the amount of some property of the indicator other than mass, e.g. heat or radiation. For definitiveness, however, m will be called mass in this chapter.

MEASUREMENT BY SUDDEN INJECTION

Consider a container (which may even be deformable) with a single inflow orifice I and a single outflow orifice E holding a mixture of indicator and fluid. The container may be part of a larger system and be composed of a network of channels as in Fig. 2.1. The cross-sectional areas of these channels are assumed to be so small and the action occurring at them is such that the variation of fluid properties across the section can be neglected.

The concentration at these orifices must first be defined in a convenient way. Let A represent any orifice of the container. Let $m_A(t)$ denote the mass of indicator and $V_A(t)$ denote the total volume of fluid that has passed A by time t. Let $(t, t + \Delta t)$ denote the interval of time consisting of all the times between t and $t + \Delta t$. During this interval, the mass $\Delta m_A = m_A(t + \Delta t) - m_A(t)$ of indicator and the volume $\Delta V_A = V_A(t + \Delta t) - V_A(t)$ of fluid pass through A. If Δt is small, then the concentration $C_A(t)$ at A is approximated by $\Delta m_A / \Delta V_A$. In fact,

$$C_A = \lim_{t \to 0} \frac{\Delta m_A}{\Delta V_A} = \lim_{t \to 0} \frac{\Delta m_A / \Delta t}{\Delta V_A / \Delta t} = \frac{(dm_A/dt)(t)}{(dV_A/dt)(t)}. \qquad (2.2)$$

Here $(dm_A/dt)(t)$ and $(dV_A/dt)(t)$ are the time rates at which the mass of indicator and the volume of fluid pass through A at time t. Then

$$Q_A(t) = \frac{dV_A}{dt}(t), \qquad (2.3)$$

TIME

Figure 2.1

where Q_A, which is the time rate of change of V_A, is termed the *flux* at A. (In most of the indicator-dilution theory, Q_A is called the *flow* at A. *Flux* is used in much of the fluid dynamics literature since it allows *flow* to be used unambiguously as a synonym for the general motion of the fluid.)

The following three assumptions on the system are understood to hold throughout this chapter.

1. The motion of the indicator is representative of the fluid. This means that the indicator must have mechanical properties compatible with those of the fluid and that the indicator must be injected at I so that it is homogeneously distributed over the volume element of fluid at I. This assumption implies that, if a given fraction of incoming fluid traverses the container (network) by a certain route, then the same fraction of indicator will traverse the container by the same route, provided the indicator enters the same time the fluid enters. This also means that, if a given fraction of incoming fluid takes time τ to traverse the container, then the same fraction of indicator will have the same transit time τ, provided it enters at the same time.

2. The amount of indicator is so small that it does not disturb the flow pattern that would obtain were it absent. (This assumption is responsible for the mathematical linearity of the resulting equations. This linearity is the source of much of the mathematical simplicity that the theory enjoys.)

3. All of the indicator injected at I eventually leaves through E. (This does not preclude its reappearance at I by recirculation.)

COMPUTATION OF FLUX

Suppose that at a given instant, say $t = 0$, when there is no indicator in the container, a mass m of indicator is injected at I. (This represents an idealization of an injection over a very short time interval. In later sections injections at an arbitrary rate are discussed.) The exit concentration $C_E(t)$ that results from this sudden injection is denoted $c(t)$. A typical form of c as a function of t is shown in Fig. 2.2, assuming that there is no recirculation. Although virtually all of the indicator can be expected to leave in a short period of time, it is mathematically convenient to hedge this statement and to assert merely that all the indicator will leave after an infinite amount of

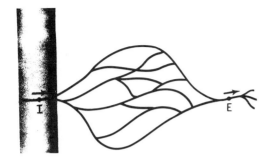

Figure 2.2

time. Thus the mass $m_E(\infty)$ of indicator that has left after an infinite amount of time must equal (by assumption 3) the mass m. Now $m_E(\infty)$ is the sum of the masses that leave in every small time interval, i.e. by (2.2), it is a sum of terms like $c(t)\Delta V_E$. In short, it is the integral

$$m_E(\infty) = \int_0^\infty c(t)\frac{dV_E}{dt}(t)\,dt = \int_0^\infty c(t)Q_E(t)\,dt. \qquad (2.4)$$

The equality of m and $m_E(\infty)$ thus yields the following equation relating c, Q_E, and m:

$$m = \int_0^\infty c(t)Q_E(t)\,dt. \qquad (2.5)$$

If it is known that Q_E is constant as a function of time, in which case we denote Q_E by Q, then we can remove it from under the integral sign in (2.5) and obtain

$$Q = \frac{m}{\int_0^\infty c(t)\,dt}. \qquad (2.6)$$

The denominator of the right side of (2.6) is just the area under the curve of c as a function of t (Fig. 2.2). Thus (2.6) permits the computation of the constant flux Q from the known mass of indicator injected at $t = 0$ and from the concentration curve c that is found experimentally.

If Q_E is not constant, as in pulsatile flows, then there is not enough information available for (2.5) to be solved uniquely for Q_E. (In fact, given m and $c(t)$, there are an infinite number of functions of t, which, when substituted into the Q_E slot of (2.5), reduce (2.5) to an identity.) If, however, Q_E has the form $Q_E(t) = Af(t)$, where A is an unknown constant but f is a known function, (e.g. $f(t) = \sin t$), then (2.5) can be used to find A. Indeed, the replacement of $Q_E(t)$ in (2.5) by $Af(t)$ leads to the expression

$$A = \frac{m}{\int_0^\infty c(t)f(t)dt}.$$
(2.7)

THE DISTRIBUTION FUNCTION

The distribution function is the fraction of injected indicator leaving the container per unit time, at time t. The concept of a distribution function (denoted by h) plays a central and unifying role in this presentation of indicator-dilution theory.

The fraction of indicator that has left the container by time t is

$$\frac{m_E(t)}{m}.$$

The time rate of change of this fraction is denoted by $h(t)$. Since m is a constant,

$$h(t) = \frac{d}{dt}\left[\frac{m_E(t)}{m}\right] = \frac{(d/dt)m_E(t)}{m}.$$
(2.8)

From (2.2) and (2.5), it follows that

$$h(t) = \frac{Q_E(t)c(t)}{m} = \frac{Q_E(t)c(t)}{\int_0^\infty Q_E(t)c(t)dt};$$
(2.9)

hence

$$\int_0^\infty h(t)dt = 1.$$
(2.10)

These equivalent characterizations are especially useful for the computation of the volume. The definition (2.8) related the distribution function to the dye. The alternative representations to be developed relate the distribution function to the fluid itself.

The reasonable requirement that $Q_F(t) \geq 0$ implies by (2.9) that $h(t) \geq 0$. If Q_E is a constant, then (2.9) reduces to

$$h(t) = \frac{c(t)}{\int_0^\infty c(t)dt}.$$
(2.11)

This says that h will have the same shape as c, but will have its amplitude adjusted so that (2.10) is satisfied. Thus the form of h is shown in Fig. 2.2.

Let V_I be the volume of fluid dyed by the indicator at time $t = 0$. (That a positive volume of fluid can be dyed instantaneously is an idealization of the same sort that permits a positive mass of dye to be injected instantaneously. In practice, it will take a short time interval to inject the mass m of dye, and during this time interval the volume V_I of fluid will enter I and be dyed. This idealization is a (mathematically subtle) limiting case of the latter situation as the time interval of injection goes to zero while the amount of dye remains constant.) That the flow of indicator is representative of the flow of the fluid itself (assumption 1) implies that

$$\frac{V_E(t)}{V_I} = \frac{m_E(t)}{m}. \tag{2.12}$$

From (2.8) it then follows that

$$h(t) = \frac{d}{dt}\left[\frac{V_E(t)}{V_I}\right] = \frac{(d/dt)V_E(t)}{V_I}. \tag{2.13}$$

Thus, h may be alternatively characterized as the time rate of change of the fraction of dyed fluid that has left by time t or, equivalently, as the fraction of dyed fluid leaving the container per unit time at time t. The utility of the distribution function, h, is a consequence of this dual characterization. Note that the words "dyed fluid," appearing in this paragraph can be replaced by "fluid which entered at time $t = 0$." The dye only aids the observation of the flow by identifying the particles entering at time $t = 0$; it does not affect the flow (assumption 2).

Now consider an arbitrary point of the system. A succession of fluid particles flow through this point, and, in general, the velocities of the particles as they pass through it are different. If these velocities are the same for each fixed point of the system, the flow is said to be *stationary* or *steady*, and it follows, therefore, that Q_I and Q_E must be constant. If the fluid is incompressible, as liquids are, then $Q_I = Q_E$ and their common value is denoted by Q.

For stationary flows, a useful modification of the last characterization of h can be obtained. In this case, the instant at which V_I passes I is immaterial. Thus, $h(\tau)$ is the time rate of change of the fraction of fluid that has transit time less than τ or, equivalently, is the fraction of fluid leaving the container per unit time at a time τ units later than when it entered. (In these last remarks t has been changed to τ for future mathematical convenience.) It is important to note that, although a flow is steady, the concentrations need not be; they will depend upon the manner of injection of indicator. On the other hand, h characterizes the flow and is independent of the manner of dye injection. Equation (2.11) merely gives a convenient formula for its computation.

In the remainder of the discussion, the following is assumed.

4. The flow is stationary and $Q_I = Q_E = Q$ (constant). Now (2.13) can be approximated by

$$h(\tau)\Delta\tau = \frac{\Delta V_E}{V_I} = \frac{V_E(\tau + \Delta\tau) - V_E(\tau)}{V_I}. \tag{2.14}$$

Thus, $h(\tau)\Delta\tau$ is approximately the fraction of fluid with transit times between τ and $\tau + \Delta\tau$. These results will be essential for the computation of volume.

COMPUTATION OF VOLUME IN THE ABSENCE OF RECIRCULATION

The volume of fluid in the container is the sum of the *movable* volume, consisting of those fluid particles that have entered through I and will ultimately leave through E, and of the *stagnant* volume, consisting of those particles trapped forever in a part of the container called a *stagnant pool*. Indicator-dilution techniques cannot detect the stagnant volume because dyed particles cannot enter it. The slight inconvenience of distinguishing the movable volume is dispensed with by the device of regarding that part of the container holding the movable volume as a new container. This is equivalent to the following assumption.

5. The container has no stagnant pools. Because the flow is stationary, the volume in the container at time t is the same as that at any other time. If the fluid particles in the container at time t are divided into classes identified by their transit times, those particles with transit times between τ and $\tau + \Delta\tau$ will certainly exit between times t and $t + \tau + \Delta\tau$. Since the flux is constant, the total volume of fluid that leaves the container in the time interval $(t, t + \tau + \Delta\tau)$ is just Q $(\tau + \Delta\tau)$. If $\Delta\tau$ is small, this last expression can be approximated by $Q\tau$. From (2.14) the fraction of fluid with transit times between τ and $\tau + \Delta\tau$ is $h(\tau)\Delta\tau$. Thus, the volume of those particles with transit times between τ and $\tau + \Delta\tau$ is approximately the product of this fraction with $Q\tau$, namely

$Q\tau h(\tau)\Delta\tau.$ (2.15)

The total volume V held in the container at time t is the sum of contributions of the form (2.15), i.e. it is the integral

$$V = Q \int_0^\infty th(t)dt.$$ (2.16)

This result is completely general for stationary flows because h characterizes the flow and not just the mode of injection. Thus, if h is determined by some means other than by (2.9), relation (2.16) is still valid. When (2.9) is used, (2.16) reduces to

$$V = \frac{Q\int_0^\infty tc(t)dt}{\int_0^\infty c(t)dt}.$$ (2.17)

This gives a rule for the computation of V from the concentration curve $c(t)$ obtained from a sudden injection experiment. The integral in the denominator is just the area under the curve of c versus t, and the integral in the numerator is only the area under the curve of tc versus t. Equation (2.17) has a simple interpretation when it is observed that the ratio of integrals in it is only the definition of the mean transit time, which is denoted $\bar{t}$. Thus (2.17) can be written as $V = Q\bar{t}$.

MEASUREMENTS BY CUMULATIVE (STEP) INJECTION

Measurement of Flux

This section extends the results of the last section to account for arbitrary rates of dye injection. Assumptions 1 through 5 are to hold here.

Let indicator be injected into I at the rate dm_I/dt, denoted $\dot{m}_I$. This rate is not necessarily constant and may include contributions from recirculated indicator. To study the efflux of an indicator at an arbitrary time t, it is useful to study the influx at time $t - s$, which is s units of time earlier. The amount of indicator injected in the time interval $(t - s - \Delta s, t - s)$ is approximately

$\dot{m}_I(t - s)\Delta s.$ (2.18)

Now if the mass m of dye is suddenly injected at time 0, relation (2.8) implies that

$$\frac{d}{dt}m_E(t) = h(t)m,$$ (2.19)

whereas if m is injected at time $t - s$, then

$$\frac{d}{dt}m_E(t) = h(s)m.$$ (2.20)

(That the form of h does not change with time is a consequence of stationarity.) If Δs is small, then (2.18) represents a nearly instantaneous injection of mass at time $t - s$. The substitution of (2.18) for m in (2.20) gives the approximate contribution to $(d/dt)m_E$ from the indicator injected in the interval $(t - s - \Delta s, t - s)$:

$$h(s)\dot{m}_I(t - s)\Delta s.$$ (2.21)

The total rate of indicator mass outflow at time t is the sum of the contributions of the form (2.21), i.e.

$$\frac{dm_E}{dt}(t) = \int_0^\infty h(s)\dot{m}_I(t - s)ds.$$ (2.22)

The use of relations (2.2) and (2.3) yields useful alternatives to (2.22):

$$QC_E(t) = \int_0^\infty h(s)\dot{m}_I(t - s)ds$$ (2.23)

and

$$C_E(t) = \int_0^\infty h(s)C_I(t - s)ds.$$ (2.24)

Note the following mathematical identities obtained by a change of variables:

$$\int_0^\infty h(s)\dot{m}_I(t - s)ds = \int_0^\infty h(t - \sigma)\dot{m}_I(\sigma)d\sigma$$
$$= \int_0^\infty h(t - s)\dot{m}_I(s)ds.$$ (2.25)

The most important special cases of (2.22) through (2.24) are obtained when $\dot{m}_I(t) = 0$ for $t < 0$, i.e. when injection begins at time 0. This means that $\dot{m}_I(t - s) = 0$ when $s > t$. In this case, contributions to the integrals of (2.22) through (2.24) are zero when $s > t$ so that we can replace the upper limit of integration ∞ by t:

$$\frac{dm_E}{dt}(t) = \int_0^t h(s)\dot{m}_I(t - s)ds.$$ (2.26)

These relations will now be used to compute the flux and volume when there is no recirculation. The absence of recirculated indicator means that $\dot{m}_I$ is known. The special case in which $\dot{m}_I(t) = 0$ for $t < 0$ and $\dot{m}_I(t) = \mu$ (constant) for $t > 0$ is treated. This is somewhat misleadingly termed "constant injection" in the indicator-dilution literature and could better be termed a *step* injection. If $C(t)$ denotes the exit concentration $C_E(t)$ due to a step injection, (2.23) reduces to

$$QC(t) = \mu \int_0^t h(s)\,ds. \tag{2.27}$$

A typical form of $C(t)$ is shown in Fig. 2.3. By virtue of the requirement that $\int_0^\infty h(t)\,dt = 1$, the limit of the right side of (2.27) as $t \to \infty$

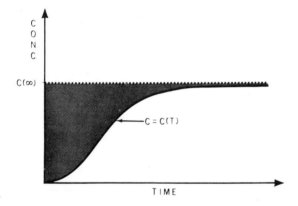

Figure 2.3

is just μ. Thus the exit concentration $C(t)$ has a limiting value as $t \to \infty$, which is just μ/Q. Since h is positive, this limit is greater than the value of C at any previous time. Denoting this limit by $C(\infty)$, we have

$$\lim_{t \to \infty} C(t) = \mu/Q = C(\infty). \tag{2.28}$$

In practice, C will attain its limiting value $C(\infty)$ (to within a minute error) in a finite length of time so that $C(\infty)$ is readily accessible to measurement. The flux is given by the formula:

$$Q = \mu/C(\infty). \tag{2.29}$$

Measurement of Volume

To find the volume of the fluid in the container at any given time, it suffices to determine h and substitute it into the formula (2.16):

$V = Q \int_0^\infty t h(t) dt$. The substitution of (2.29) into (2.27) and the differentiation of the resulting expression with respect to time yields

$$h(t) = \frac{(d/dt)C(t)}{C(\infty)}.$$ (2.30)

Thus (2.16) becomes

$$V = \frac{Q}{C(\infty)} \int_0^\infty t \frac{d}{dt} C(t) dt.$$ (2.31)

A more useful version of this formula is obtained by integrating it by parts and observing that $(d/dt)C = (d/dt) [C - C(\infty)]$. Then,

$$V = \frac{Q}{C(\infty)} \int_0^\infty [C(\infty) - C(t)] dt = Q \int_0^\infty \left[1 - \frac{C(t)}{C(\infty)} \right] dt.$$ (2.32)

The integral $\int_0^\infty [C(\infty) - C(t)] dt$ appearing in (2.32) is the area between the curve of $C(t)$ and the horizontal line with ordinate $C(\infty)$. This region is hatched in Fig. 2.3. An alternative development of (2.32) is given by Zierler (1962a, b).

If h is known, then Q can be found from (2.23) for arbitrary $\dot{m}_I$ because all of the other variables in (2.23) are known. Equation (2.16) can then be used to obtain V. If there is an effective way to measure C_I, then (2.24) could be solved for h. In theory, this solution can be obtained by means of the Laplace transform, but in practice it would have to be constructed numerically; it is a relatively simple problem on a computer. Some other ways of finding h are discussed in the next section.

Finally, the relationship between the exit concentration for sudden injection c and that for step injection C can be obtained by comparing the first equation of (2.9) with (2.30). Since it is the distribution function h that characterizes the flow and not the mode of dye injection, there is no theoretical advantage in using one method over the other (or even in using some other method, such as pulsatile injection).

MEASUREMENTS FOR RECIRCULATING SYSTEMS

There is both experimental and theoretical evidence indicating that the exit concentration due to a sudden injection should decay exponentially to zero in the absence of recirculation. If there is recirculation, the exit concentration shows an upward distortion due to the reappearance of the indicator some time after the injection takes place.

If this distortion does not appear until after the decay of the concentration curve has been established, the effects of the recirculated indicator can be removed by an extrapolation of that part of the concentration curve obtained before the distortion is manifested. Because of exponential decay, this extrapolation is best effected on semi-logarithmic paper which converts exponential curves to straight lines. Once the concentration curve has thus been "purified of contamination" by recirculated indicator, the previous results can be applied. If another mode of injection is used, the transient behavior decays exponentially, and an extrapolation can again be employed. Note that these extrapolation procedures are unaffected by losses of indicator outside the container. If, however, the recirculated indicator appears before the decay of the concentration curve has been established, a meaningful extrapolation may be impossible, and a more intricate analysis is required. This section describes several techniques for dealing with such a problem.

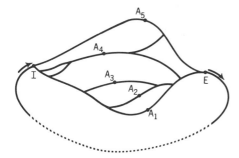

Figure 2.4

Assume now that the container, with entrance I and exit E, is part of a closed, circulating system. In addition to assumptions 1 through 5, there is a further set of assumptions that (6) no indicator can leave the system, (7) every fluid particle ultimately traverses the system, and (8) every fluid particle must pass I (and therefore E) in such a traversal. A set, A, of points of the system is called a *section* if any fluid particle that traverses the system once (from I to I, say) must pass through exactly one point of the section. In Fig. 2.4, A, which consists of the points, $A_1, \ldots, A_5$, is a section. I is a section and E is a section, but the pair of points, I, E, is not a section because a fluid particle traversing the system must pass through both of these points. In principle, the concentration at any section can be determined.

The relations among concentrations at different sections of the system will now be found. These relations will then be used in the determination of flux and volume of the container.

In Fig. 2.5, let A, B, D be three sections of the system with the property that fluid passing A must pass B before passing D and returning to A. Starting at time 0, dye is added to the flow at B at the rate $\dot{m}_B$. The concentration C_D at D consists of a contribution from the dye passed through A plus a contribution from the dye injected at B (that has not yet passed through A). If h_{AD} and h_{BD} denote the

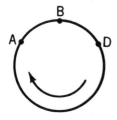

Figure 2.5

distribution functions for the containers AD and BD, then (2.23) and (2.24) imply

$$C_D(t) = \int_0^t h_{AD}(s)C_A(t - s)ds + \int_0^t h_{BD}(s)\frac{\dot{m}_B(t - s)}{Q} ds. \qquad (2.33)$$

This formula determines the general relations between concentrations at different sections of the system.

Several specializations of this formula will now be obtained. First, I of Fig. 2.4 is identified with B of Fig. 2.5. Let D move clockwise about the circuit until it approaches A. In this process, h_{AD} approaches the distribution function for the entire system, which can be denoted by g. Note that g does not depend upon the particular section, in this case section A, at which the cycle is started. Next, let A and D approach I (that is, B) clockwise. Then h_{ID} also approaches g. C_I^* denotes the limit of C_A and C_D as A and D approach I clockwise. This limit process and the concomitant use of the asterisk are employed to render precise the fact that C_I^* does not include the contribution of the dye being injected at I. In practice, C_I^* is measured at a section near I through which the fluid must pass before reaching I. In this case (2.33) reduces to

$$C_I^*(t) = \int_0^t g(s)C_I^*(t - s)ds + \int_0^t g(s)\frac{\dot{m}_I(t - s)}{Q} ds, \qquad (2.34)$$

an integral equation relating C_I^* and g. If g is known, then (2.34) is the well known renewal equation for the unknown C_I^*.

To establish further relations between Figs. 2.4 and 2.5, D is now identified with E, and B retains its identity with I. Let A approach I clockwise. Then h_{AD} and h_{ID} each reduce to h and (2.33) reduces to

$$C_E(t) = \int_0^t h(s)C_I^*(t-s)ds + \int_0^t h(s)\frac{\dot{m}_I(t-s)}{Q}\,ds. \qquad (2.35)$$

This relates the exit concentration to the concentration just before the entrance and the concentration due to injected dye. It is a generalization of (2.24) to account for recirculation. The renewal equation satisfied by C_E itself is obtained by keeping the identification of B with I and D with E and by moving A counterclockwise about the circuit until it approaches E. Then h_{AE} approaches g and h_{IE} is h. In this case, (2.33) becomes

$$C_E(t) = \int_0^t g(s)C_E(t-s)ds + \int_0^t h(s)\frac{\dot{m}_I(t-s)}{Q}\,ds. \qquad (2.36)$$

These relations are simplified for sudden and for step injections. For sudden injection of mass m of indicator at time 0, the second integral of (2.34) becomes

$$\frac{mg(t)}{Q}. \qquad (2.37)$$

This follows by analogy with the first part of (2.9). (To see this, note that the second integral of (2.34) represents the concentration at I^* for the first passage of dye injected at I. Equation (2.9) likewise gives the concentration c due to the sudden injection of the dye by the formula $c = mh(t)/Q$. Because there is no recirculation accounted for in (2.9), this represents the first and only passage of the dye. By replacing h for the distribution function g for the entire circuit, (2.37) is obtained from the above representation for c. Alternatively, the Dirac delta can be used.)

For the step injection at the rate μ, the second integral of (2.34) becomes

$$\frac{\mu}{Q}\int_0^t g(s)ds. \qquad (2.38)$$

The second integrals in (2.35) and (2.36) are treated analogously.[1]

The same sort of approach that gave the flux and volume for the simpler problems treated above will now be extended to the recirculation problem. As before, the unknown Q, which appears in (2.34) through (2.36), can be isolated when the equations are evaluated for a large time. For this purpose, knowledge of the behavior of the distribution functions h, g, and their indefinite integrals as $t \to \infty$ can be exploited. Such information about the distribution functions will yield results that do not depend on the specific form of these functions.

As a preliminary result, the volume of the whole system, denoted V_T, is obtained. Consider (2.34) when the injection is sudden, so that (2.37) replaces the second integral on the right. It is specifically required that the distribution functions h and g have the shape shown in Fig. 2.2, but an explicit mathematical characterization of this shape will not be given here. It can then be shown that the solution C_I^* of (2.34) in this case has the limit

$$\lim_{t \to \infty} C_I^*(t) = \frac{m}{Q \int_0^\infty sg(s)\,ds} = \frac{m}{V_T}. \tag{2.39}$$

This result is obtained by an asymptotic analysis of the inverse Laplace transform of the solution of (2.34). The techniques for this are discussed in books on asymptotic expansions. Some asymptotic results for the renewal equation are given by Bellman and Cooke (1963). It is emphasized that this result and analogous ones that follow are rigorous consequences of equations (2.34) to (2.36); they are intuitive, but they do not rely on intuitive arguments for their justification as Zierler's (1962a, b) surveys may lead one to believe. Laplace transform techniques can give formal solutions of the equations of the theory, but these are of little practical significance, except for their utility in describing the long-time asymptotic behavior of the solutions. If explicit solutions are required, a numerical analysis is simpler and quicker.

In (2.39), m/V_T is just the average concentration of the entire system. Thus, the concentration at I approaches the expected uniform

[1]Zierler (1962a, b) neglected to include the contributions of the injected dye in the equations corresponding to (2.34) and (2.35). Thus, the equations (27), (28), (30), and (31) of 1962a and the equation preceding (18) and equations (18), (19), (20), and (21) of 1962b are wrong. Since his arguments were intuitive rather than analytic, he recovered the correct asymptotic forms of these equations in (31) of 1962a and (21) of 1962b. "Asymptotic" here refers to approximations valid for large time.

concentration of the system. If C_I^* is monitored, then (2.39) allows V_T to be determined by the formula $V_T = m/\lim C_I^*$.

The asymptotic behavior of (2.34) for a step injection will now be examined. In this case, (2.38) replaces the second integral of (2.34). It can be shown that the solution of (2.34) then has the asymptotic form

$$C_I^*(t) \sim \frac{\mu t}{V_T} + a \qquad (2.40)$$

for large t, where a is a constant that does not have to be determined. This behavior is illustrated in Fig. 2.6. (The tilde should be read "be-

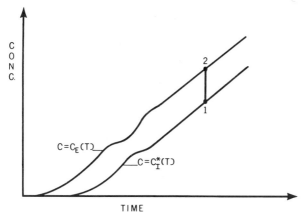

Figure 2.6 A plot of $C_E(t)$ and $C_I^*(t)$. For sufficiently large time t, the curves will be straight lines as a consequence of equations (2.43) and (2.45). The difference in the ordinates for this large time is the right side of equation (2.46) and is represented by the length of the segment 1 to 2.

haves like.") This sort of steady growth of concentration is to be expected because the dye is being injected at a constant rate. The asymptotic behavior of C_E for large time is then determined from (2.35) when the step injection is used. In this case, the second integral of (2.35) is just

$$\frac{\mu}{Q} \int_0^t h(s)\,ds \qquad (2.41)$$

and this approaches μ/Q as $t \to \infty$ by virtue of (2.10). The asymptotic behavior of the first integral of (2.35) is obtained by replacing C_I^* by its asymptotic representation (2.40). (This step is readily justified. By a change of variables (cf. 2.25), the first integral of (2.35) can be written as $\int_0^t h(t-s)C_I^*(s)ds$. Now $h(u)$ rapidly approaches 0 as u gets large. Thus $h(t-s)$ is small for large t and small s. Therefore, most of the contribution to this integral comes from where $h(t-s)$ is relatively large, i.e. from where $t-s$ is small. If t is large and $t-s$ is small, then s must be large. And for large s, $C_I^*(s)$ behaves like $\mu s/V_T + a$. A negligible error is therefore committed by using this asymptotic form for C_I^*. The direct evaluation of the resulting expression yields (2.42).) Thus the first integral of (2.35) behaves like

$$\frac{\mu t}{V_T}\int_0^t h(s)ds - \frac{\mu}{V_T}\int_0^t sh(s)ds + a\int_0^t h(s)ds. \tag{2.42}$$

Since h decays very rapidly for large time, an insignificant error is committed by replacing the upper limit of integration t by ∞. The use of (2.10) permits the resulting expression to be written as

$$\frac{\mu t}{V_T} - \frac{\mu \bar{t}}{V_T} + a, \tag{2.43}$$

where $\bar{t}$ is the mean transit time:

$$\bar{t} = \int_0^\infty sh(s)ds. \tag{2.44}$$

(Compare comments following 2.17). The long-time asymptotic behavior of C_E is obtained by using (2.43) in (2.35):

$$C_E(t) \sim \frac{\mu}{Q} + \frac{\mu t}{V_T} - \frac{\mu \bar{t}}{V_T} + a. \tag{2.45}$$

The subtraction of (2.40) from (2.45) eliminates the unwanted constant a and yields

$$C_E(t) - C_I^*(t) \sim \frac{\mu}{Q} - \frac{\mu \bar{t}}{V_T}. \tag{2.46}$$

Since C_E and C_I^* are easily monitored, (2.46) can be used to find the unknown right side in terms of the known data on the left. Fig. 2.6 shows how this should be done. Since V_T can be determined by the independent calculation described above, there are only two unknowns on the right side of (2.46). These are Q and $\bar{t}$. (Note that the container volume V is found from $\bar{t}$ by (2.16).) A second relation to supplement (2.46) is therefore needed so that both unknowns can be found.

For this purpose, a second experiment must be performed in which the indicator is injected at a step rate at a section J outside of the container IE. This is a further specialization of (2.33). D is identified with I and B with J. Let A move counterclockwise about the circuit (Fig. 2.5) until it approaches I. Let the rate of this injection be denoted λ, and let the resulting concentrations be denoted by K. Then (2.33) yields

$$K_I(t) = \int_0^t g(s)K_I(t-s)ds + \frac{\lambda}{Q}\int_0^t h_{JI}(s)ds. \tag{2.47}$$

Next, identify D with I and A with E. Since there is no injection of dye between I and E, (2.33) becomes

$$K_E(t) = \int_0^t h(s)K_I(t-s)ds. \tag{2.48}$$

Equation (2.47) is completely analogous to (2.34) so that the asymptotic behavior of the solution K_I is completely analogous to (2.40), namely

$$K_I(t) \sim \frac{\lambda t}{V_T} + b, \tag{2.49}$$

where b is a constant. As before, the long-time asymptotic behavior of K_E is obtained by substituting (2.49) into (2.48) and then letting the upper limits of integration go to ∞. Then

$$K_E(t) \sim \frac{\lambda t}{V_T} - \frac{\lambda \bar{t}}{V_T} + b, \tag{2.50}$$

whence

$$K_E(t) - K_I(t) \sim -\frac{\lambda \bar{t}}{V_T}. \tag{2.51}$$

This gives an equation for the unknown mean transit time $\bar{t}$. Then (2.46) can be solved for Q. V is then obtained from (2.16), i.e. from $V = Q\bar{t}$. The treatment of (2.51) is the same as that for (2.46) and is described in Fig. 2.6. A prescription has thus been provided for determining flux and volume in the presence of recirculation when extrapolation techniques are inadequate.

CONCLUSION

This chapter was planned to present an easily understandable mathematical basis for dye curve theory, rather than an exhaustive review of the literature on this aspect. The principle papers on which this chapter is based are those of Zierler (1962a, b) and Stephenson (1948). No effort was made to survey the wealth of special results, but a few articles can be specifically cited. In that regard, an alternative treatment for recirculation is offered by Zierler (1962), diffusion is treated by Harris and Newman (1970) and by Grodins (1962), and an abstraction of the mathematical models is given by Gonzáles-Fernández (1962). The renewal equation and its asymptotic properties are discussed in Bellman and Cooke (1963). A simple but careful treatment of the Dirac delta is presented by Stakgold (1967).

REFERENCES

Bellman, R., and K. Cooke. 1963. Differential-Difference Equations. Academic Press, New York.

González-Fernández, J. M. 1962. Theory of the measurement of the dispersion of an indicator in indicator-dilution studies. Circ. Res. 10:409–428.

Grodins, F. S. 1962. Basic concepts in the determination of vascular volumes by indicator-dilution methods. Circ. Res. 10:429–446.

Harris, T. R., and E. V. Newman. 1970. An analysis of mathematical models of circulatory indicator-dilution curves. J. Appl. Physiol. 28: 840–850.

Stakgold, I. 1967. Boundary Value Problems of Mathematical Physics, Vol. 1. The MacMillan Company, New York.

Stephenson, J. L. 1948. Theory of the measurement of blood flow by the dilution of an indicator. Bull. Math. Biophys. 10:117–121.

Zierler, K. L. 1962a. Circulation times and the theory of indicator-dilution methods for determining blood flow and volume. *In* Handbook of Physiology, Vol. 1, p. 585–615. American Physiological Society, Washington, D.C.

Zierler, K. L. 1962b. Theoretical basis of indicator-dilution methods for measuring flow and volume. Circ. Res. 10:393–407.

chapter 3/ A Method for Performing an Indicator-Dilution Curve to Measure Cardiac Output

Dennis A. Bloomfield
Division of Cardiology/Maimonides Medical Center/
Brooklyn, New York

There are numerous methods of introducing indicator into the circulation and withdrawing diluted samples from the blood to record dilution curves and to calculate cardiac output. A comparative analysis of these methods is presented in succeeding chapters, each devoted to a single facet of indicator-dilution performance. However, as a practical guide to the method of recording indicator curves, one such procedure is presented here in detail. This technique has the virtue of simplicity without compromising accuracy, and its use is standard practice in the Cardiac Catheterization Laboratory of Maimonides Medical Center.

EQUIPMENT, INSTRUMENTS, AND MATERIALS

Prerequisites to performing and recording an indicator-dilution curve include the appropriate equipment and instruments, materials, and routes of access to two sites in the vascular system.

A. Equipment and instruments
   1. Cuvette densitometer
   2. Blood-withdrawal pump
   3. Analog recorder
   4. Injection syringe, reproducible volume type (Cornwall; Becton, Dickinson & Co., Rutherford, New Jersey)
   5. 100-cc beaker (2)
   6. 30-cc glass syringe
   7. 20-cc syringe, plastic or glass (2) with 18-gauge needle (2)
   8. Three-way stopcock
   9. Connecting tubing
B. Materials
   1. Cardiogreen (Hynson, Westcott and Dunning, Inc., Baltimore, Maryland)
   2. Heparin, 50 mg/cc (0.5 cc)
   3. Normal saline

The Cardiogreen is dissolved in the total volume of solute provided in the package, as recommended by the manufacturer, by using a 20-cc syringe and needle. This solution is prepared in advance of requirement, since mixing invariably causes frothing in the bottle. The froth must settle and disappear before use if the concentration of the solution is to remain constant throughout the study. However, because Cardiogreen so prepared is mildly unstable in light and contains no bacteriostatic properties, it should be used within 1 hour or so of preparation and as close as possible to calibration. When not being used, the rubber top of the bottle is covered with an alcohol swab. Heparin (0.5 cc) is placed in the beaker selected for collection of withdrawn blood.

The withdrawal equipment is set up as shown in Fig. 3.1.

The injection syringe is set to a volume roughly equivalent to 2 cc more than the dead space of the catheter through which the indicator is to be injected into the circulation. Table 3.1 indicates the approximate dead-space volumes for most cardiac catheters used for introduction of indicator.

CATHETERIZATION SITES

Simultaneous access to two sites in the vascular system is necessary when using nondiffusable indicators. The sites must be separated by at least one "mixing chamber." This criterion is met by the capillary

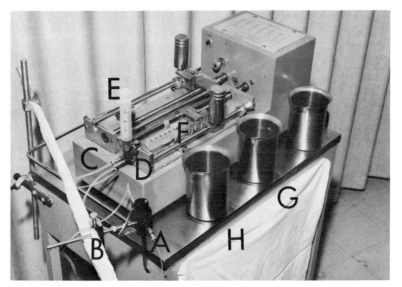

Figure 3.1 The cuvette (A), which, for use with human subjects is gas-sterilized, is attached to a universal-position frame covered by a sterile sleeve drape (B) allowing its close apposition to the blood withdrawal site. The densitometer is connected by a large bore (PE190) tube (C) to a three-way stopcock (D) mounted on the 30-cc withdrawal syringe (F) in the withdrawal pump. To the side arm of the stopcock is connected the hand-operated waste syringe (E). Beakers for waste (G) and collected blood samples (H) stand beside the withdrawal pump assembly on the wheeled trolley. The withdrawal pump is set to speed 1, which, with the 30-cc syringe, affords a withdrawal rate of approximately 25 cc per minute.

bed of the lungs, the right side of the heart, or either the left atrium or left ventricle. Usually, the withdrawal site is a peripheral artery (brachial or femoral), whereas the injection site may commonly be the pulmonary artery or either left-sided chamber. To allow rapid introduction of dye, the injection catheter must be a size 6 French or greater, whereas the withdrawal site must provide a bore of an 18-gauge needle or larger, to allow adequate withdrawal rates.

A very satisfactory femoral arterial withdrawal site is obtained with a Longdwel (Becton, Dickinson and Co., Rutherford, N.J.) Teflon needle introduced by the Seldinger technique. The needle provides, as well, minimal vessel wall damage, stability with patient movement, and the versatility to be replaced by a regular catheter for retrograde aortic catheterization.

Table 3.1. Catheter internal volume chart

Size	Length	Description	Approximate volume
	cm		*ml*
5F	125	Cournand (standard wall)	0.5
6F	125	Cournand (standard wall)	0.8
7F	125	Cournand (standard wall)	1.3
8F	125	Cournand (standard wall)	2.0
5F–8F	125	Goodale-Lubin (standard wall)	(same as above)
5F	125	Lehman (thin wall)	0.7
6F	125	Lehman (thin wall)	1.3
7F	125	Lehman (thin wall)	2.1
8F	125	Lehman (thin wall)	3.0
5F–8F	125	NIH (thin wall)	(same as above)
6F	125	Eppendorf (thin wall)	1.3
7F	125	Eppendorf (thin wall)	2.1
8F	125	Eppendorf (thin wall)	3.0
5F	100	Lehman ventriculography (thin wall)	0.5
6F	100	Lehman ventriculography (thin wall)	1.1
7F	100	Lehman ventriculography (thin wall)	1.7
8F	100	Lehman ventriculography (thin wall)	2.4
8.5F	69	Brockenbrough transseptal adult left heart Teflon catheter	1.6
7F	55	Brockenbrough transseptal child's left heart Teflon catheter	0.8
8.5F	70	Brockenbrough transseptal adult right heart Teflon catheter	1.6
7F	55	Brockenbrough transseptal child's right heart Teflon catheter	0.8
7F	100	Judkins right coronary	1.2
8F	100	Judkins right coronary	1.7
7F	110	Pigtail	1.2
8F	110	Pigtail	1.8
5F	110	Swan Ganz	1.1

METHOD

To ensure amplifier stability, both densitometer and recorder are switched on at least 30 minutes before inscription of the curve.

The densitometer output is connected to a DC input on the recorder, and, with densitometer balance and sensitivity controls set at zero, the recorder channel output signal is positioned to a suitable base line. The appropriate sensitivity of the conjoined densitometer-DC amplifier channel is selected, and the recording paper speed is set at 10 mm/second and with 1-second time markings.

All of the prerequisites are now satisfied, and the stage is set for inscription of the curve. This requires simultaneous maneuvers at four stations: injection, withdrawal, densitometer, and recorder. For greater understanding of this coordination, the sequence of steps is presented in Table 3.2.

Step 1

Injection
The injection catheter must be well flushed at all times to minimize any clot formation within the lumen. The calibration technique to be described with this method requires exactly reproducing the indicator volume injected, and any variation in the intraluminal volume from clot formation between curve inscription and calibration will be manifested as inaccuracy in the cardiac output calculation.

Attachment of the filled injection syringe to the catheter is the last act of Step 1.

Withdrawal
The short connecting tube from the densitometer is attached to the withdrawal catheter or needle, and freedom of flow through the system is checked by withdrawing blood manually into the vertical syringe. There should be no resistance, and the plunger should actually rise from the transmitted intra-arterial blood pressure alone. A few seconds of suction, however, will reveal any loose connections or punctured tubes by air bubbles entering the system.

All of the saline or air previously filling the connecting tubing is withdrawn into the vertical syringe, and the withdrawal is stopped as soon as this is achieved to minimize blood wastage.

Densitometer
The hand withdrawal and filling of the densitometer cuvette with blood is utilized to reset a base line zero on this instrument. This setting is influenced by the velocity of withdrawal through the cuvette

Table 3.2. Sequence and coordination of dye curve inscription

Steps	Injection	Withdrawal	Densitometer	Recorder
Preliminary steps	Mix indicator Fill injection syringe	Assemble syringe system	Balance zero Select sensitivity	Select base line Set paper speed
Step 1	Flush injection catheter Attach injection syringe	Attach withdrawal line Hand-withdraw to fill lines with blood	Set zero	—[a]
Step 2	—	Run withdrawal pump	Reset zero	Run recorder
Step 3	Inject dye	—	—	—
Step 4	—	Stop pump	—	Stop recorder
Step 5	Detach syringe Flush catheter	Empty syringe Refill lines with saline	—	Note sensitivity Note injection and withdrawal sites

[a]It is essential that no manipulation or readjustment be made at the indicated stations during steps designated by a dash.

and can only be accurately selected when the withdrawal pump is providing a uniform and sufficiently rapid flow. However, in the interests of blood conservation, it is only necessary, at this step, to set the base line roughly with blood drawn through the densitometer with the hand withdrawal.

The sequence of Steps 2 to 4 follow each other closely, and the operation flows more smoothly and efficiently if under the control of one coordinator. This office is best served by the densitometer operator whose responsibility it is to make the final adjustment before the inscription of the curve. This coordinator gives the signals to "withdraw" blood with the pump in Step 2, to "inject" the indicator in Step 3, and to "stop" the withdrawal pump after the curve is recorded in Step 4.

Step 2

Withdrawal
On instruction from the coordinator, the withdrawal pump is started, and the operator observes that the first few milliliters of blood fill freely into the withdrawal syringe. This satisfactory condition of the withdrawal system is reported to the coordinator, and the action proceeds. The operator continues to observe the satisfactory smooth withdrawal until Step 4.

If withdrawal is not free or air bubbles are observed, the withdrawal pump is stopped, and the curve inscription is suspended until the fault is located and rectified.

Densitometer
As soon as satisfactory withdrawal is reported, the coordinator makes the final base line adjustment to the densitometer and instructs that the recorder be started.

Recorder
The recorder is started on instruction and, if required, final base line adjustment is made. The monitor screen or write-out is displayed in a manner in which the coordinator may observe the curve.

Step 3

Injection
As soon as the coordinator is satisfied that the base line is stable (not rising or falling) and artifact-free, instruction is given to inject the indicator. The indicator syringe plunger is smoothly and quickly depressed and maintained in the fully depressed position until the

record is terminated. It is usual to record the beginning and duration of the injection with a marker activated simultaneously with the injection.

Step 4

Withdrawal and Recorder
When the coordinator has observed satisfactory recording of the curve, instructions to stop the withdrawal pump and recorder are given. Unless special interest is attached to the later phases of recirculation, the termination of the curve recording is made when the curve amplitude has diminished to between one-third and one-quarter of the peak amplitude.

Step 5

Injection
The injection syringe is detached, and the injection catheter is flushed. The injection syringe should not be flushed if further dye curves are anticipated.

Withdrawal
The vertical waste syringe is emptied into the waste beaker, and the blood in the withdrawal syringe is transferred, via the vertical syringe, to the heparinized beaker. The volume of blood is noted. If further curves are to be inscribed within 1 minute or so, blood may be left in the connecting tubes and cuvette. If a greater time interval is contemplated, the withdrawal system is disconnected from the withdrawal site and flushed by drawing saline through the system into the waste syringe.

Recorder
A record is made of the sensitivity at which the curve was recorded and of the sites of injection and sampling.

After the procedure, the injection syringe (its volume setting unchanged), the injection catheter, the bottle containing the remaining indicator solution, and the heparinized withdrawn blood are recovered for use in calibration.

SPLITTING THE CURVE

Even with considerable experience, an indicator-dilution team may be unable to estimate, with any degree of certainty, the sensitivity settings on their equipment which will enable the curve to be recorded with optimal amplitude in any given patient. To reduce this uncertainty, to increase the percentage of "usable" curves, and to minimize the blood loss from the patient, a simple signal-splitting patch cord is recommended (Fig. 3.2). The output from the densitometer is connected in parallel to patch cords communicating with two DC

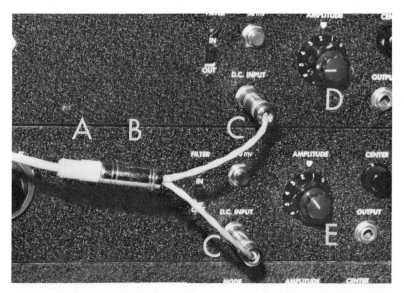

Figure 3.2 Splitting the signal. The signal from the densitometer at the terminals of male jack (A) is connected in parallel to two patch cords in female jack (B). Two further jacks (C) transmit the identical signals to inputs on two DC amplifiers (D and E) set at different sensitivities.

amplifiers set at different sensitivities, one somewhat higher and the other somewhat lower than the estimated optimal sensitivity. Curves with peaks too high to be recorded or too low to be accurately quantitated are thereby endowed with a second chance of utilization. Calibration is performed with the same double sensitivity connections (Fig. 3.3).

MARKING INJECTION TIME

When it is important to know the time interval between injection of the indicator and observation of the initial segment of the curve (the

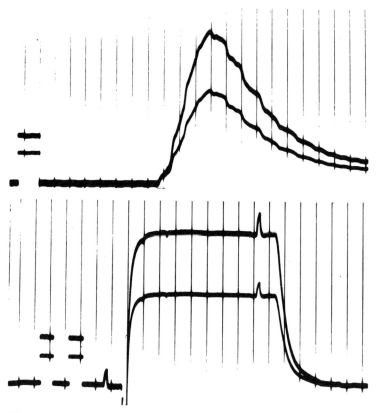

Figure 3.3 The *upper panel* shows a single dilution curve recorded simultaneously at two sensitivities with the signal splitting patch cord. The *lower panel* shows the calibration measurement similarly recorded. The step-function markings at the *left* of each recording are the sensitivity confirmation signals. The equal deflections on *upper* and *lower panels* indicate that the calibration has been recorded at the same sensitivity as the dilution curve.

appearance time), it is necessary to record the instant and duration of the injection. This is usually provided by a footswitch marker depressed by the injection operator at the start of the injection and released at the end of the injection.

The densitometer amplifiers in use at Maimonides Medical Center have been modified to incorporate a variable 2K ohm resistance which can be superimposed momentarily on the base line signal, providing a square-wave deflection used for confirming sensitivity settings. The instant of the return of this calibrating signal to base line is used as the instant of beginning the indicator injection, obviating the need for a separate marker system (Fig. 3.3).

CALCULATION OF CURVE AREA

The estimating of cardiac output requires the calculation of the area encompassed between the base line and the dilution curve. A resumé of methods of calculating curve area are presented in Chapter 5, and all of the methods discussed are suitable for use with this technique of recording the dilution curve.

CALIBRATION

The curve calibration designed for this method of curve recording is designated the "Vanderbilt" technique and is treated in detail in Chapter 6.

ARTIFACTS IN RECORDING DILUTION CURVES

A number of identifiable artifacts are capable of interfering with the satisfactory recording and interpretation of dilution curves. Since the majority of them become apparent during the inscription of the base line, rapid recognition and correction minimizes wastage of blood, indicator, and time, and increases the yield of meaningful data. These artifacts, together with their cause and remedy, are illustrated and identified in Table 3.3.

Table 3.3. Artifacts in dilution curves: causes and correction

Problem	Diagnosis	Cause	Correction
	Air leak proximal to cuvette	1. Loose connection 2. Punctured or cut tubing	Tighten or replace
Bubbles in syringe	Air leak distal to cuvette	1. Loose connection 2. Punctured or cut tubing	Tighten or replace
	Pulsatile flow through densitometer	1. Air in syringe 2. Plunger loose in holder 3. Slow withdrawal rate 4. Very high phasic pressure shift at the withdrawal site (e.g. arterial site in aortic insufficiency, left ventricular site in aortic stenosis)	1. Empty syringe into waste 2. Pack the plunger-holder gap with paper or wood wedge (match stick) 3. Increase withdrawal rate 4. Increase withdrawal rate
	Nonuniform withdrawal rate (N.B. difference from combined right-to-left and left-to-right shunting)	1. Partly obstructed withdrawal needle or catheter 2. "Stuck" withdrawal syringe 3. Slipping clutch on withdrawal pump	1. Flush or reposition 2. Wash withdrawal syringe 3. Tighten clutch

Curve	Appearance	Cause	Remedy
	Very slow but smooth rise and descent— recirculation contamination		Reduce injection-to-sampling path length
	Irregular time-interval lines	1. Variation in recorder paper speed 2. End of paper roll	1. Check paper drive 2. Replace paper
	Drifting (usually falling) base line Descending limb approaches or passes paper base line	1. Densitometer not warmed up 2. Nonreplacement of wedge filter (Gilford densitometer) when flushing cuvette with saline	1. Warm up densitometer (30 minutes from cold) 2. Replace wedge filter
	Rapid, short dip in base line before curve	Excess flushing saline	Reduce volume of flush
	Reversed polarity of curve	Reversed output terminals from densitometer to recorder	Reverse output terminals
	Flat-topped curve	Saturation of cuvette	Reduce amount of injected dye

chapter 4/ Indicator Injection and Sampling and the Problem of Mixing

Dennis A. Bloomfield
Division of Cardiology/Maimonides Medical Center/
Brooklyn, New York

In 1921, Stewart described two practical methods for the measurement of fluid flow in a system of tubes. In the first method, an indicator is injected at a constant rate into the fluid flowing through the tubes, and the concentration of indicator at a distal sampling site is determined when an equilibrium concentration is obtained. In the second method, the total amount of indicator is injected as rapidly as possible, and its mean concentration is determined at the sampling site. Provided that the indicator mixes uniformly with the fluid in the tubes and, additionally for the second method, that the flow is constant during the sampling procedure, both methods give accurate and comparable values for volume of the flow.

The adaption of these methods to present day technology, their limitations, and the conditions which must be met for validity are discussed in this chapter.

INJECTION

There are a number of general requirements regarding the injection of indicator and, although they have been considered in Chapter 2, they will be restated here because of their fundamental nature. To fulfill the conditions of the indicator-dilution theory, indicator must mix completely and uniformly with the native fluid, if not immediately upon injection, at least before sampling. The addition of indicator to the native fluid should not physiologically alter the flow or volume of the native fluid. For the calculation of flow, the exact amount of indicator injected (or amount per unit time for constant injection) must be known before or measurable after delivery. For curves with minimal recirculation artifact after bolus injection, all of the indicator should arrive at the central mixing chamber over the shortest possible time duration.

The practical points for achieving these requirements are described for the numerous bolus and continuous injection methods.

BOLUS (SUDDEN) INJECTION

The most common method of introducing indicator into the circulation to measure cardiac output is by a sudden, complete (or bolus) injection. The various ways in which this is achieved differ only in the methods by which the amount of delivered indicator can be measured before or after injection.

Nonflush Methods

A simple technique has been briefly described in the preceding chapter. A reproducible-volume syringe containing a roughly calculated amount of indicator is attached directly to the injection catheter. (Although the exact amount of indicator is not known, it can be exactly reproduced by refilling the syringe.) The syringe is emptied by rapid injection into the saline or blood-filled catheter and not followed by any flush. The important requirements for using this technique are the following.

1. The syringe is held closed (empty) after the injection or is disconnected from the catheter lumen by a stopcock, preventing slight withdrawal and reinjection of dye contained in the lumen of the catheter from introducing the error of a second injection of an unknown amount of indicator. This further ensures that the syringe is not contaminated with blood or saline which may alter the concentration or light transmission properties of the dye subsequently drawn into the syringe for repeated determinations or calibration.

2. The injection catheter is meticulously flushed and cleaned of indicator before subsequent injections. Any indicator remaining in the catheter or its stopcock will be introduced into the circulation on the next injection, adding an error of unknown magnitude to the amount of indicator.

3. The volume set roughly on the syringe is determined by the sum of the actual amount of indicator intended to be delivered into the circulation and the internal volume of the injecting catheter.

Table 3.1 in Chapter 3 lists the internal volumes of the more commonly used catheters.

The information in this table is also necessary for determining the "flush" volume in other methods of bolus injection. A variation of this method requires completely filling the catheter with indicator before injection. An accurately calibrated or reproducible volume syringe, also filled with indicator, is attached to the catheter, and a volume equal to that in the syringe is injected by displacement. No compensation for, or knowledge of, the catheter dead space is necessary for calibration.

Disadvantages of these methods relate to the accuracy of assessing the delivered volume of indicator. The ability to reproduce a volume in these syringes depends on careful filling technique and the elimination of air bubbles, a problem common to all methods of indicator dilution. The effect of loss of indicator from the catheter tip after the injection is minimized by small holes at the catheter tip (this also aids mixing of the injectate) and the use of a "delivered volume" which is large in relation to the internal volume of the catheter. This may be achieved by reducing the concentration of the indicator.

Flush Methods

These methods of injection are performed by introducing into the injection catheter an accurately known mass of indicator and ensuring its total entry into the circulation by immediately flushing the catheter with saline or blood. There are various techniques of increasing sophistication for performing these methods of indicator injection. The simplest is illustrated in Fig. 4.1.

The smaller syringe is loaded with an accurately known amount of dye; the flush syringe is loaded with saline. The injection of dye is followed as quickly as possible by the saline flush of between 5 and 10 cc, depending on the dead space of the catheter system. To obtain a single bolus injection, two persons need to combine operations, one working each syringe.

A second system, utilizing an open-ended, dye-filled tube with the flushing syringe in series is illustrated in Fig. 4.2. The tube is com-

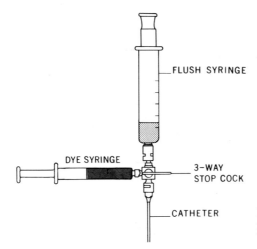

FLUSH SYRINGE

DYE SYRINGE

3-WAY
STOP COCK

CATHETER

Figure 4.1 A simple system
for the "flush" method of
bolus injection.

pletely filled with indicator, which is then contained by the stopcock.
The injection is made by attaching the flushing syringe to the open
end of the dye tube. With the stopcock open to the catheter, the
syringe is depressed, delivering an appropriate amount of saline
which clears the dye from both the tube and the catheter. The
amount of dye injected is found by accurately (pre-) measuring the
internal volume of the dye tube. The advantage of this system is that
it guarantees instantaneous flushing of dye. The disadvantage centers
on the necessity of meticulously clearing the saline from the tube
before refilling with dye for subsequent injections; this can best be
achieved by flushing through with dye. Only then can the injected
amount of dye be accurately reproduced. More sophisticated systems
can be devised, using reservoirs of flushing and indicator solutions,
connected via one-way valves to reproducible-volume syringes. The
ease of repeatability with these systems is outweighed by the complexity of construction, the potential errors in accuracy arising from

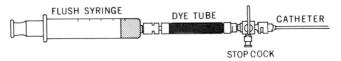

FLUSH SYRINGE   DYE TUBE   CATHETER

STOP COCK

Figure 4.2 A system for the "flush" method of bolus injection requiring a
single operator and providing a single indicator concentration peak.

the greater tubing lengths and from possible component failure in the one-way valves.

The flushing method of injection can be modified for indicator delivery in predetermined phases of the cardiac cycle by replacing the flushing syringe in the second system with an electrocardiographically programmed pressure injector. Injection systems which are fully automatic, remote controlled (Rosenhamer, 1968), and programmed (Hansen and Pace, 1962) have been described. These systems automatically mark the time and duration of the injection, which is important in the interpretation of curves in congenital heart disease and in the measurement of mean transit time for regional blood volume determination. Benson and co-workers (1964) have described a hand-operated syringe with electronic marking of injection. In less automated systems, a footswitch marker is activated by the operator performing the injection. The mark is continued to the end of the injection, and the midpoint of the mark is considered the actual moment of injection.

A single exception to the practice of making bolus injections as close to instantaneous as possible arises in the method of quantitating aortic valvular insufficiency by simultaneous upstream and downstream sampling, as described in detail in Chapter 9. In this instance, the injection is made over the course of one or more cardiac cycles (Armelin *et al.*, 1963).

CUMULATIVE (CONTINUOUS) INFUSION

The injection of indicator at a constant rate into the blood stream was the first method described by Stewart in 1897 to estimate cardiac output. The technique is simple, and the results can be quickly calculated. The only measurement required is the dye concentration at equilibrium, and the only assumptions that have to be made are that uniform mixing occurs and that, at equilibrium, the plateau concentration includes no recirculated indicator.

The technique requires a constant, known rate of indicator infusion provided by a constant infusion pump or weight-and-pulley-driven syringe (Shillingford, Bruce, and Gabe, 1962). Special injection catheters have been designed (Kountz, Dempster, and Shillingford, 1964) to provide a fine spray of indicator, considered necessary to enhance mixing at the low flow infusion rates. Fig. 4.3 represents

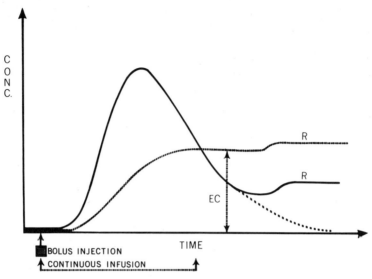

Figure 4.3 A simultaneous comparison of bolus and continuous infusion curves. *EC* indicates the equilibrium concentration or plateau of the continuous-infusion curve, and *R* identifies the recirculation in both methods. The *dotted line* suggests the downslope concentration of the primary circulation of the bolus injection.

a comparison of the two methods by recording simultaneous indicator-dilution curves with different indicators injected at the same point in the pulmonary artery and sampled from the same point in the radial artery. Theoretically, there is no basic difference between bolus and cumulative injection techniques for measuring cardiac output. Numerous authorities have compared the two methods under controlled conditions and found the variability between them to be no greater than the variability of replicate estimations with the bolus technique alone (Rashkind and Morton, 1949; Shepherd, Bowers, and Wood, 1955).

In practice, however, the cumulative infusion method has a number of limitations. The plateau is limited on one side by the rise time (the ascending limb) and on the other by the recirculation time. The rise time characteristics are determined by complex functions of the flow to be measured, the flow of indicator into the stream of blood, the length of the sampling catheter, and the response time of the densitometer. The recirculation time is also characterized by the flow to be measured and the circulatory path length between injection and sampling sites. Coronary circulation is believed to occupy only 8 to 10 seconds. Mean recirculation time in healthy patients at rest and during upright exercise has been measured at 8 to 15 seconds (Sowton *et al.*, 1968). Consequently, from peripheral venous injection sites and arterial sampling sites where build-up time for the constant infusion curve may approach or even exceed 8 seconds, no plateau at all may be recognized. Hamilton and Remington (1947) believed that, only when the site of injection and sampling are such that a "complete passage" of the indicator can occur before recirculation, will continuous infusion utilizing those sites give valid results. The value of this statement depends obviously on how long a duration the "plateau" needs to persist to be clearly identified, and numerous workers have demonstrated valid plateaus and obtained comparable results in experimental and clinical work (Holt, 1944; Rashkind and Morton, 1949). Howard, Hamilton, and Dow (1953) reinvestigated the validity of plateaus and found that they were not frequently encountered. On the other hand, "spurious plateaus," unrelated to any valid estimate of flow were frequently observed, and one cause was shown to be fluctuation in venous inflow related to respiration. Constant infusion methods have been used to measure regional flows (Andres *et al.*, 1954; Grace *et al.*, 1957; Kountz *et al.*, 1964) and valvular insufficiency (Frank *et al.*, 1966), but are not in common use for the measurement of cardiac output.

INJECTION SITE

The influence of the injection site on the dilution curve has been extensively studied. Hetzel, Swan, and Wood (1954) compared curves drawn from peripheral venous and pulmonary artery injection sites and noted delay in appearance, prolonged passage, lowered maximal concentration, and reduced slope of the logarithm of the declining concentration from the peripheral site. Thorburn (1961) described changes in curve symmetry which are more or less important to cardiac output measurement depending on the method used to calculate curve area (Chapter 5).

Left ventricular and pulmonary arterial injection sites have been compared in determination of cardiac output (Shepherd, Higgs, and Glancy, 1972). The two sites gave nearly identical and equally reproducible values. Physiologically identical outputs were obtained from left atrial and pulmonary artery injection sites (Samet, Bernstein, and Medow, 1965). In normal subjects, no significant difference between cardiac output from left ventricular and aortic root injection was observed (Rahimtoola and Swan, 1965). However, in the dog, injection sites beyond the aortic root produced progressive differences in paired output determinations (Krovetz and Benson, 1965).

Newman *et al.* (1951) postulated that the slope of the logarithm of the declining concentration of indicator is determined entirely by the washout of indicator from the largest volume in the central vascular system, that is, the lungs. These workers felt that the slope was independent of the site of injection, but this argument was based on the assumption of instantaneous injection which, in relation to the central mixing chambers, is achieved less frequently as the injection becomes more peripheral.

In all indicator-dilution methods, the errors of greatest magnitude arise from the inability to separate the curve of the first passage of dye from the contamination by recirculated dye. The recirculation time is a function of the circulation to be measured and wholly independent of sites of indicator injection or sampling. On the other hand, longitudinal dispersion of indicator is in part determined by the distance of the circulatory path between injection and sampling sites. At one extreme with peripheral venous and distal arterial sites (where the dye must traverse the whole circulatory pathway except for the systemic capillary bed), the time course of a curve resulting from a slow injection may be nearly equal to total body recirculation time. Recirculated dye will be present during the inscription of the descending limb of the curve, invalidating the measurement of cardiac output. At the other (optimal) end of the scale, with the injection and sampling sites moved centrally and separated only by a mixing chamber, complete passage of the dye past the sampling site may occur in a shorter interval than the shortest recirculation pathway. In practice, this is achieved by selecting injection sampling site pairs in the combinations of pulmonary artery and left atrium, left atrium and aorta, or, if flow is not greatly depressed, pulmonary artery and systemic artery.

In shock, with very low cardiac outputs, the difficulties and invalidating factors noted in the above paragraph are exaggerated. There is no obvious recirculation peak; semilogarithmic extrapolation of the downslope does not exclude recirculated indicator which has merged with the primary curve. The area is always overestimated, and the flow is always underestimated very significantly. In these circumstances, the remedy detailed above, i.e. the movement of injection and sampling sites closer together, is essential. Theye and Kirklin (1963) used this approach postoperatively, leaving a left atrial catheter in place (for injection) after repair of septal defects. Because of higher flow and shorter transit times, the femoral artery is preferred to the brachial artery for sampling. The practical aspects of providing short, low-volume sampling systems with high sampling flow rates to increase curve accuracy in shock has been discussed by Bassingthwaite, Sturm, and Wood (1970).

In special cases, where other information besides cardiac output is required, special sites are selected to define abnormal anatomic pathways or measure valvular incompetence, as detailed in Chapters 8 and 9.

MIXING

A basic assumption of classical dilution theory is that complete mixing of the indicator occurs with the native fluid into which it is injected, but it is universally accepted that this theoretical goal is never reached in man. Much work has been undertaken to define the extent of "non-mixing." Silver, Kirklin, and Wood (1956) demonstrated the preferential flow of blood from the inferior vena cava and the right pulmonary veins through experimental atrial septal defects, and further evidence of streaming in the right heart and left atrium in congenital defects in man is cited by Rowe (1962). Maseri and Enson (1968) confirmed the existence of systematic regional differences in blood turnover in the right ventricle and demonstrated sequential emptying of this chamber with a significant fraction of the stroke volume passing through the ventricle without mixing, or only partial mixing, with the residual volume. However, flow can be estimated by indicator-dilution methods with "reasonable" accuracy, even in the absence of a localized "perfect" mixing site between injection and sampling provided that the indicator attains a "sufficiently" homogeneous dispersion in blood by the time it reaches the sampling point (Andres et al., 1954). That these circumstances can be attained on the right side of the heart is attested to by the close agreement between flow values from atrial injection-pulmonary artery sampling dilution curves and those obtained with reference techniques (Fritts *et al.,* 1957; Fox and Wood, 1957; Hamlin *et al.,* 1962).

Complete mixing of blood and injected saline during one cycle in the left ventricle is rare (Irisawa, Wilson, and Rushmer, 1960). Mixing can be demonstrated across the greater part of the diameter in the ascending aorta and proximal part of the arch, but support for this point of view (Peterson *et al.,* 1954) is challenged by Pavek, Pavek, and Boska (1970). Grace *et al.* (1957), using high energy jet injection of indicator, obtained evidence of uniform mixing in the thoracic aorta, but Freis and Heath (1964) suggested a tendency to streamlining in the descending segment. Pavek *et al.* (1970), using multiple thermistors in the arch of the aorta, showed a temperature fluctuation across the vessel in the range of $\pm 13.3\%$ (S.D.) when indicator was injected into the left ventricle. The cross-mixing was further improved to $\pm 6.1\%$ (S.D.) when the indicator was infused into the left atrium. Neither the method of injection into the left atrium, the addition of the mixing and dispersion effect of the right heart and pulmonary circulation (from a right atrial injection), nor the alteration of heart rate significantly improved this mixing.

Within the framework of these requirements, it is well established that cardiac output determined from a right atrial injection shows no physiological difference, whether sampled from the pulmonary artery or a systemic artery. Alterations in the morphology of the curves, however, are pronounced and are detailed in Chapter 5. Shepherd *et al.* (1972) demonstrated that mixing adequate for reproducible cardiac output determinations occurred by the time the indicator reached the innominate artery after left ventricular injection.

SAMPLING

The theoretical requirements of a sampling system are that it allows "representative" sampling and minimizes distortion of the sampled concentration of indicator with respect to time.

The requirement of representative or uniformly mixed sampling is met by withdrawing dyed blood from as near to the site of concentration change as possible (yet far enough away to insure completion of the concentration change), and by insuring that the withdrawing catheter or needle is not sampling from an area of nonrepresentative laminar flow (such as against a vessel wall or beside a tributary).

The second requirement can be met by reducing the internal volume of the sampling system, while increasing its internal cross-sectional area and flow rate. These factors are further discussed in Chapter 7.

The practical necessity of conservation of patients' blood may dictate compromise in the above optimal conditions. A minimal flow rate of withdrawn blood will be required by the characteristics of the densitometer. This figure ranges from approximately 10 to 20 ml/minute with the densitometers commercially available, and withdrawal rates should not exceed a small margin above the densitometer flow required for instrument stability. The total volume of blood removed during a study on one patient can be reduced by pooling the blood withdrawn during curve inscription and using that, rather than pre-injection blank blood, for calibration purposes.

WITHDRAWAL SYSTEMS

Withdrawal needles smaller than 19 gauge and withdrawal catheters 70 to 100 cm in length and less than French size 7 are, in some or most cases, unable to provide optimal withdrawal flows. Withdrawal pumps are for the most part constant-volume syringe-reservoir instruments. Roller pumps without reservoirs, to sample and reinfuse blood, have been designed specially for pediatric work, where blood volume depletion is a serious handicap to widespread indicator-dilution use. Specific details of the commercially available pumps are provided in Chapter 20. All of the systems feature load-independent, constant-speed motors to provide steady withdrawal rates. Sampling at constant rates (time averaging) rather than the more correct rates proportional to the instantaneous flow past the sampling site (volume averaging) produce inherent errors under conditions of pulsatile flow. Hand withdrawal is not used.

MIXING AND SAMPLING ERRORS

The indicator-dilution method gives considerable variation in flow measurements ($\pm 20\%$) when compared to the Fick method (Smulyan, 1961, 1962) and to electromagnetic flow metering (Hamilton et al., 1967). The dilution technique itself shows variation in man up to $\pm 12\%$ when estimated simultaneously from two different arteries (Sleeper et al., 1962). These variations occur when the requirements for validity of "complete mixing" and "representative sampling," the classical assumptions of the theory, cannot be met in practice. Ideally, an accurate description of the concentration of indicator could be recorded if either the indicator is distributed proportionally to the volume flow rate or the rate of sampling is proportional to the instantaneous flow at the sampling site. Mixing errors occur when the flux of indicator across a cross-section of a vascular segment is not proportional to the blood flow through each element of the cross-section. Sampling errors occur when the observations of concentration at the sampling site are not in proportion to the flow past the sampling site. Variations in flow measurements detailed at the beginning of this section are inherent in the practice of dye dilution in man, because complete mixing is rarely achieved and sampling is time proportional rather than flow proportional. Pavek *et al.* (1970) and Bassingthwaighte, Knopp, and Anderson (1970) have shown that these errors are more profound with bolus injection methods than with cumulative infusion methods. Despite these observations, the difficulties of recovering first passage information from recirculation has rendered the constant-infusion technique of indicator dilution less popular.

REFERENCES

Andres, R., K. L. Zierler, H. M. Anderson, W. N. Stainsby, G. Cader, A. S. Ghrayyib, and J. L. Lilienthal, Jr. 1954. Measurement of blood flow and volume in the forearm of man: with notes on the theory of indicator-dilution and on the production of turbulence, hemolysis and vasodilation by intra-vascular injection. J. Clin. Invest. 33:482–504.

Armelin, E., L. Michaels, H. W. Marshall, D. E. Donald, R. J. Cheesman, and E. H. Wood. 1963. Detection and measurement of experimentally produced aortic regurgitation by means of indicator-dilution curves recorded from the left ventricle. Circ. Res. 12:269–290.

Bassingthwaighte, J. B., T. J. Knopp, and D. U. Anderson. 1970. Flow estimation by indicator dilution (bolus injection): reduction of errors due to time-averaged sampling during unsteady flow. Circ. Res. 27:277–291.

Bassingthwaighte, J. B., R. E. Sturm, and E. H. Wood. 1970. Advances in indicator dilution techniques applicable to studies of the acutely ill patient. Mayo Clin. Proc. 45:563–572.

Benson, R. W., L. J. Krovetz, and G. L. Schiebler. 1964. A new type of syringe holder for performance of indicator dilution curves. J. Appl. Physiol. 19:1022–1023.

Fox, I. J., and E. H. Wood. 1957. Applications of dilution curves recorded from the right side of the heart or venous circulation with the aid of a new indicator dye. Proc. Staff Meet. Mayo Clin. 32:541–550.

Frank, M. J., P. Casanegra, M. Nadimi, A. J. Migliori, and G. E. Levinson. 1966. Measurement of aortic regurgitation by upstream sampling with continuous infusion of indicator. Circulation 33:545–557.

Freis, E. D., and W. C. Heath. 1964. Hydrodynamics of aortic blood flow. Circ. Res. 14:105–116.

Fritts, H. W., Jr., P. Harris, C. A. Chidsey III, R. H. Clauss, and A. Cournand. 1957. Validation of a method for measuring the output of the right ventricle in man by inscription of dye-dilution curves from the pulmonary artery. J. Appl. Physiol. 11:362–364.

Grace, J. B., I. J. Fox, F. S. Rodich, and E. H. Wood. 1957. Thoracic aorta flow in man. J. Appl. Physiol. 11:405–418.

Hamilton, F. N., J. C. Minzel, and R. M. Schlobohm. 1967. Measurement of cardiac output by two methods in dogs. J. Appl. Physiol. 22:362–364.

Hamilton, W. F., and J. W. Remington. 1947. Comparison of the time concentration curves in arterial blood of diffusible and non-diffusible substances when injected at constant rate and when injected instantaneously. Amer. J. Physiol. 148:35–39.

Hamlin, R. L., W. P. Marsland, C. R. Smith, and L. A. Sapirstein. 1962. Fractional distribution of right ventricular output in the lungs of dogs. Circ. Res. 10:763–766.

Hansen, J. T., and N. Pace. 1962. Apparatus for automatic dye dilution measurement of cardiac output. J. Appl. Physiol. 17:163–166.

Hetzel, P. S., H. J. C. Swan, and E. H. Wood. 1954. Influence of injection site on arterial dilution curves of T-1824. J. Appl. Physiol. 7:66–72.

Holt, J. P. 1944. The effect of positive and negative intrathoracic pressure on cardiac output and venous pressure in the dog. Amer. J. Physiol. 142:594–603.

Howard, A. R., W. F. Hamilton, and P. Dow. 1953. Limitations of the continuous infusion method for measuring cardiac output by dye dilution. Amer. J. Physiol. 175:173–177.

Irisawa, H., M. F. Wilson, and R. F. Rushmer. 1960. Left ventricle as a mixing chamber. Circ. Res. 8:183–187.

Kountz, S. L., W. J. Dempster, and J. P. Shillingford. 1964. Application of a constant indicator dilution method to the measurement of local venous flow. Circ. Res. 14:377–386.

Krovetz, L. J., and R. W. Benson. 1965. Mixing of dye and blood in the canine aorta. J. Appl. Physiol. 20:922–926.

Maseri, A., and Y. Enson. 1968. Mixing in the right ventricle and pulmonary artery in man: evaluation of ventricular volume measurements from indicator washout curves. J. Clin. Invest. 47:848–859.

Newman, E. V., M. Merrell, A. Genecin, C. Monge, W. R. Milnor, and W. P. McKeever. 1951. The dye dilution method for describing the central circulation: an analysis of factors shaping the time-concentration curves. Circulation 4:735–746.

Pavek, E., K. Pavek, and D. Boska. 1970. Mixing and observational errors in indicator-dilution studies. J. Appl. Physiol. 28:733–740.

Peterson, L. H., M. Helrich, L. Greene, C. Taylor, and G. Choquette. 1954. Measurement of left ventricular output. J. Appl. Physiol. 7:258–270.

Rahimtoola, S. H., and H. J. C. Swan. 1965. Calculation of cardiac output from indicator dilution curves in the presence of mitral regurgitation. Circulation 31:711–718.

Rashkind, W. T., and J. H. Morton. 1949. Comparison of the constant and instantaneous injection techniques for determining cardiac output. Amer. J. Physiol. 159:389–393.

Rosenhamer, G. 1968. Remote-controlled injection systems for repeated dye dilution studies. Scand. J. Clin. Lab. Invest. 21:347–350.

Rowe, G. G. 1962. The myth of mixing. Amer. Heart J. 63:425–427.

Samet, P., W. H. Bernstein, and A. Medow. 1965. Effect of site of injection upon left ventricular indicator-dilution output. Amer. Heart J. 69:241–244.

Shepherd, J. T., D. Bowers, and E. H. Wood. 1955. Measurement of cardiac output in man by injection of dye at a constant rate into the right ventricle or pulmonary artery. J. Appl. Physiol. 7:629–638.

Shepherd, R. L., L. M. Higgs, and D. L. Glancy. 1972. Comparison of left ventricular and pulmonary arterial injection sites in determination of cardiac output by indicator dilution technique. Chest 62:175–178.

Shillingford, J., T. Bruce, and I. Gabe. 1962. The measurement of segmental venous flow by an indicator dilution method. Brit. Heart J. 24:157–165.

Silver, A. W., J. W. Kirklin, and E. H. Wood. 1956. Demonstration of preferential flow of blood from inferior vena cava and from right pulmonary veins through experimental atrial septal defects in dogs. Circ. Res. 4:413–418.

Sleeper, J. C., H. K. Thompson, Jr., H. D. McIntosh, and R. C. Elston. 1962. Reproducibility of results obtained with indicator-dilution technique for estimating cardiac output in man. Circ. Res. 11:712–720.

Smulyan, H. 1961. Reliability of the indicator-dilution technique. Amer. Heart J. 62:140–141.

Smulyan, H., R. P. Cuddy, and R. Eich. 1962. An evaluation of indicator-dilution technique in the dog. J. Appl. Physiol. 17:729–734.

Sowton, E., D. Bloomfield, N. L. Jones, B. E. Higgs, and E. J. M. Campbell. 1968. Recirculation time during exercise. Cardiovasc. Res. 4: 341–345.

Theye, R. A., and J. W. Kirklin. 1963. Physiologic studies following surgical correction of ventricular septal defect. Circulation 27:530–540.

Thorburn, G. D. 1961. Estimates of cardiac output from forward part of indicator dilution curves. J. Appl. Physiol. 16:891–895.

chapter 5/ The Calculation of Curve Areas

Dennis A. Bloomfield
Division of Cardiology/Maimonides Medical Center/
Brooklyn, New York

Indicator-dilution curves present a great variety of shapes which are dependent not only on the cardiac output, but on the anatomic sites of indicator injection and blood withdrawal, the optical characteristics of the densitometer, the mechanical characteristics of the withdrawal system, and last, but of principal importance, the circulatory pathology. The characteristic distortions to the morphology of the curve in such conditions as cardiac valvular incompetence and intracardiac shunts will be discussed in detail in the appropriate chapters of Part II of this book. However, regardless of these considerations, when a bolus of indicator is injected suddenly into the circulation, its concentration monitored downstream from the injection site demonstrates an increase to a peak value, followed by a decrease which continues until recirculation of indicator on its second passage supervenes. The "area under the primary circulation curve," regardless of the shape of the curve, is inversely proportional to the greatest flow encountered by that indicator during its primary circulation.

When Hamilton and co-workers described their method of measuring cardiac output in 1928, it was apparent that recirculation of the indicator was interfering with the analysis of the primary curve and that indicator was being "counted twice." From comparative analysis of open-ended and recirculating models, they were able, during the next year, to define the logarithmic nature of the descending limb of the primary curve and apply this knowledge to the extraction of primary circulation information from the recirculating curve. All the methods of curve area calculation to be discussed in this chapter have been devised principally to compensate for, or eliminate, the artifact of recirculation.

Originally, calculation of output was based on the mean concentration of indicator in the pooled samples of blood collected during the inscription of the curve and the actual measurement of the curve was purely to determine the time course of the primary circulation. These measurements then provided the denominator (concentration × time) in the cardiac output formula.

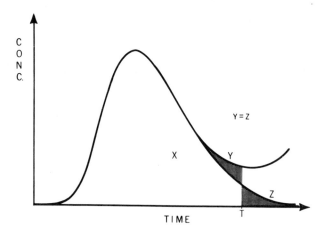

Figure 5.1

A more recent adaption of this approximation was described by Levinson *et al.* (1962). This was based on the argument that, at a critical time interval from appearance of the curve, the pooled sample, including some recirculated indicator, would contain the same mean concentration of indicator as the primary circulation in its entirety. Fig. 5.1 represents an actual indicator-dilution curve with recirculation apparent along the line Y. The curve representing only the first passage of the indicator follows the line Z. At the critical time interval T, from the beginning of the curve when the shaded area Y equals the shaded area Z, the actual pooled sample of withdrawn blood would contain the same mean concentration of indicator as the hypothetical complete first circulation sample. This time T is supposed to occur at the nadir of the curve. Flow per minute is then given by the formula

$$\frac{60 \text{ sec} \times \text{mass of indicator injected}}{\text{pooled sample mean concentration} \times T \text{ sec} \times \text{plasmacrit}} .$$

Despite the equality of area and mean concentration of $X + Y$ and $X + Z$, the true primary circulation time extends beyond T to a point where the line Z approaches zero. This technique, therefore, overlooks the fact that the measured time is shorter than the true primary circulation curve time. Furthermore, it requires an arbitrary and manual definition of the "right" time to cease collection of the pooled sample. The advantages of reduced calculation time and increased yield of utilizable curves (curve peaks extending beyond the recorder scale do not invalidate this method) are outweighed by the above disadvantages.

INTEGRATION METHODS

In 1932, Hamilton's group reported further analyses of the injection method in which manual integration of the curve was described. This method is still recognized as the standard in this field and will be described in detail.

Step 1

The height of the curve is measured at 1-second intervals to an arbitrary point on the descending limb in the region of one-third of the peak height. For purposes of illustration, Fig. 5.2 shows the height of the curve measured well into the area of obvious recirculation.

Step 2

The values obtained from the descending limb of the curve are replotted on semilogarithmic paper (Fig. 5.3). The following segments of this replot should be identified from the figure: A, the convexity of the peak of the curve; B, the exponential (straight-line) portion; and C, the concavity of the onset of recirculation.

Step 3

The straight-line segment is projected to the base line, and the values on this straight line are noted at 1-second intervals from the point of departure of the actual curve from the projected line to a point ap-

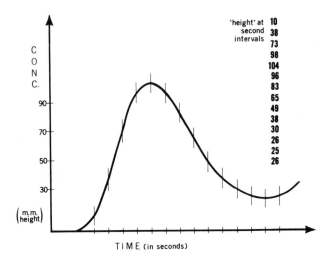

Figure 5.2

proximately 1 to 5% of the peak concentration. These values represent the corrected or primary circulation curve. Only those values on the straight-line (exponential) segment are considered, and they not only extend to lower levels than the original curve, but replace values

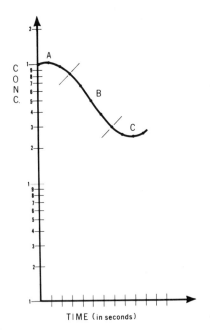

Figure 5.3

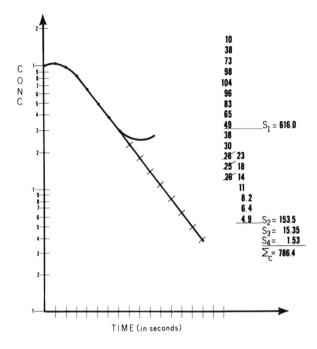

TIME (in seconds)

Figure 5.4

recognized to be on the "recirculation" segment of the curve (Fig. 5.4).

Step 4

Two values, as close to one logarithmic scale apart as the numbers will allow, are selected on the straight line. In Fig. 5.4 they are 49 and 4.9, but they could as satisfactorily be 65 and 6.4 or 83 and 8.2. The summed total of 1-second values from the outset of the curve through the peak and down to the highest of the two selected values is designated S_1 (the first sum), and the summed total of 1-second values from (but not including) the upper selected value to (and including) the lower value is designated S_2. S_3 is one-tenth of S_2, S_4 is one-tenth of S_3, and so on. The total corrected curve area Σ_c is then equal to $S_1 + S_2 + S_3 + \ldots$ until the values summed are in the range of 1% of the peak concentration.

PLANIMETRY

The use of a planimeter to measure the curve area is advocated by some (Hetzel and co-workers, 1958).

It requires an extra preliminary step, that of replotting the corrected semilogarithmic curve back onto the initial linear plot. Since all of the data required for the manual integration method have been obtained by this time, there seems to be no clear advantage to this method. In fact, errors may be introduced in the reading of the planimeter. Despite these points, curve area calculation by planimetry remains a common and popular technique.

OTHER METHODS

Among other methods, the integration-extrapolation method of Hamilton has been repeatedly demonstrated to provide the best prediction concerning the tail of the curve on the first passage of the indicator. This is true, however, only if the linearity of the exponential portion of the curve is clearly established before a sharply marked recirculation supervenes.

When these somewhat interdependent conditions are not met, the validity of the conventional extrapolation may be extremely dubious. This uncertainty, coupled with the laboriousness of the replotting procedure has stimulated the development of simplified formulae to substitute for the standard method of curve analysis. Under satisfactory conditions these alternative methods provide calculated areas which are within the biological variance and acceptable limits of reproducibility when compared to the standard method. However, none is based on established dye curve theory. They may be categorized as (1) purely empirical, (2) mathematically based, or (3) hybrid systems combining graphic and mathematic processes for the derivation of the curve area formulae. In the development of these alternative formulae, emphasis has been on the rapidity of calculation, although not all techniques substantially achieve this purpose and some do so only at the cost of introducing significant errors (Kelman, 1966).

Empirical Methods

Total Triangle Concept

After the development of the continuously recording densitometer and the substantial reduction in calculation time this afforded, Nicholson and Wood (1951) revived the early Hamilton concept of extrapolating the linear plot of the downslope to the base line of the indicator curve. They originally applied the manual integration method to their adjusted curve, but later (Warner and Wood, 1952) modified this concept to consider the extrapolated linear plot of the curve as a straight-line triangle (Fig. 5.5). The sides of the triangle are formed by the onset to peak concentration (*PC*) and peak to the base line intersection of the straight-line projection of the descending limb of the curve. The time from "appearance" (curve onset) to intersection of the descending limb with the base formed the third side of the triangle. The curve area was considered as ½ peak height (*PC*) × time. The formula was found to increase the variability of curve areas to 14% compared with 11% for the standard technique.

Fore-'n'-Aft Triangle Formula
The fore-'n'-aft triangle formula (Bradley and Barr, 1969) utilizes an overly elaborate mathematical approach visualizing two sets of congruent triangles in the linear curve (Fig. 5.6). The total area considered is equivalent to the product of the peak concentration (*PC*) and the time from one-half of the peak concentration (PC_{50}) on the ascending limb to the similar point on the descending limb (T_{50}). Since this measurement is one-half of the total base time from onset to descent intersection when the curve is considered as a single triangle, the calculation is, in essence, the same as in the preceding method.

Forward Triangle Concept
The "emergency" formula was developed by Dow (1955) in an attempt to recover data from curves with such horizontal or flattened downslopes that recirculation could not be clearly identified and extrapolation was fallacious. This formula relates appearance time

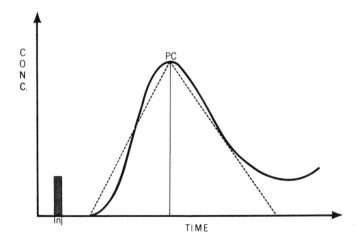

Figure 5.5

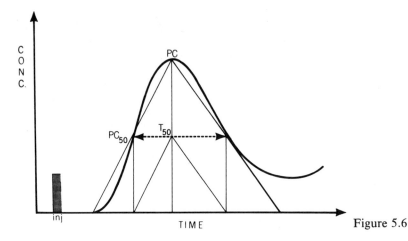

Figure 5.6

(AT), time from appearance to peak concentration (PCT), maximal dye concentration (PC), and certain empirically determined constants and is independent of the downslope of the curve (Fig. 5.7). In curves where the appearance time is less than one-third of the time for the descending limb of the curve to cross one logarithmic decade (i.e. to fall from any given value to $1/10$ of that value), or where the descending limb does not fall below 50% of the peak value before recirculation, the formula

$$\text{area} = \frac{PC \times PCT}{3.0 - 0.9 \, (PCT/AT)}$$

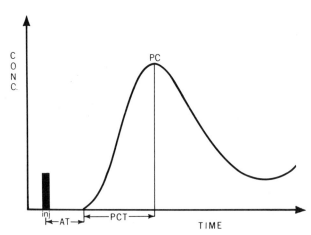

Figure 5.7

was recommended as an alternative to the conventional extrapolation. Oriol (1967) corrected this formula for the distortion due to the sampling system, making minor alterations in the constants to 2.98 and 0.91, respectively. Hetzel *et al.*, between 1955 and 1957, developed a forward triangle formula utilizing only the peak and the appearance-to-peak concentration times (Fig. 5.8). Although indepen-

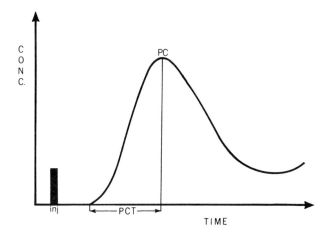

Figure 5.8

dent of the extraneous factors which may affect the appearance time of the previous formula, Hetzel's formula (1958) needed to be modified depending on peripheral or central venous sites of injection. The formula was stated as

$$\text{area} = \frac{\frac{1}{2}\, PCT \times PC}{K},$$

where $K = 0.35$ for peripheral venous injections of indicator and 0.37 for central or superior vena caval injections (Thorburn, 1961). Benchimol *et al.* (1963) found that, regardless of the site of injection, a value of $K = 0.34$ gave the closest approximation to the Hamilton formula results with the same curve. The average percentage difference between the two methods was found to be $10.42 \pm 8.25\%$, with the greater discrepancies obviously when the tail of the curve was particularly flat and prolonged.

The methods of Warner and Wood, Hetzel, and Dow were compared to planimetry by Gil-Rodriguez *et al.* (1970). Warner and Wood's formula consistently underestimated (-11.1%) cardiac output, whereas the forward triangle formulae overestimated it (+8.8 and 23.7%, respectively).

It must be recognized that all forward triangle methods fail to check for exponentiality of the concentration decay and are exceedingly sensitive to inaccuracy in the appearance-to-peak time measurement. This is an inherent and fundamental drawback to the method, because the appearance tends to be gradual (and hence the appearance time less certain) in those very curves recommended for the method where cardiac output is low and the downslope is prolonged.

The Mathematical Analysis Methods

Williams, O'Donovan, and Wood, 1966
In this analysis, illustrated in Fig. 5.9, the segment of the indicator-

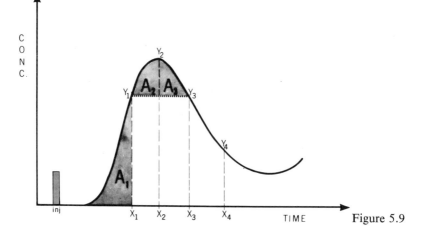

Figure 5.9

dilution curve before the exponential segment is regarded as a sequence of three parabolas (A_1, A_2, and A_3). The remaining curve is analyzed as a single exponential, and its area is calculated without replotting. The further assumption is that the exponential decay has been established at a point approximately 75% of the peak deflection and that it is still uncontaminated by recirculation at 37% of peak deflection. The geometric construction requires the selection of the point Y_3 on the downslope as any convenient integer in the vicinity of 75% of peak deflection. The other points selected are Y_1 (equal to Y_3 on upslope), Y_2 (peak deflection), Y_4 (half of Y_3 on down-slope) and X_1, X_2, X_3, and X_4, the corresponding coordinates on the time axis. The curve area is calculated from standard formulae for areas under parabolas and exponentials, and approximates to

$$\frac{2}{3} Y_2 (X_3 - X_1) + Y_1 \left[\frac{X_3}{3} + \frac{10 (X_4 - X_3)}{7} \right].$$

Although this method gives excellent correlation with the Hamilton method (1.4% standard deviation of differences) and requires less time, it does not allow for checking the exponential character of the downslope.

Jorfeldt and Wahren, 1967
In this analysis, the curve is divided into three parts, A, B, and C (Fig. 5.10). As in the previous method, part A is considered in terms

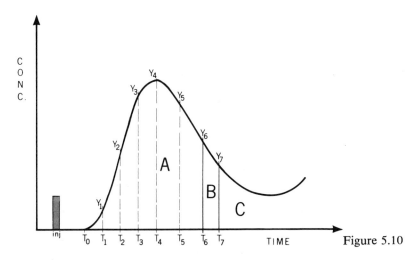

Figure 5.10

of three parabolas. Area B is calculated in terms of a constant fraction of a right-angled trapezium having, as its upper side, a straight line joining the points Y_6 and Y_7 (0.6 and 0.4 of the peak height Y_4, respectively). Area C is shown to be twice area B. The time appearance (T_0) to peak deflection (T_4) is divided into four equal intervals by T_1 to T_4 and their corresponding coordinates Y_1 to Y_4. T_5 bisects the interval T_4 to T_6 and corresponds with Y_5. The area under the curve is given as

$$A = (4Y_1 + 2Y_2 + 4Y_3 + Y_4) \times \frac{(T_4 - T_0)}{12}$$
$$+ (Y_4 + 4Y_5 + Y_6) \times \frac{(T_6 - T_4)}{6},$$

and

$$B + C = 1.48 \times Y_4(T_7 - T_6).$$

This method assumes that Y_6 and Y_7 appear before recirculation and that the curve does possess a mono-exponential component.

Both of the above methods were evaluated against the Hamilton standard by George *et al.* (1970), who found that the correlation coefficient for either formula was close to unity and that the standard deviation about each regression line was small over a wide range of values. The technique of Jorfeldt and Wahren, however, took twice as long as that of Williams *et al.* (1966) to perform and was subject to more errors in calculating from the relatively complex formula. Moreover, in some instances it could not be used as the descending limb of the curve did not fall to 0.4 of the peak height.

The Hybrid Systems

A number of systems have been devised in which a combination of methods has been used to calculate the total curve area. The initial pre-exponential segment of the curve is usually derived by standard integration or planimetry.

Gorton and Hughes (1964) described the production of a family of lines representing the shape of exponential decay curves within the limited range of time constants encountered in cardiac output measurements in their laboratory. The recorded linear indicator-dilution plot, traced on transparent rectilinear paper, was placed over the exponential curves and moved horizontally until one line in the family was found which matched the traced downslope. The extrapolation beyond the point of recirculation was then traced onto the rectilinear paper, and the resulting "primary circulation" curve area measured by integration or planimetry.

This concept was carried further by Killen and France (1969), who designated values to each curve in the family such that, when the "best fit" had been selected in the manner described above, the area under this exponential could be calculated from a simple formula. A still further extension of this approach was described by Boyett *et al.* (1964), who used an on-line electronic integrator to calculate the pre-exponential area and a curve-fitting technique, as above, for the remainder of the area.

These methods are quick and show a negligible difference in precision when compared to conventional methods (Boyett *et al.*, 1964). They require prior preparation of the exponential functions and transcription onto transparencies. In the last example, supplementary electronic equipment was required. They also need certain "laboratory constants" such as curve orientation, paper speed and Y-axis limits, and calibration for the sensitivity differences between the scale of the exponential functions and the individual curves.

Other hybrid methods include that of Lilienfield and Kovach (1956) combining integration with a simple graphic technique applied to the dilution curve washout replotted against an optical density scale. The method of Clarke and Stoik (1964) combines integration with a graphic construction applied to a linear plot of the washout segment of the curve and assumes the presence of an exponential function.

ERRORS IN CALCULATING CURVE AREAS

The errors attendant upon utilizing the various numerical integration formulae have been analyzed by Kelman (1966).

The Hamilton method makes use of the trapezoid rule, by which the curve is considered as a series of trapezoids, each with a short base on the time axis. The formula has only a very small error but in the translation to cardiac output curves, the numerical integration is usually performed by simple summation of successive ordinates. This may lead to considerable inaccuracy unless the ordinates are very closely spaced, or the first and last ordinates are zero or small. In most instances, the time interval of 1 second between ordinates is usually sufficiently close, but in high flows, shunts, and pediatric indicator curves, 0.5-second intervals may be required. It is standard practice for the first ordinate to have zero value and the last counted ordinate to be approximately 1% of the peak value.

A more accurate mathematical treatment of the curves, and that utilized in the formulae of Williams *et al.* and Jorfeldt and Wahren, is represented by ordinates at equally spaced time intervals between which the curve is approximated by a series of arcs of parabolas. This requires the numerical analysis known as Simpson's rule. An alternative and attractive method is to determine the decay constant of the exponential fall by measuring the gradient of the straight line in the semilogarithmic plot. This is used in some of the hybrid methods described above.

No significant difference was noted between the accuracy of the trapezoid rule and Simpson's rule techniques, provided that the qualifications regarding summation substitution in the trapezoid method were adhered to. However, considerable variability (1.2 to 7.8% coefficient of variation) was found in estimates of decay constants made by different investigators and even between duplicate estimates made by the same investigator. A curve-fitting technique would reduce this variation. A significant error is introduced into the hybrid systems from summation of only the initial segment of the curve. Under these circumstances, the last ordinate is not zero but has a considerable dimension, and it is inappropriate to use this method of integration. The trapezoid rule is sufficiently accurate to eliminate this error.

The most significant errors in curve area determination, however, result from contamination of the primary circulation by earlier-than-anticipated recirculation. This problem can only be reduced by minimizing the injection-to-sampling time interval. This, in turn, is achieved by approximating the injection and sampling sites within the circulation, as closely as is compatible with fulfilling the criteria of adequate mixing and representative sampling, as defined in Chapter 4. The shortest recirculation time in maximally exercising adults in the absence of shunts is approximately 9 seconds (Sowton *et al.,* 1968), and curves with the washout segment well inscribed within this interval from the appearance time will be practically free of recirculation. Injection in the left ventricle with femoral artery sampling, or superior vena caval injection with pulmonary artery sampling, will fulfill these requirements.

REFERENCES

Benchimol, A., E. G. Dimond, F. R. Carvalho, and M. W. Roberts. 1963. New method: the forward triangle formula for calculations of cardiac output, the indicator-dilution technic. Amer. J. Cardiol. 12:119–125.

Boyett, J. D., D. E. Stowe, L. H. Becker, and W. E. Britz, Jr. 1964. Rapid method for determination of cardiac output by indicator dilution techniques. J. Lab. Clin. Med. 64:160–167.

Bradley, E. C., and J. W. Barr. 1969. Fore-'n-aft triangle formula for rapid estimation of area. Amer. Heart J. 78: 643–648.

Clarke, R. G., and M. W. Stoik. 1964. A method of rapid computation of indicator dilution downslope areas. J. Appl. Physiol. 19:526–527.

Dow, P. 1955. Dimensional relationships in dye dilution curves from humans and dogs, with an empirical formula for certain troublesome curves. J. Appl. Physiol. 7:399–408.

George, M., B. W. Lassers, A. L. Muir, and D. G. Julian. 1970. A comparison of two formulae for measuring the area of indicator dilution curves. Cardiovasc. Res. 4:127–131.

Gil-Rodriguez, J. A., D. W. Hill, J. T. Horny, S. Lundburg, and A. H. Wilcock. 1970. A comparison of some methods for estimating the area under a dye dilution curve. Brit. J. Anaesth. 42:981–987.

Gorten, R. J., and H. M. Hughes. 1964. Reliable extrapolation of indicator dilution curves without replotting. Amer. Heart J. 67:383–387.

Hamilton, W. F., J. W. Moore, J. M. Kinsman, and R. G. Spurling. 1928. Simultaneous determinations of the pulmonary and systemic circulation times in man and of a figure related to the cardiac output. Amer. J. Physiol. 84:338–344.

Hamilton, W. F., J. W. Moore, J. M. Kinsman, and R. G. Spurling. 1932. Studies on the circulation. IV. Amer. J. Physiol. 99:534–551.

Hetzel, P. S., A. A. Ramirez DeArellano, and E. H. Wood. 1955. Estimation of cardiac output from initial portion of arterial indicator dilution curves. Fed. Proc. 14:72.

Hetzel, P. S., H. J. C. Swan, A. Ramirez DeArellano, and E. H. Wood. 1958. Estimation of cardiac output from first part of the arterial dye dilution curves. J. Appl. Physiol. 13:92–96.

Jorfeldt, L., and J. Wahren. 1967. A simplified procedure for calculation of cardiac output from dye dilution curves. Acta Med. Scand. (Suppl.) 472:75–80.

Kelman, G. R. 1966. Errors in the processing of dye dilution curves. Circ. Res. 18:543–549.

Killen, D. A., and R. France. 1969. A simple "curve-fit" method of calculating area under the terminal downslope of indicator dilution curves. Cardiovasc. Res. 3:107–111.

Levinson, G. E., P. H. Lehan, P. J. Coleman, and H. K. Hellems. 1962. A short method for measurement of cardiac output by indicator dilution. Amer. Heart J. 64:489–497.

Lilienfield, L. S., and R. D. Kovach. 1956. Simplified method for calculating flow, mean circulation time and downslope from indicator dilution curves. Proc. Soc. Exp. Biol. Med. 91:595–598.

Nicholson, J. W., and E. H. Wood. 1951. Estimation of cardiac output and Evans blue space in man, using an oximeter. J. Lab. Clin. Med. 38: 588–603.

Oriol, A. 1967. Determination of cardiac output, using Dow's formula. J. Appl. Physiol. 22:588–590.

Sowton, E., D. Bloomfield, N. L. Jones, B. E. Higgs, and E. J. M. Campbell. 1968. Recirculation time during exercise. Cardiovasc. Res. 4:341–345.

Thorburn, G. D. 1961. Estimates of cardiac output from forward part of indicator dilution curves. J. Appl. Physiol. 16:891–895.

Warner, H. R., and E. H. Wood. 1952. Simplified calculation of cardiac output from dye dilution curves recorded by oximeter. J. Appl. Physiol. 5:111–116.

Williams, J. C. P., T. P. B. O'Donovan, and E. H. Wood. 1966. A method for the calculation of areas under indicator dilution curves. J. Appl. Physiol. 21:695–699.

chapter 6/ The Calibration of Indicator-Dilution Curves and Computation of Cardiac Output

Dennis A. Bloomfield
Division of Cardiology/Maimonides Medical Center/
Brooklyn, New York

Indicator-dilution studies were initially carried out by obtaining multiple, timed blood samples and assessing each sample individually for its indicator concentration. The curve was then plotted directly in terms of concentration and time. With the introduction of the continuously recording optical densitometer, however, the curve was drawn directly in terms of millimeters of paper deflection and time, and calibration of this deflection in terms of concentration was required.

The first procedures applicable to continuously recorded curves were developed in the early 1950s and utilize either serial dilutions of a measured volume of dye unrelated in volume to the injectate (Friedlich, Heimbecker, and Bing, 1950; Nicholson and Wood, 1951; Shadle *et al.*, 1953) or spectrophotometric analysis of a low and falling dye concentration after the curve inscription (Theilen, Paull, and Gregg, 1954).

In this chapter, the calibrating techniques in current usage are described and compared. The utilization of the calibration factor for the final computation of cardiac output, for both bolus and continuous injections of indicator, will be detailed.

THE NICHOLSON TECHNIQUE

Of the initial techniques, the method described by Nicholson and Wood has been most widely used, both in its original and modified forms.

The calibration samples are prepared by diluting 1 ml of a 5-mg/cc solution of indicator in 7 cc of distilled water. Three samples of progressively less concentration are prepared by diluting 4 cc from the previous sample with an equal amount of distilled water. A 5-cc amount of the patient's blood, withdrawn before the inscription of the curve, is pipetted into each of four beakers, and 0.2 cc of indicator from each concentration is pipetted into its corresponding beaker of blood. The samples are sequentially drawn through the densitometer used during, and set at the same sensitivity as, the inscription of the output curve. The paper deflection of each progressive concentration is plotted on a linear scale (Fig. 6.1), and a straight line best fitting the points is constructed. The calibration factor is then designated as

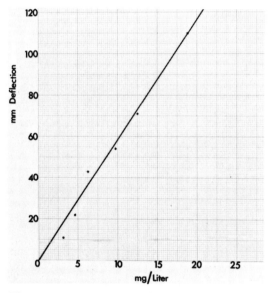

Figure 6.1 Nicholson's technique. The relationship of successive concentrations of indicator to the output voltage of the densitometer (recorded deflection) is shown with the *straight line* which best fits the points. In this example, 17 mg of indicator per liter causes 100 mm of deflection. The calibration factor for associated indicator curves is 0.17 mg/liter/mm.

the ratio of the deflection (in millimeters of paper) to the concentration of indicator (in milligrams per liter). The use of this relationship allows the dye curve area, measured in millimeters per second to be expressed in milligrams (of dye) per liter per second.

The Nicholson technique has a primary disadvantage in terms of accuracy, in requiring multiple pipetted samples and subsequent dilutions. The amount of dye is unrelated to that used for the inscription of the curve and, being small, permits slight inaccuracy in measurement to assume significant proportions. Initial errors in concentration are perpetuated throughout the samples with no ability to be recognized or adjusted. The method is time-consuming and laborious—factors which may in themselves potentiate inaccuracies in the necessarily meticulous dilutions. This is, in part, compensated for by multiple point determinations providing the ability to mean-out minor variations in dilution accuracy. The majority of cardiac catheterization laboratories utilize a modification of the Nicholson technique, in which only one or two dyed samples are prepared. Although this reduces the amount of blood required, it does not materially lessen the labor, and, although still permitting initial dilution inaccuracy to go undetected, it eliminates the "multiple point" ability to correct for subsequent variations in dilution.

The complete technique as recommended, however, requires at least 25 cc of blood for a single set of dilutions. Duplicate analysis requires as much again. The requirement for blood for replicate calibrations and the contingencies of spillage or recognized errors in technique may become prohibitive, particularly in infants and small children.

THE VANDERBILT TECHNIQUE

The technique practiced in Newman's laboratory at Vanderbilt University since 1961, requires the delivery into a 25-cc volumetric flask of the exact amount of indicator injected into the patient during the inscription of the output curve. This is accomplished by using a Cornwall (Becton, Dickinson & Co., Rutherford, N.J.) automatic syringe attached to the injection catheter during the cardiac catheterization and reinjecting through the self-same system after the procedure (Fig. 6.2).

This is diluted to the 25-cc mark with plasma, and 1 cc of this dilution is mixed with 39 cc of the patient's blood, giving a 1/1000

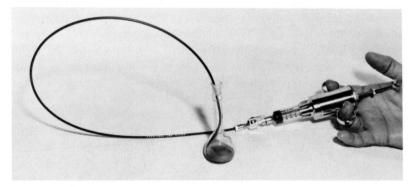

Figure 6.2 Vanderbilt technique. The calibrating volume of dye is delivered into the volumetric flask by repeating the Cornwall syringe injection through the injecting catheter used for the cardiac output study.

dilution. A "blank" is also prepared by diluting 1 cc of plasma in 39 cc of the patient's blood. Blank and diluted dye samples are alternately drawn through the densitometer, and the deflection is recorded. The samples must be constantly stirred during this operation. This can easily be effected by the construction of a simple stirring system, as shown in Fig. 6.3A, or by a pair of magnetic stirrers. In present usage, it has been found more practical to prepare a final dilution of 1/500 by adding 1 cc of the diluted dye to 19 cc of blood. This alteration is then corrected by halving the recorded deflection. The paper deflection (Fig. 6.3B) is equivalent to the concentration of the amount of dye per liter injected into the patient. This gives a value both for the amount of dye injected and for the calibration of the curve area, without ever needing to measure directly the amount of injectate. The capability of the Cornwall syringe to reproduce constant volumes allows for the reproducibility of the volume delivered at the catheter tip, if the syringe volume is larger than the catheter dead space. In this laboratory, injecting catheters range in interluminal volume from 1.0 to 2.0 cc, and the approximate figure of dead space for each catheter type is available to the operator at the time of dye curve inscription. The Cornwall syringe is set to a volume of approximately 2 cc plus catheter dead space and is locked at this volume.

The Vanderbilt technique offers advantages over the previously described method. The accuracy of syringe setting and knowledge of absolute amount injected are not necessary with this technique, as will be shown later. The method is rapid, requiring only a blank and a dye sample to be prepared. Dilution volumes are relatively large (25 cc, 40 cc) and consequently minimize the significance of measurement errors. The samples allow multiple measurements of densitometer deflection, permitting averaging of this value and thereby further minimizing errors. The total variation in replicate studies, incorporating possible inaccuracies in delivery of indicator dilutions, stability of densitometer, and measurement of deflection, amounted to a standard deviation of only 0.93%.

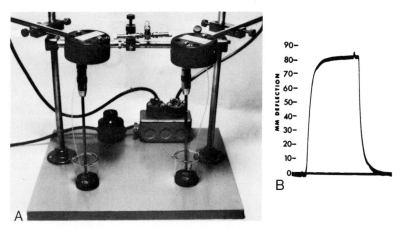

Figure 6.3 Vanderbilt calibration. *A*, a simple stirring system for calibration samples. Thin polyethylene tubes with weighted ends lying in the beakers connect to a three-way stopcock and thence to a cuvette densitometer (*not shown*). *B*, calibration signals produced by rapidly turning stopcock from "blank" to sample while withdrawing through a densitometer.

The disadvantages of this technique, however, must be borne in mind. The single point calibration does not allow for recognition of either volumetric errors in the calibration technique or nonlinearity in the densitometer. Even in the modified 1:500 form, the blood requirement is greater than for the Nicholson calibration.

THE INTEGRATED SAMPLE TECHNIQUE

The integrated sample technique requires the collection of the blood containing the dye which has inscribed the curve. This is achieved by a special withdrawal system (Fig. 6.4), the design of which was controlled by two major requirements:

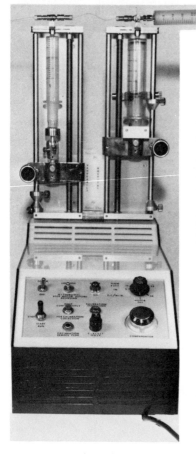

Figure 6.4 Vanderbilt Hospital integrated withdrawal pump. The sample syringe (small syringe) is on the left. The universal joint plunger attachment is to eliminate play in the plunger movement. The instrument panel contains an on-off switch, withdrawal-infuse switch, power light, flow rate (in cc/minute) regulator, stop-start switch, a button which when depressed stops the withdrawal in the large syringe and starts the collection sample in the small syringe without alteration in the withdrawal rate (the reverse happens when the button is released). The machine also has a remote calibration plug, a motor circuit fuse, a calibration indicator that marks the dilution curve on the recorder simultaneously with depression of the button, and a variable time delay trigger that compensates for the various size tubing between the densitometer photocell and small syringe. (This instrument was designed by Mr. R. E. White of Vanderbilt University.)

1. The maintenance of constant flow rates over a range of selected values.

2. The ability to segment the blood sample during withdrawal into a collection fraction and a calibration fraction, which is selected over a short period of time near the peak of the concentration curve.

A finite and predictable time is required for the blood to travel from cuvette to calibration syringe. A time delay control is added to synchronize the concentration curve plotted from the densitometer with the collection of the calibration fraction in the withdrawal syringe. The point at which the calibration fraction is selected is decided by the operator monitoring the concentration curve. At an appropriate time, near the peak of the concentration curve, a button is depressed

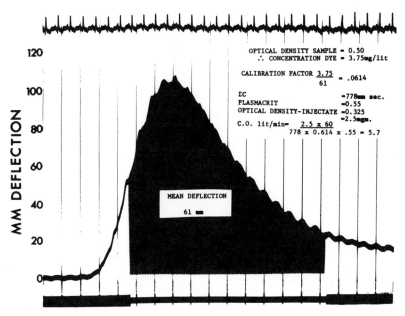

Figure 6.5 Integrated sample calibration. The *cross-hatched section* represents that curve area inscribed during the collection of the calibrating sample, as indicated by the narrowing of the marker signal recorded below the curve. The indicator concentration in this sample was 3.75 mg/liter, corresponding to a mean deflection of 61 mm, and giving a calibration factor of 0.0614 mg/liter/mm. For explanation of cardiac output computation, see Table 6.1.

to trigger the time delay, which in turn switches the collection into the calibration syringe (Fig. 6.5). The reverse sequence occurs when the button is released as the peak concentration is passed. A table of time delay per inch of catheter as a function of flow rate for each of the most used catheter sizes has been prepared, and the appropriate delay for a particular flow rate and catheter size is set before the withdrawal run. The calibration sample is centrifuged, and the plasma containing the dye is separated. The optical density of the sample at a wave length of 800 nm is assessed in a spectrophotometer, and the concentration of dye in the sample is read from a graph equating serial concentrations of dye in plasma to optical density at 800 nm. For the calculation of the cardiac output, the amount of injectate can also be measured by diluting it in 1 liter of distilled water and measuring its concentration per liter via its optical density in the same manner as the sample.

The integrated sample technique is of theoretical merit in that the calibration is performed with the sample collected during the inscription of the curve. In this way, inaccuracy due to instability in densitometer sensitivity is avoided. Although considerable labor is required to meticulously compile the correlate of optical density and concentration of indicator, the graph holds good for all samples of dye from the same batch. The principal inaccuracies in this method are related to technical problems concerned with the collection period and the phase difference between dye passing through the densitometer and dye appearing in the collection syringe. In unsophisticated systems, the collection period extends from the point before indicator injection until the curve has been inscribed and the lowest possible recirculating dye concentration is present. This results in dilution of the indicator so that only low concentrations are recorded in the spectrophotometer. The internal volume of the connection between cuvette and collection syringe also constitutes an error, in that indicator present in the blood in the tubing has been represented in the inscribed curve but not represented in the collection. The higher the dye concentration in the tube, the greater will be the error. Therefore, attempts to improve accuracy by increasing dye concentration by stopping the collection during the downslope of the curve are negated by the greater errors incorporated in the higher concentrations left in the tubing.

A special withdrawal pump (Fig. 6.4) built by Mr. R. E. White at Vanderbilt University has eliminated both these problems. The system permits collection of a selectable high dye concentration portion of the curve without interruption or alteration in flow. This blood sample provides concentrations high enough to be accurately estimated, and the exact segment of the curve relating to the sample is marked on the curve.

There is one remaining and inherent problem with this technique. Only the first study, in which the curve base line represents an absolute rather than a relative zero concentration of dye can be used. Should this be incorrect or be artifactual, or should the peak of the curve exceed the maximal deflection of the recording sensitivity, a delay in the procedure must occur until the indicator has been eliminated or has reached an insignificant concentration in the blood.

THE DYNAMIC TECHNIQUE

The dynamic technique was introduced by Sparling in 1960 and requires setting up, in series with the densitometer, a miniature circulation with a known flow. Calibration is performed by injecting a known amount of indicator into the flowing system and recording a dilution curve. With three factors known in the expression,

$$\text{flow (cc/sec)} = \frac{\text{indicator (mg)}}{\text{curve area (mm/sec)}} \times \text{calibration factor,}$$

the equation can be solved for the calibration factor. The flow must be constant and known exactly, the injectate must be accurately measured, and adequate mixing must take place between injection and sampling sites. In practice, this last requirement has been met with a cylinder filled with glass beads in the models of Sparling (1960), Emanuel (1966), and Scheel (1969) and their co-workers, and a plastic helical coil in the model of Wexler and Frater (1969).

A further development of this coil calibrating chamber has been tested in this laboratory. The coil (Fig. 6.6) is constructed from glass wound into nine loops with an internal diameter of 14 mm and an internal volume of 1.2 cc. It is attached directly to the cuvette of the densitometer, and blood from a constantly stirred reservoir is drawn through it at a rate of 24.7 cc/minute by a Harvard pump. The calibrating dye is introduced through a short latex tube at the inflow of the coil. This system has been shown to provide an exponential washout of a bolus injection of indicator. The output of the densitometer is fed into a Lexington cardiac output computer, from which is obtained both a calibrating deflection and a zero base line check. The system is designed for a reproducible but unmeasured amount of indicator, as in the Vanderbilt technique. The calibrating sample is determined by delivering, into a volumetric flask, the exact amount of indicator, I, used for each injection into the patient. This is diluted to 25 cc with plasma, and 0.25 cc of the dilution (that is, 1/100 of the patient injection) is used for the calibrating injection. The curve area produced by this injection into the calibrating circuit flowing at 24.7 cc/minute will, therefore, be equal to the area of a patient's curve produced by the full amount of indicator in a circulation flowing at 100×24.7 cc/minute or 2.47 liters/minute. Consequently, the computer is calibrated for any patient curve produced

GLASS CALIBRATION COIL

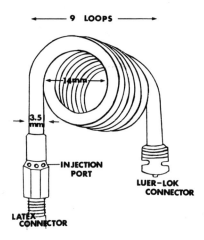

Figure 6.6 Glass calibration coil. The direction of flow is from the injection latex to the Luer-Lok to which point the densitometer is connected. For details, see text.

by the volume I of indicator by being set to 2.47 liters/minute during the calibrating curve. Potential errors owing to lack of uniform cross-section in the syringe or variation in withdrawal rate are minimized by carrying out three calibrations during a single withdrawal of the syringe. Drift in computer base line is further monitored by simultaneously recording the integrating circuit (Fig. 6.7).

The dynamic method of calibration, although described nearly 10 years ago, has failed to gain enthusiastic usage in this country. The method involving the application of the area of a curve obtained from a small calibration system built into the sampling line is particularly suited to "computer" analysis of indicator curves. The accuracy of this method depends on a constant and known sampling rate and the exact measurement of the amount of dye injected both into the subject and the calibration circuit. Despite the advantages of requiring little blood and no spectrophotometric analysis, the technique has significant practical disadvantages. The initially described calibration loops contained small silicone glass beads, and inaccuracy was reported from dye adhering to fibrin deposited in the mixing chamber (Völlm and Rolett, 1969). A variant designed by Wexler and Frater (1969), consisting of a tightly wound polyethylene helix, has too

COIL CALIBRATION

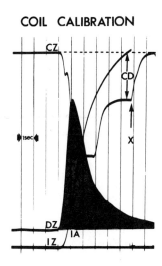

Figure 6.7 Coil calibration. The calibrating curve (*shaded area*) originates from the densitometer zero (*DZ*) line. The computed calibration output originates from the computer zero (*CZ*) and adopts a horizontal configuration during the exponential portion of the shaded curve. The calibration deflection (*CD*), which corresponds to a cardiac output of 2.47 liters/minute, is the vertical displacement from the computer zero to the horizontal segment. True computation of curve area is assured by noting no area integration (*IA*) above the integrator zero (*IZ*) before inscription of the shaded curve. At point *X*, the computer is re-zeroed, and the calibrating measurement is repeated without stopping the withdrawal pump.

small a mixing volume and, at withdrawal rates sufficient to produce mixing, has a transit time of only 1 or 2 seconds. Errors associated with the densitometer response time may then attain significant proportions. The withdrawal rate must be fast enough to render insignificant those changes due to variations in syringe cross-section area and to ensure adequate mixing. As this rate is crucial, measurement of withdrawal with a particular syringe against a constant and reproducible load is essential. Consequently, it appears that a delicate balance between withdrawal rate and blood volume in the calibration circuit is necessary to render this system accurate from a practical viewpoint.

THE END-TAIL TECHNIQUE

After the inscription of the indicator-dilution curve, the withdrawal is continued until the level of recirculated indicator in the blood is steady, as judged by a steady output from the densitometer. This output is recorded, and, simultaneously, a separate arterial blood sample of 10 cc is drawn. This sample is centrifuged, and the plasma is subjected to spectrophotometric analysis of indicator concentration, as for the integrated sampling technique. The calibration factor is determined as the ratio of the millimeter paper deflection of the steady recording above the prerecorded base line to the spectrophotometric indicator concentration.

The end-tail method calibration utilizes the deflection obtained from the systemic recirculation and the spectrophotometric analysis of indicator concentration in blood drawn during the recirculation to provide the calibration data. Both the concentration of dye and its deflection are relatively low during the recirculation phase, and measurements are more subject to significant errors under these circumstances than with high concentrations and large deflections. The calibration can only be applied to the first study. Further, the method provides only a single point calibration and, in this regard, is subject to the same objections cited for the Vanderbilt method. However, the major inaccuracy results from instability of the densitometer system. Small shifts in base line at 20 to 40 seconds after the curve is inscribed will lead to significant errors in the measurement of calibration deflection. Because of these unavoidable problems, this technique is not recommended and is important only in an historic sense.

CALIBRATION OF EARPIECE DENSITOMETERS

The calibration of indicator-dilution curves sensed by earpiece densitometers is usually undertaken by the end-tail technique and is subject to all of the associated inaccuracies. An alternative method was proposed by Barr and Bradley (1968), in which a polyethylene tube cuvette was constructed to interpose between the arms of the earpiece and convert it to a "flow-through" densitometer. All calibration methods for flow-through densitometers could then be applied to earpiece densitometers. The optical density variation between polyethylene and the ear pinna, however, requires that this modification be only used with dichromatic earpiece densitometers, where compensatory adjustments for these nonspecific changes can be effected.

CALIBRATION ARTIFACTS

Indicator-dilution curves are obtained from the sudden injection and rapid mixing of indicator in a stream of fresh, 37°C, non-anticoagulated whole blood, with a widely varying and unknown oxygen saturation and a low (but rising, if multiple curves are drawn) prior concentration of indicator. A totally critical approach would require that calibration be performed under identical conditions. The practicalities of these requirements, however, have stimulated investigation into the extent to which these nonspecific variables influence the accuracy of calibration. The nonspecific factors include hematocrit, tonicity, temperature, pH, and flow rate of the calibration blood, together with the presence of hemolysis or anticoagulants and other chemicals.

The effect of *hematocrit* on response characteristics of Indocyanine green densitometers has been investigated by Edwards *et al.* (1963), Sekelj *et al.* (1967), and Cropp (1969). Not only do variations in hematocrit affect the absolute optical density of indicator-free blood at the isosbestic point, but they also affect the sensitivity of the cuvettes to a given concentration of indicator in whole blood. Maximum sensitivity was noted with hematocrits in the normal range, with reduction of sensitivity both above and below this range. The effect was more marked in monochromatic than in dichromatic densitometers.

With increasing *tonicity* of the surrounding fluid, erythrocytes become increasingly crenated, and light transmission through the blood is reduced. Although many chemicals, including glucose and sodium chloride, can affect the accuracy of the dye curve analysis by their hypertonic action at the time of curve inscription or by inclusion in the calibrating sample, the most important is angiographic contrast material, by virtue of the high tonicity and the large volumes regularly employed in clinical practice. This material may also cause red cell agglutination and Castaneda *et al.* (1966) showed that an intravenous infusion of 20% NaCl simultaneously caused crenation (producing decreased light transmission) and agglutination (producing increased transmission).

Temperature variation may effect subtle changes in the volume of erythrocytes with concomitant alteration in light transmission. If extreme temperatures result in hemolysis, however, increase in light transmission may be significant.

Sudden increase in *pH* is known to cause red cells to assume a more spherical shape with associated increase in light transmission. The opposite occurs with sudden reduction of pH.

Variations in the *rate of blood withdrawal* through the densitometer cause relatively large changes in light transmission (Sutterer, 1967). Particularly marked in monochromatic instruments, the sharp decrease in transmission as the flow abruptly ceases and the asymptotic increase in transmission with stationary blood are produced by the shape and depth of the cuvette lumen and the orientation of blood cells in the cuvette.

Hemolysis may occur from the sudden admixture of low osmolarity indicator solution but is more likely to accompany those calibration procedures requiring centrifugation for plasma separation. The hemolyzed products remain with the plasma fraction and affect optical density estimation, particularly when this is measured at 800 nm (Simmons and Shephard, 1971).

Anticoagulants, such as potassium oxalate and pure heparin, have not been shown to effect optical density (Simmons and Shephard, 1971). However, sodium bisulfite, sometimes used as a stabilizing agent in heparin preparations, significantly reduces optical density of Indocyanine green in blood, when present in concentrations of 1 mg/100 ml *in vitro*. Concentrations of 7.5 mg/kg *in vivo* produce a significant but transient reduction in optical density, being most marked at 2 minutes after injection but absent at 15 minutes (Manning *et al.,* 1972).

Saunders *et al.* (1970) described a further source of error in which Indocyanine molecules remain in a macroaggregated state for a short, but consequential period. This is of particular importance *in vitro* where the flowing stream may not contain albumin to bind the indicator.

Ideally, the circumstances of dye curve inscription and subsequent calibration should be identically matched for each of the factors cited above as potential sources of inaccuracy. Variation in optical density, however, does not necessarily imply variation in densitometer sensitivity to fixed increments of indicator concentration, and, in practice, it is considered adequate and sufficient to calibrate with only the following limitations.

1. Calibration blood should be drawn from the subject in whom the curves are obtained.

2. The calibrating sample must be free from saline or other diluting fluids and contain only a minimal amount of anticoagulant.

3. Calibrating blood should not be obtained following the administration of angiographic contrast material to the subject.

4. The withdrawal flow rate through the densitometer should be the same for calibration and curve inscription.

It is not common practice to calibrate with blood at body temperature nor to compensate for pH, osmolarity, or hemolysis. Except in prolonged studies or those concerning the hemodynamic adjustment to shock, blood volume expansion by plasma, effects of transfusion, and the like (when calibration must be hematocrit-matched to each phase of the study), the above qualifications suffice to provide meaningful and generally acceptable calibrations.

A factor far more significant than any previously discussed, however, is that of individual variability in the responses and linearity of densitometers. Differences exist not only between mono- and dichromatic instruments, but also between instruments of the same model and manufacture. Every investigator is advised to closely examine the behavior of his densitometer and recording equipment to changes in linearity, sensitivity, and response to nonspecific factors.

Instrument variation is so marked and unpredictable that Hazelwood, Gumpert, and Rivera, using a monochromatic densitometer, found only insignificant difference between patients' fresh and stored blood and bank blood in regard to calibration factor (unpublished data). It was further demonstrated that, for this densitometer, changes in pH, temperature, and protein content produced in the collection and storage of calibration blood did not effect the calibration measurement. Consequently, although it is not recommended as sound dilution practice, bank blood could be used for calibration purposes with the appropriate densitometer when the patient's size or condition required the reinfusion of blood sampled during the curve inscription.

COMPUTATION OF CARDIAC OUTPUT

The way in which the "calibration factor" relates to the calculation of cardiac output adds a further dimension for comparison between calibration techniques. The formulae for cardiac output calculation for each technique are shown in Table 6.1.

Table 6.1 Cardiac output calculation[a]

Calibration technique	Formula for cardiac output (liters/min)
Nicholson and end-tail	$\dfrac{I \times 60}{\Sigma_c \times \text{calibration factor}}$
Vanderbilt	$\dfrac{\text{calibration deflection} \times 60}{\Sigma_c}$
Integrated sample	$\dfrac{I \times 60}{\Sigma_c \times \text{calibration factor} \times \text{plasmacrit}}$
Dynamic	
Standard	$\dfrac{I \times Q\text{cal} \times \Sigma_c\,\text{cal}}{I\text{cal} \times \Sigma_c}$
Coil modification	$\dfrac{(\text{computer}) \ \text{calibration curve area} \times 2.47}{(\text{computer}) \ \Sigma_c}$

[a]The formula for each technique is designated, where I is the amount of indicator injected, Σ_c is the total curve area, and Qcal, $\Sigma_c$cal, and Ical are the flow, curve area, and injected amount of dye, respectively, in the dynamic calibrating circuit.

In the Nicholson and end-tail techniques, the calibration factor is the ratio of millimeters of deflection to milligrams of dye per liter; by multiplying the curve area by this factor, the "deflection × time" area is converted to a "concentration × time" product. The amount of indicator must be independently known.

The Vanderbilt technique provides a calibration deflection that represents not only a ratio of deflection per unit of concentration, but also represents the injectate in terms of deflection. The indicator, unknown in regard to amount, is diluted into 1 liter and drawn through the densitometer. The calibration ratio is then $(I\text{ mg/liter})/(\text{mm deflection})$, where I is the amount of injectate. Substituting in the standard formula:

$$\text{cardiac output (liters/min)} = \frac{I\text{ mg}}{\Sigma_c} \times \frac{60}{I\text{ mg/mm deflection}}$$

where Σ_c is the curve area for a single circulation (see Chapter 5). Canceling of the unknown I gives

$$\frac{\text{mm deflection} \times 60}{\Sigma_c}.$$

The simplicity of this application and the elimination of errors inherent in the measurement of the standard I invests this technique with ease, rapidity, and accuracy.

The integrated technique provides a blood sample from which a concentration of indicator in plasma is obtained by spectrophotometry. The calibration factor ratio is then provided by relating this concentration to the mean deflection in the segment of the curve marked during the calibration-specimen collection. This factor must then be corrected for the plasma volume/whole blood ratio, which must also be estimated at the time of the study. The indicator amount, as in the Nicholson technique, must be independently estimated. This may be directly measured for each injection, or the technique can be extended to utilize repeatable unknown volumes as for the Vanderbilt method. This is facilitated by diluting a duplicate injection into 1 liter of distilled water and reading the concentration from the spectrophotometer. A graph relating dye concentration in water to optical density is prepared for each batch of dye in the same manner as for the blood concentration graph.

The standard dynamic technique requires the most measurements for calculation of output. The amount of indicator and curve area are required, as in the previous techniques; in addition, the amount of indicator used in the calibrating curve, the flow rate, and the calibrating curve area must be calculated. The coil modification described here requires only the ratio of the computer-generated calibration deflection and curve deflection multiplied by 2.47 liters/minute.

CARDIAC OUTPUT BY CONSTANT-RATE INJECTION

The constant-rate or "step" injection method of measuring cardiac output has been considered from a theoretical viewpoint in Chapter 2. In this original method of Stewart, volume flow is determined by dividing the quantity of indicator injected per unit time by the concentration of the indicator in the sampled blood at equilibrium. Since the volume of the injectate adds to the flow and must be accounted for, the cardiac output is computed from the formula

$$\text{cardiac output} = \left(\frac{C}{c} \times f\right) - f = f\left(\frac{C}{c} - 1\right),$$

where C is the concentration of the indicator injected, c is the concentration of the indicator in the sample, and f is the rate of indicator injection. Only the equilibrium sample concentration needs to be measured, as the other values are arbitrarily chosen. This concentration can be determined by the Nicholson method with the inscribed dilution curve providing the equilibrium deflection or by direct spectrophotometric analysis, when the inscribed curve is used only to determine the time of equilibrium.

COMPARISON OF TECHNIQUES

Numerous studies have demonstrated satisfactory agreement between the various techniques of calibration of indicator-dilution curves. Emanuel et al. (1966) demonstrated a standard deviation of 6.1% between paired cardiac output calculated by Hamilton's multiple arterial sampling technique and the dynamic method. Völlm and Rolett (1969) found the Nicholson and dynamic methods to be interchangeable with regard to calculated cardiac outputs. Emanuel et al. (1957), working with a prototype, manually switched, integrated withdrawal system, found a standard deviation of ±2.16% in cardiac output when compared with a Hamilton technique of calibration. In this laboratory, the Vanderbilt technique has been compared to the Fick output with a 4.8% variation. In practice, therefore, the factors which determine the preference for one technique over another are the following.

1. Ease and time necessary to carry out the method.
2. Number of variables to be measured. The accuracy of the method is, for the most part, inversely related to the number of measurements.
3. Cost, space, and operational requirements of special equipment necessary for the calibration measurement.
4. Limitation imposed by the method on the contingency of not obtaining meaningful data on the first dye curve of the study. The integrated (and end-tail) methods are sensitive to and rendered inaccurate by the presence of a circulating basal concentration of indicator.
5. Adaptability of the technique to on-line computer applications.

For the latter requirement, the dynamic technique is most suited. The integrated technique will provide a method of extreme accuracy but the Vanderbilt technique is satisfactory to replace the Nicholson method in all general catheter room applications of indicator-dilution calibrations.

ACKNOWLEDGMENTS

The author is greatly indebted to Mr. Marcos Rivera, C. P. T., Chief Technician of the Cardiopulmonary Laboratory, Patricia Hazelwood, B. S., C. P. T., and Carol Gumpert, R. N., who assisted with this work. Further acknowledgments are due to Dr. E. V. Newman and Mr. T. G. Arnold, Jr., who kindly loaned the special withdrawal pump for the integrated sample technique study.

REFERENCES

Barr, J. W., and E. C. Bradley. 1968. A calibration device for the earpiece dichromatic densitometer. J. Appl. Physiol. 25:633–635.

Castaneda, A. R., K. C. Weber, G. W. Lyons, and I. J. Fox. 1966. Effect of agglutinating substances on optical density of blood in vitro and in vivo. J. Appl. Physiol. 21:928–932.

Cropp, G. J. A. 1969. Effect of hematocrit on response characteristics of Indocyanine green densitometers. J. Appl. Physiol. 27:132–136.

Edwards, A. W. T., J. Isaacson, W. F. Sutterer, J. B. Bassingthwaighte, and E. H. Wood. 1963. Indocyanine green densitometry in flowing blood compensated for background dye. J. Appl. Physiol. 18:1294–1304.

Emanuel, R. W., J. Hamer, B. N. Chiang, J. Norman, and J. Manders. 1966. A dynamic method for the calibration of dye dilution curves in a physiological system. Brit. Heart J. 28:143–146.

Emanuel, R. W., W. W. Lacy, and E. V. Newman. 1957. An improved method for the calibration of continuously recorded dye dilution curves. Circ. Res. 5:527–530.

Friedlich, A., R. Heimbecker, and R. J. Bing. 1950. Device for continuous recording of concentration of Evans blue dye in whole blood and its application to determination of cardiac output. J. Appl. Physiol. 3: 12–20.

Manning, R. D., Jr., R. A. Norman, Jr., and T. G. Coleman. 1972. Decrements in the optical density of Indocyanine green due to interaction with sodium bisulfite. Clin. Res. 20:29.

Nicholson, J. W., III, and E. H. Wood. 1951. Estimation of cardiac output and Evans blue space in man, using an oximeter. J. Lab. Clin. Med. 38:588–603.

Saunders, K. B., J. I. E. Hoffman, M. I. M. Noble, and R. J. Domenech. 1970. A source of error in measuring flow with Indocyanine green. J. Appl. Physiol. 28:190–198.

Scheel, K. W., D. G. Watson, and P. H. Lehan. 1969. An improved on-line calibrator for dye dilution curves. J. Appl. Physiol. 26(5):667–669.

Sekelj, P., A. Oriol, N. M. Anderson, J. Morch, and M. McGregor. 1967. Measurement of Indocyanine green dye with a cuvette oximeter. J. Appl. Physiol. 23:114–120.

Shadle, O. W., T. B. Ferguson, D. Gregg, and S. R. Gilford. 1953. Evaluation of a new cuvette densitometer for determination of cardiac output. Circ. Res. 1:200–205.

Simmons, R., and R. J. Shephard. 1971. Does Indocyanine green obey Beer's law? J. Appl. Physiol. 30:502–507.

Sparling, C. M., G. A. Mook, J. Nieveen, L. B. Van der Slikke, and W. G. Zijlstra. 1960. Calibration of dye dilution curves for calculating cardiac output and central blood volume. Proc. 3rd Eur. Congr. Cardiol. Rome. Excerpta Med. Int. Congr. Ser. 2:595–598.

Sutterer, W. F. 1967. Optical artifacts in whole blood densitometry. Pediat. Res. 1:66–75.

Theilen, E. O., M. H. Paull, and D. E. Gregg. 1954. Cardiac output determinations with the cuvette densitometer. Techniques for blood of reduced oxygen saturation. Amer. J. Physiol. 179:679.

Völlm, K. R., and E. L. Rolett. 1969. Calibration of dye dilution curves by a dynamic method. J. Appl. Physiol. 26:147–150.

Wexler, H., and R. W. M. Frater. 1969. A simplified method for calibration of dye dilution curves. Read at New York Heart Association Scientific Session of Research, N. Y. Academy of Science (April 24th).

White, R. E. 1968. The design and fabrication of a blood withdrawal system to measure cardiac output. Master of Science thesis, Vanderbilt University, Nashville, Tennessee.

chapter 7/ Distortion and Correction of Indicator-Dilution Curves

Ephraim Glassman
Department of Medicine/
New York University Medical Center/New York, New York

The ideal indicator-dilution curve is one which records the actual time-indicator concentration relationship as it exists at a chosen, intravascular point; however, it has been long known that the properties of the sampling system influence the shape of such curves (Sheppard, 1957). In fact, every indicator-dilution curve is distorted to some degree by the physical characteristics of the sensing and recording systems used to obtain it. This distortion occurs because of the inability of the sensing and recording systems to respond instantaneously to changing indicator concentrations. For example, consider a heated stylus which requires 0.1 second to traverse its full-scale deflection. If a true step function of indicator were detected by some sensing device, and transmitted with absolute fidelity to the recorder, the resultant, inscribed wave would no longer be a square wave. Its leading edge would require 0.1 second to attain the final height (Fig. 7.1). Furthermore, if the sensing device had a similar prolonged response, the distorting effect on the final curve would be additive. Clearly, the slower the response of the components of the measuring system, the greater is the distortion introduced into the recorded curve. As well as such delayed response distortions in the curve, the

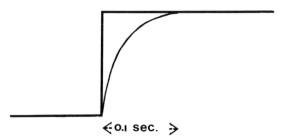

<-0.1 sec. ->

Figure 7.1 Curvilinear recording obtained with a heated stylus in response to the introduction of a step input of indicator dye.

necessity for withdrawing blood and measuring the dye concentration in an external cuvette rather than at an intravascular site produces further alterations in curve configuration. The greater the time required for the cuvette to register a change after a variation in intravascular dye concentration, the less faithful will be the curve. It is apparent that the rapidity with which blood is transported to the cuvette will affect the curve configuration. The speed of blood withdrawal, the volume of tubing interposed between the sampling site and cuvette, and the volume of the cuvette itself will each have an effect on the rate at which blood traverses the sampling system. In addition, the rate at which the cuvette and its associated amplifiers respond to changes in dye concentration will also influence the overall system response. Finally, as in the example of Fig. 7.1, the recorder itself may further distort the final curve by adding its own delayed response.

TIME-DEPENDENT FACTORS

Several investigators (Lacy *et al.,* 1957; Milnor and Jose, 1960) have attempted to quantify the effects of varying blood withdrawal rates as well as using different lengths and diameters of tubing for connection to the densitometer cuvette. Lacy and co-workers found that these variables were related and that the resulting distortion of the curve could be minimized by increasing the flow rate, decreasing the length, or decreasing the diameter of the sampling system. Generally, the purpose of these investigations has been to determine the effect on various specific curve parameters, e.g. appearance time, build-up time, or mean transit time, and to determine applicable correction factors. Parenthetically, all of these corrections apply to time-dependent factors. The measurement of cardiac output is independent of the time-distorting response of the system. Application of these correction factors results in a corrected estimation of the individual factors cited.

One method of improving the curve fidelity is based upon rapid blood sampling rates and, concomitantly, minimizing the volume of the withdrawal system. Sampling rates of 6 ml/second in dogs (Holt, 1956) and 1 to 2.2 ml/second in humans have been achieved for this purpose (Wilcken, 1965; Freis et al., 1960; Levinson et al., 1967). Curves recorded in this fashion have demonstrated discrete downsteps of dye concentration when withdrawal has been made from the pulmonary artery or aorta. The method has the disadvantages of requiring special withdrawal pumps, not available in the average catheterization laboratory, and of removing large volumes of blood.

Sherman and co-workers (1959) considered the distortion to be predictable from theoretical considerations based on laminar flow. Deductions regarding this distortion were compared favorably to the experimental data of Lacy over a wide range of catheter and flow parameters. It was felt that the fidelity of the sampling systems could be improved by placing the sampling catheter tip as close to the source of concentration change (in this case, the ascending aorta) as possible at the expense of increasing the internal volume of the catheter. At the same time that the volume should be minimized, radius and the flow rate should be maximized within the limits set by the necessity to conserve blood. A figure of merit for catheter sampling systems was devised, requiring for acceptable distortion that the ratio of the catheter volume and twice the withdrawal rate should be less than the time in which significant changes in indicator concentration occur, that is, the cardiac cycle interval.

A satisfactory withdrawal system, then, is one in which

$$\frac{\text{catheter volume (cc)}}{2 \times \text{withdrawal rate (cc/sec)}} < \text{R–R interval/sec.}$$

Deviations from the parabolic velocity gradients with zero wall velocity characteristics of laminar flow occur with low flows and high hematocrits, and in these circumstances redefining the figure of merit is required.

THE TRANSFER FUNCTION

Another technique for correction of dye curve distortion has been reported by Glassman, Blesser, and Mitzner (1969). According to their analysis, the actual recorded curve may be considered as having been determined by a distorting factor, introduced by the sampling and recording systems, operating on the true *in vivo* curve (Fig. 7.2).

Figure 7.2 The distortion factors operate upon the true curve and result in the production of the recorded, distorted curve.

If the actual distorting effect or transfer function of the system is determined, application of it, in an inverse function to the recorded curve, should result in determination of the original undistorted curve. In equation form this is written

$$F(s) = G(s) \cdot H(s), \tag{7.1}$$

where $F(s) =$ Laplace transform of the recorded curve of dye concentration with respect to time (output function), $G(s) =$ Laplace transform of the intravascular concentration of dye with respect to time (input function), and $H(s) =$ transfer function of the sampling and recording system. Once $H(s)$ has been determined for a given system, and $F(s)$ is recorded as the standard dye curve, then $G(s)$, the intravascular curve, can be calculated. In actual practice, a simple analogue device was constructed by the authors, which performed the calculation on line, and allowed the simultaneous recording of both the distorted and undistorted curves (Fig. 7.3).

To determine the transfer function, $H(s)$, the authors utilized a system which allowed undyed blood to be withdrawn, at a set rate, through a given type catheter and then through a cuvette densitometer. A stopcock was then rotated, as rapidly as possible, switching to a source of dyed blood. In effect, this introduced a step function input, $G(s)$, of dye into the system. The output of the densitometer, $F(s)$, was recorded. Upon analysis, the recorded curve was found not to be a step function, but instead had the form of a first order exponential curve (Fig. 7.4).

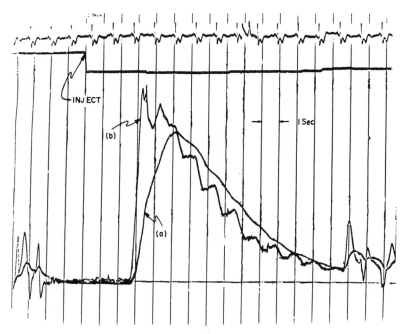

Figure 7.3 Indocyanine green injection into the left ventricle and sampling in the ascending aorta. The original, distorted curve (*a*) and the curve obtained by use of a compensating network (*b*).

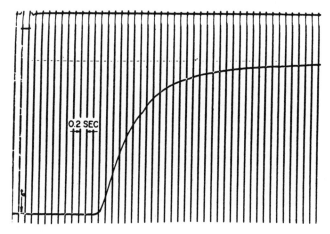

Figure 7.4 The curve recorded when a step function input of dyed blood is withdrawn through a size 8 French, 70-cm catheter, and a cuvette densitometer.

The time constant, τ, of the exponential is determined by measurement of the slope of the line obtained by replotting the curve on semilogarithmic axes. Subsequent mathematic analysis (see Appendix) demonstrated that it is possible to reconstruct the original undistorted curve by adding to the distorted curve τ times its first derivative. This may be written

$$g(t) = f(t) + \tau f'(t), \tag{7.2}$$

where $g(t) = $ input function (undistorted dye curve), $f(t) = $ output function (distorted or uncompensated dye curve), $f'(t) = $ first derivative of output function, and $\tau = $ time constant of system. This equation indicates that the input function is equal to the output function plus a portion of the first derivative of the output function. The amount of $f'(t)$ which must be added to obtain compensation is dependent upon the time constant of the particular system being used. It will vary with changes in catheter length or diameter, the volume of the interconnecting tubing, the volume of the cuvette, the response time of the densitometer and recording system, and the sampling rate. Each of these, however, may be kept reasonably constant without difficulty. The time constant, τ, must be determined for each catheter type which is to be used.

To obtain compensated curves directly, without the need of applying any manual calculation, a simple analog computing device was introduced into the recording system. This automatically performed the differentiation, multiplication, and addition described in (7.2). The output of the computer was recorded as $g(t)$, the input function or undistorted dye curve.

The ability to calculate ejection fractions from the compensated curves was evaluated by using a pulsatile pump with variable residual volumes. Clearly defined concentration steps, corresponding to individual pump strokes, were easily determined. By using the method of Holt (1956), excellent estimation of the ejection fraction was possible. The method was also applied clinically. Dye was injected into the left ventricle and sampled just above the aortic valve. A representative curve is reproduced in Fig. 7.5. Ejection fractions ranging from 0.18 to 0.61 were obtained in a diverse group of patients. Comparison with other methods of obtaining intravascular indicator-dilution curves by using fiberoptic catheters or thermodilution methods is being performed.

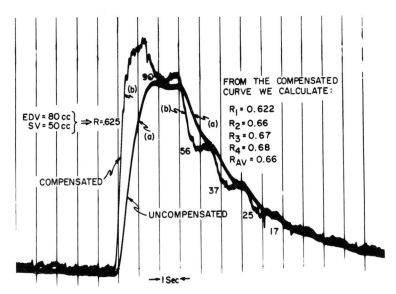

Figure 7.5 Calculation of the left ventricular ejection fraction from a compensated curve (b); the uncompensated curve (a) is simultaneously recorded.

APPENDIX

It is desired to recreate the true input to the system. To accomplish this, an expression for the system transfer function must first be found.

Assuming a step input, $f(t)$, and a delayed exponential response output, $y(t)$, one may then write

$$f(t) = u(t) \text{ (unit step)},$$
$$y(t) = (1 - e^{-(t-T)/\tau})u(t - T),$$

where $\tau = $ time constant and $T = $ time delay. Taking Laplace transforms of input and output yields

$$F(s) = 1/s$$
$$Y(s) = \left[\frac{1}{s} - \frac{1}{s + 1/\tau}\right] e^{-sT} = \left[\frac{1/\tau}{s(s + 1/\tau)}\right] e^{-sT}$$

where $F(s)$ and $Y(s)$ are the Laplace transforms of $f(t)$ and $y(t)$, respectively. With $H(s) =$ the system transfer function, $Y(s) = H(s) F(s)$,

$$\frac{1/\tau}{s(s + 1/\tau)} e^{-sT} = H(s)1/s,$$

$$H(s) = \left[\frac{1/\tau}{s + 1/\tau}\right] e^{-sT}.$$

Neglecting the time delay (e^{-sT}), the true input to the system may be found in the following manner. Let $G(s)$ be the transform of the unknown input and $F(s)$ the transform of the recorded output. Then,

$$F(s) = H(s)G(s) = G(s)\frac{1/\tau}{s + 1/\tau},$$

$$G(s) = \tau sF(s) + F(s),$$

and converting back to the time domain yields

$$g(t) = f(t) + \tau\frac{df(t)}{dt}.$$

Hence, the actual dye curve input to the system may be recreated by taking the recorded curve and adding to it τ times its first derivative.

REFERENCES

Freis, E. D., G. L. Rivara, and B. L. Gilmore. 1960. Estimation of residual and end-diastolic volumes of the right ventricle of men without heart disease, using the dye-dilution method. Amer. Heart J. 60:898–906.

Glassman, E., W. Blesser, and W. Mitzner. 1969. Correction of distortion in dye dilution curves due to sampling systems. Cardiovasc. Res. 3: 92–99.

Holt, J. P. 1956. Estimation of the residual volume of the ventricle of the dog's heart by two indicator dilution techniques. Circ. Res. 4:187–195.

Lacy, W. W., R. W. Emanuel, and E. V. Newman. 1957. Effect of the sampling system on the shape of indicator dilution curves. Circ. Res. 5:568–572.

Levinson, G. E., M. J. Frank, M. Nadimi, and M. Braunstein. 1967. Studies of cardiopulmonary blood volume. Measurement of left ventricular volume by dye dilution. Circulation 35:1038–1048.

Milnor, W. R., and A. D. Jose. 1960. Distortion of indicator-dilution curves by sampling systems. J. Appl. Physiol. 15:177–180.

Sheppard, C. W. 1957. Catheter artefacts in indicator dilution experiments. Fed. Proc. 16:118.

Sherman, H., R. C. Schlant, W. L. Kraus, and C. B. Moore. 1959. A figure of merit for catheter sampling systems. Circ. Res. 7:303–313.

Wilcken, D. E. L. 1965. The measurement of the end-diastolic and end-systolic, or residual volumes of the left ventricle in man, using a dye-dilution method. Clin. Sci. 28:131–146.

part 2/ SPECIAL APPLICATIONS

chapter 8/ Detection and Quantification of Intracardiac and Great Vessel Shunts

L. Jerome Krovetz
Division of Pediatric Cardiology, and the Departments of Pediatrics and Biomedical Engineering/The Johns Hopkins University School of Medicine/The Johns Hopkins Hospital/Baltimore, Maryland

One of the first (and still very important) applications of dye-dilution techniques to cardiac catheterization was the use of the abnormal curve morphology to localize the site of a right-to-left shunt (Swan and Wood, 1953). Since then, the diagnostic as well as the investigative value of indicator-dilution techniques in congenital heart disease has received progressively widespread recognition (Braunwald *et al.,* 1957; Crane *et al.,* 1957; McDonald, 1959; Swan, 1954; Wood, 1960). The development of methods to describe both uni- and bidirectional shunts has been extensive, and a major factor of this increased utilization has been the availability of suitable instrumentation and the use of foreign gases as indicators in these techniques (Marshall *et al.,* 1962). The nitrous oxide test was introduced in 1958 (Morrow, Sanders, and Braunwald), followed by radioactive gases, krypton-85 (Sanders, 1958), and iodine-131 (ethylodide) (Amplatz and Marvin, 1959). Hydrogen (Clark and Bargeron, 1959) has also been used for the detection of left-to-right shunts, as well as thermal dilution (Paul *et al.,* 1958) and external scanning techniques (Huff *et al.,* 1957). The various methods of utilizing Indocyanine green and Freon in the diagnosis and quantitation of shunts are detailed in this chapter.

DETECTION AND QUANTIFICATION
OF LEFT-TO-RIGHT SHUNTS

In the presence of a left-to-right shunt, injection of indicator into the central venous circulation with peripheral arterial sampling results in an "early recirculation." In other words, dyed blood flows to the

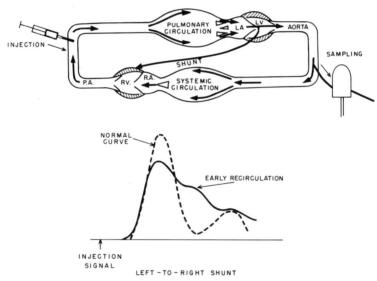

Figure 8.1 Schema of the circulation in a patient with a left-to-right shunt; for example, a ventricular septal defect. When the dye-blood mixture reaches the site of the defect, it divides into two streams, one through the normal channels and out the aorta and the other through the shunt. If the dye and blood are adequately mixed, the proportion of the dye traversing the shunt and the normal circulation should be representative of the magnitude of the shunt. Methods for extracting this information from the composite dye curve are discussed in the text.

site of the shunt, and then a portion of its passes through the shunt and recirculates through the lungs. Since pulmonary circulation time is significantly shorter than systemic circulation time, the shunted, dyed blood appears between the normal peak and the normal recirculation peak (Fig. 8.1). Left-to-right shunts, in excess of 25% of pulmonary blood flow, will usually distort the disappearance slope of the indicator-dilution curve.

Although the detection and localization of left-to-right shunts may be sufficient for some purposes, an estimate of the magnitude of the shunt is generally desired, and a number of methods of providing such estimates have been described.

The method of Carter and co-workers (1960) is based on the rate of disappearance of dye from the circulation. The formula for dilution curves recorded at the radial artery and validated by them for right-sided injections in older children and adults is

$$141 \times \frac{C(2BT)}{C(BT)} - 42,$$

$$135 \times \frac{C(3BT)}{C(BT)} - 14,$$

(8.1)

where C refers to concentration after each multiple of the build-up time (BT). The values obtained from these two calculations are averaged and expressed as per cent of pulmonary blood flow contributed by the left-to-right shunt and is illustrated in Fig. 8.2. The empirical constants in Carter's formula were obtained from 100 right heart injections with arterial sampling in 47 subjects ranging from 3 to 56 years of age. Generally excellent agreement with oximetric calculations were obtained when the left-to-right shunt exceeded 35% of pulmonary blood flow. Below that value, the Fick method becomes increasingly inaccurate (Wood et al., 1963), and the correlation of left-to-right shunts by dye and Fick becomes increasingly poor. That this is due to the limitations of oximetry is suggested by the excellent correlation $(r = 0.997)$ obtained for six subjects in whom the slope disappearance ratio method was compared to values obtained from the areas of two simultaneous dye curves sampled from venous and arterial sites (Wood and associates, 1958).

Carter's formula was not proposed and has never been validated for left heart injections. Dye curves obtained after left heart injection generally show a distinct notch after the inscription of peak concentrations when a left-to-right shunt is present (Fig. 8.3). Attempts to apply this formula to left heart injections frequently resulted in the calculation of two grossly disparate values, with the first value as much as 50% lower than the second. The use of an average value obtained from two such disparate calculations is of doubtful validity.

Krovetz and Gessner (1965) proposed a method for calculating left-to-right shunts from left heart dye curves as follows:

L-to-R shunt (as per cent of pulmonary blood flow)

$$= \frac{Cs(Ts - PCT)}{Cs(Ts - PCT) + CnTn,} \qquad (8.2)$$

where Cn is the initial (normal) concentration peak in milligrams per liter (if the densitometer used has a linear calibration curve, centimeters of galvanometer deflection may be used instead of milligrams

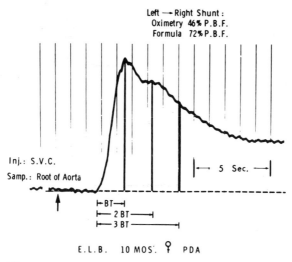

Figure 8.2 Shunt calculation by the formula of Carter *et al.*: indicator-dilution curve in patent ductus arteriosus, from injection in superior vena cava and sampling in the root of aorta. The *heavy vertical lines* indicate the dye concentration at each multiple of the build-up time (BT). The *arrow*, in this and subsequent figures, refers to the midpoint of the injection of dye, corrected for transit of arterial blood through the sampling system. (From Krovetz, L. J., and I. H. Gessner, 1965. A new method utilizing indicator-dilution techniques for estimation of left-to-right shunts in infants. Circulation 32:772–777. With permission.)

per liter for Cn and Cs), Tn is the time from the initial appearance of dye to Cn, Cs is the second (shunt) concentration peak, Ts is the time from initial appearance of dye to Cs, PCT is the pulmonary circulation time, estimated in turn by the difference in appearance times of dye curves obtained after right heart and left heart injections of

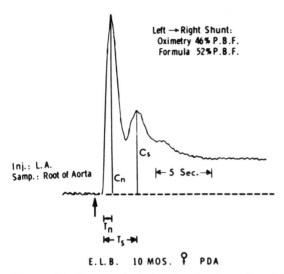

Figure 8.3 Shunt calculation by the formula of Krovetz and Gessner: indicator-dilution curve recorded after left atrial injection in the same patient as that of Fig. 8.2. Note the shorter appearance time and the distinct notch after the peak concentration. Cn and Cs are the dye concentrations of the primary and shunt peaks; Tn and Ts are the time intervals between appearance and the respective peaks. (Reprinted, as Fig. 8.2, with permission.)

dye, in the absence of right-to-left shunts. It is preferable to minimize the time between the performance of such a pair of dye curves.

Formula (8.2) requires the estimation of pulmonary circulation time to obtain the appearance time for shunted dye. In infants and small children, the time of appearance of dye shunted through pulmonary flow pathways occurs before the inscription of the concentration peak. This results in a high disappearance ratio and an erroneously high estimation of the magnitude of the shunt. Typical curves obtained after injection into the right and left atria in a normal newborn are shown in Fig. 8.4. From the difference in appearance times between the pair of curves, the *solid straight line* was added to indicate the estimated time shunted, dyed blood lags behind dyed blood flowing in normal circulatory pathways. With right atrial injection, shunted dye appears on the upstroke of the recorded dye curve, whereas after left atrial injection, shunted dye appears after peak concentration is reached. In 30 neonates and infants studied, pul-

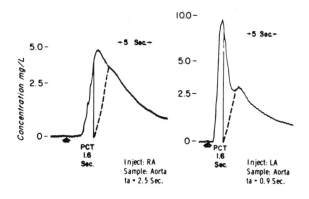

Baby Sc   38 Minutes After Birth

Figure 8.4 Estimated time of appearance of recirculated ductal flow in relation to the primary curves obtained after right and left atrial injections with sampling in the aorta. The *solid line* indicates the start of estimated pulmonary recirculation of shunted blood. Note that pulmonary circulation time (*PCT*) is longer than the time from initial appearance of dye to peak concentration after left heart injection. *PCT* was always less than time from initial appearance to peak concentration after right atrial injection. (Reprinted, as Fig. 8.2, with permission.)

monary circulation time was always greater than the time from initial appearance of dye to peak concentration after left heart injection. Pulmonary circulation time was always less than the time from initial appearance to peak concentration after right atrial injection.

A typical dye curve obtained after injection into the right atrium of an infant with a left-to-right shunt is shown in Fig. 8.2 and after left atrial injection in Fig. 8.3. As noted, calculations based on right heart injection (Carter's formula) overestimated the degree of left-to-right shunt, whereas calculations based on left atrial injection (Krovetz and Gessner formula) showed excellent agreement with that calculated from oximetric data. Comparison of the two methods of shunt estimation, with oximetry as the standard, are shown in Figs. 8.5 and 8.6. It is obvious that the estimated left-to-right shunt after right heart injection of dye in 11 infants shows a large error, generally an overestimation (Fig. 8.5). Only 16 of 36 values were within 20% of the oximetric calculations. Estimated shunts based on left heart injections in these same infants were all within 20% of the oximetric calculations (Fig. 8.6). The Krovetz and Gessner formula has also

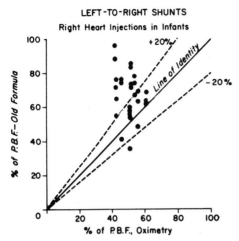

Figure 8.5 Correlation of left-to-right shunts calculated after right heart injections using Carter's formula and left-to-right shunt calculated by oximetry. Included in this chart are 37 curves from 11 infants. The line of identity is *solid;* the *dotted lines* delimit a zone of agreement within 20% of the oximetric value. (Reprinted, as Fig. 8.2, with permission.)

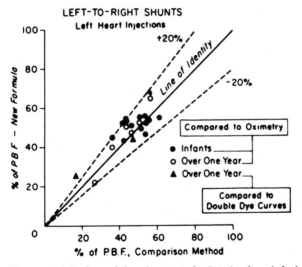

Figure 8.6 Left-to-right shunts calculated after left heart injections by the formula of Krovetz and Gessner. Included in this chart are 16 sets of data from the same 11 infants and 8 sets from 7 older subjects. (Reprinted, as Fig. 8.2, with permission.)

been tested in older patients ranging in age from 28 months to 39 years. Essentially similar results were noted, namely eight of nine left heart injections were within 20% of the value established by oximetry. This formula, although necessary in infants, is thus also applicable to older and larger subjects.

Oximetry failed to detect one of the shunts in this group. In this child, dye was injected into the left ventricle and sampled simultaneously through densitometers attached to catheters in the main pulmonary artery and the femoral artery. For this small shunt, directly calculated at 16% of pulmonary blood flow, the dye curves obtained after right-sided injection of dye were normal. Injection into the left ventricle showed an obvious second concentration peak similar to that illustrated in Fig. 8.4. The left-to-right shunt calculated by the formula was 25% of the pulmonary blood flow.

Phinney et al. (1964) proposed the use of a spread-to-appearance time ratio to take account of alterations in the contour of dye curves. Such alterations have been shown experimentally to be due to mixing alone, independent of cardiac output or the physical volume between injection and sampling sites (Cliffe and co-workers, 1959). The spread of the curve is defined as the time in seconds measured on the semilogarithmic plot at one-tenth of the curve's peak concentration. They obtained a slightly better correlation with left-to-right shunts calculated by the Fick method than the disappearance slope ratio method proposed by Carter. However, as shown in Table 8.1, the correlation coefficients decrease from 0.81 for spread/appearance time measured after injections into the pulmonary artery to 0.71 for right ventricular injections and to 0.65 for right atrial injections.

In addition, the slopes and intercepts recalculated from the data of Phinney et al. were the highest of the formulas compared (Table 8.1). Furthermore, there was a marked deviation from the results obtained by Carter with a slope 0.46 and a high intercept value when applying the slope disappearance ratio method to Phinney's data. This strongly suggests that much of the lack of correlation lies in the oximetry data obtained by Phinney.

A method for quantitation of left-to-right shunts using an earpiece dilution curve was presented by Nakamura and associates (1967). The principle consists of an estimation of concentration area of the abnormal recirculation hump on the downslope and its comparison with the area on the first circulated dye curve. The downslope of the first dye curve was determined by a modification of the forward tri-

Table 8.1 Comparison of formulas for quantification
of left-to-right shunts with calculations based on oximetry

Author and year	Reference for data (per column 1)	n	Slope	Y Intercept	Correlation coefficient (r)
Carter et al.,	a	54	0.73	15.70	0.850
1960 (a)	a	6[a]	0.98	6.56	0.997
	c	32	0.46	24.00	0.524
	d	33[b]	−0.13	60.33	−0.299
Mook and Zijlstra, 1961 (b)	b	33	0.82	5.05	0.920
Phinney et al.,	c	32 PA	2.32	25.89	0.806
1964 (c)	c	28 RV	2.74	27.17	0.709
	c	30 RA	3.30	24.42	0.647
Krovetz and Gessner, 1965 (d)	d	23	0.93	0.71	0.801
Nakamura et al.,	e	40	N.A.[c]	N.A.	0.83
1967 (e)	f	34	0.99	−1.95	0.903
Sato et al., 1969 (f)	f	34	1.01	−1.82	0.845

[a]Based on two simultaneous dye curves sampled from venous and arterial sites (Wood, Swan, and Marshall, 1958).

[b]Infants; see text for discussion.

[c]N.A., data not available.

angle method. Estimates using this method yield a reasonably high agreement with shunts calculated by oximetry (Table 8.1). The effect of injection sites on the calculated value of the shunt was shown to be insignificant. However, their method is extremely complicated, especially in the calculation process of ts, which is a time constant of the exponential downslope, being the time interval in seconds required for the dye concentration to decay to $1/e$ (37%) of the peak concentration.

Sato and colleagues (1969) simplified the method proposed by Nakamura. Ts time is estimated from the log of the ratio of peak concentration to build-up time; the time at which the downslope is decaying exponentially is also estimated. They reported no significant difference in correlation obtained with the Fick method for various sites of injection, including peripheral veins and left atrium. However, the method does not seem to be significantly simpler than that of Nakamura.

The most complex, but perhaps the most accurate method of estimating left-to-right shunts was described by Mook and Zijlstra (1961). This method is a direct independent method and does not depend upon correlations between curve parameters and the magnitude of shunts calculated from oxygen saturation data. The basis of this method is a semilogarithmic extrapolation of the downslope of the dye curve as in the usual method for obtaining cardiac output. If the extrapolated portion is now subtracted from the original dye curve, a second curve results which represents the left-to-right shunt. The downslope of this curve is then again extrapolated by using a semilogarithmic replot, and the area representing the shunted dye is measured by using planimetry. This method has the advantage of being very sensitive to small left-to-right shunts. The authors studied eight patients in whom oximetry had failed to show a left-to-right shunt, whereas, with the dye-dilution curves, very small shunts could be clearly demonstrated. This method has a further advantage of requiring only a single venous dye curve for calculation purposes and would not appear to be affected by rapid circulation times that occur in infants and invalidate the downslope ratio method.

For extremely large left-to-right shunts in excess of 70% of pulmonary blood flow, or a pulmonary-systemic flow ratio greater than about 3:1, the indicator-dilution methods show an increasingly poor correlation with the Fick method. It has been pointed out, however, that in this range the errors involved in the estimation of left-to-right shunts using oximetry clearly invalidate that method (Schostal and co-workers, 1972). In addition, most of the dye-dilution methods depend upon the ability to locate a distinct break on the downslope representing shunted dye or to be able to extrapolate the semilogarithmic downslope of the first circulation of dye. In the presence of large left-to-right shunts, it may not be possible to apply these techniques. This would seem to leave the downslope disappearance ratio method of Carter as the only useable one for large left-to-right shunts. From a practical viewpoint, it would appear to make little difference whether the Qp/Qs was 4:1 or 5:1 or 6:1. It is our custom to report shunts of this magnitude as simply Qp/Qs greater than 4:1.

TECHNIQUES FOR SMALL LEFT-TO-RIGHT SHUNTS

A normal peripheral arterial indicator-dilution curve may be present when a small shunt exists, i.e. one that is less than 25% of pulmonary blood flow. More sensitive methods are needed to detect these small left-to-right shunts. Indicator may be injected into left heart, preferably into a pulmonary vein, and sampled from the pulmonary artery. This will detect the presence of a small left-to-right shunt at or distal (downstream) to the injection site by causing indicator to appear quickly at the pulmonary artery sampling site. By appropriate selection of injection sites, one can localize the site of the left-to-right shunt (Marshall, Helmholz, and Wood, 1962). A schematic representation of the dye curves obtained by this method is shown in Fig. 8.7.

INJECTION INTO DISTAL PULM. A. OR PULM. V.

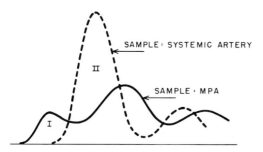

Figure 8.7 A technique for small left-to-right shunts. Dye curves obtained by injecting into the distal pulmonary artery or a pulmonary vein and sampling simultaneously from catheters in the main pulmonary artery (*MPA*) and a systemic artery. Area *I* is proportional to the left-to-right shunt, and area *II* is proportional to the pulmonary blood flow.

After correction for any variations in sensitivity of the two densitometers, area I, the area under the initial portion of the main pulmonary artery (*MPA*) curve, is proportional to the left-to-right shunt and area II, the area under the initial portion of the femoral artery curve, to pulmonary blood flow. The downslopes in both instances are extrapolated by semilogarithmic replotting in the usual manner. The left-to-right shunt may then be calculated by a simple ratio between areas I and II.

The main problem with this technique, namely lack of adequate mixing, is illustrated by one of our patients. All of the clinical and catheterization data in this child were consistent with a small ventricular septal defect, except for one pair of dye curves obtained after injection of dye into the left ventricle with simultaneous sampling from the main pulmonary artery and the femoral artery. In this case, the shunt was estimated at 45%, whereas the other dye curves showed the shunt to be 16 to 25% of pulmonary blood flow. It is probable that the aberrant values were obtained with the tip of the injection catheter pointing either toward or actually within the limits of the ventricular septal defect. This illustrates the value of multiple injection sites. An upstream injection site, i.e. pulmonary vein or left atrium, should be used whenever possible to provide greater opportunity for mixing of dye and blood. In common with all indicator-dilution techniques, mixing of dye and blood is a necessary assumption. Less opportunity for this exists when the injection site is distal to the pulmonary circulation (Krovetz and Benson, 1965; Swan and Beck, 1960). It is probable that this method will give erroneous values for left-to-right shunts through an atrial septal defect, even when averaged for injections into both right and left pulmonary veins.

In the "double catheter" modification of the above technique, one venous catheter is placed in peripheral pulmonary artery (but not the wedge position), and a second venous catheter is placed in the main pulmonary artery (Marshall et al., 1962; Wood et al., 1958). Indicator is injected into the distal pulmonary artery and sampled from the main pulmonary artery. If a left-to-right shunt exists, the indicator will appear earlier than normal at the pulmonary artery sampling site, as above. A more exact definition of "earlier" may be made by comparing a suspected abnormal curve with one obtained after sampling in a location proximal to the shunt.

Another double catheter technique uses one venous catheter in the main pulmonary artery and a second venous catheter in a peripheral vein (Braunwald and associates, 1959). Indicator is injected into the peripheral vein and sampled from pulmonary artery. If a small left-to-right shunt exists, there will be a distortion of the downslope of the dye curve because any indicator shunted will appear earlier than the normally recirculating dye. In this technique, distortion of the disappearance slope of the pulmonary artery curve indicates a left-to-right shunt. This is a useful but less sensitive technique than those comparing appearance times.

Detection and localization of small left-to-right shunts may also be accomplished by the hydrogen inhalation technique (Clark and

Bargeron, 1959). A catheter with a platinum electrode attached to the tip is advanced into the pulmonary artery. Hydrogen is administered with an anesthetist's bag to a three-way valve and face mask. The mask is held over the face while the patient breathes room air. Turning the valve provides a single breath of hydrogen. When hydrogen in blood reaches the platinum tip, an electrical current is generated and a suitable recording can then be made. Thus, in essence, an indicator is placed into the pulmonary capillaries similar to a distal pulmonary artery injection. If hydrogen appears at the sampling site early, a left-to-right shunt exists at or proximal to the sampling site.

If there is some doubt about the "early" appearance time, the electrode catheter may be placed into a pulmonary artery wedge position. This will give almost instantaneous appearance of hydrogen. An alternate method is to repeat the hydrogen curves in a more proximal chamber. If the appearance time is significantly longer in this chamber, the left-to-right shunt is localized to the more distal chamber. Unfortunately, hydrogen curves are qualitative only. The use of gaseous hydrogen has been popular but is potentially explosive when combined with oxygen and hazardous to use.

A simple test for the detection of small left-to-right shunts utilizes Freon (registered trademark for DuPont fluorinated hydrocarbons) as an indicator and a commercially available halogen leak detector as a sensing device. This method has the advantages of being easy to perform with standard cardiac catheters and safe for both patient and operating personnel. The Freon compounds are stable, noninflammable, inert gases with extremely low toxicity and are, therefore, widely used as propellants in spray cans and as refrigerants.

The patient is given a single inhalation of Freon while blood is withdrawn through the cardiac catheter at a slow, steady rate (0.2 cc/second) into a clear plastic chamber, where Freon is forced from the blood stream by a high velocity CO_2 jet. The carbon dioxide and the released Freon gases are passed from the spray chamber into a sensing and detecting unit, which produces a recordable signal. The test is described by Amplatz et al. (1969) and may be used to detect and localize the shunt but not to quantify it. In the 178 patients studied, there appeared to be only one false negative Freon test. This failure was due to faulty sampling through an end-hole catheter. In contrast, of the 62 cases in which the Freon test indicated a left-to-right shunt, 25 could not be detected by oximetry, 8 could not be detected by selective right-sided angiography, and 4 could not be detected by venous dye curves. Furthermore, in 11 patients, the shunt was detected only by the Freon method and was not found by any combination of the other methods. There appeared to be only one false positive test in a child with pulmonic regurgitation. More recent communication with one of the authors (R. V. Lucas) reveals that he has had more apparently false positive tests in patients with pulmonic valve insufficiency. This suggests limited usefulness in evaluation of postoperative tetralogy of Fallot patients. On the other hand, false positive results in unoperated patients have been extremely rare, and the test is of great usefulness in those with a precatheterization diagnosis of left-to-right shunt. If oximetry and the Freon test are both negative, the authors believe that a left-to-right shunt may be excluded.

DETECTION AND QUANTIFICATION OF RIGHT-TO-LEFT SHUNTS

When the indicator is a foreign gas (NO, ^{85}Kr, H_2) the gas must first be dissolved in saline solution. The indicator is injected into superior or inferior vena cava and sampled simultaneously in pulmonary artery and systemic artery. Since foreign gases are cleared almost entirely by one passage through the lungs, any indicator appearing at the peripheral arterial sampling site must have by-passed the lungs through a right-to-left shunt. If the gas appears at the systemic arterial and pulmonary arterial sites at about the same time, this also indicates a right-to-left shunt.

When the indicator is a dye, injection into the superior or inferior vena cava with peripheral arterial sampling will demonstrate a right-to-left shunt in a similar manner. In this case, the indicator creates an early deflection on the upstroke of the dilution curve (Fig. 8.8). When the right-to-left shunt is at the atrial level, the right ventricular and pulmonary artery curves have a longer appearance time with no early deflection. Similarly, when the shunt is at the ventricular level, injection in superior vena cava, inferior vena cava, right atrium, or right ventricle will cause an earlier deflection than injection into the pulmonary artery.

Swan et al. (1953) proposed a scheme for calculating the magnitude of right-to-left shunts by indicator-dilution curves (Fig. 8.9). In this scheme, right-to-left shunts are quantitated by assuming that the

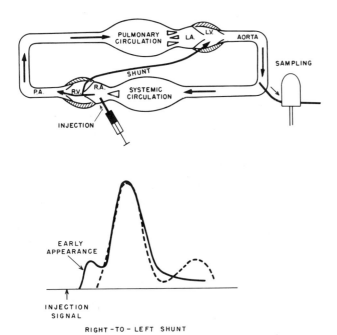

RIGHT – TO – LEFT SHUNT

Figure 8.8 Schematic representation of the circulation in a patient with a right-to-left shunt at the ventricular level. Dye injected proximal to the shunt will result in early appearance of dye at the arterial sampling site since the dye-blood mixture does not go through the pulmonary circulation. Dye injected downstream to the site of the shunt will result in an essentially normal dye curve appearance time and contour.

curve areas proportional to the shunt and the systemic blood flow can be estimated by a forward triangle method (see Chapter 5). The appearance time must be corrected for the volume of the sampling catheter, as well as the speed of blood withdrawal. Dead-space transit time (td) equals withdrawal rate in seconds/cc $\times$ dead space in cc $\times$ 0.6. The latter is an empirical constant to correct for the parabolic nature of flow through the lumen. Dead-space transit time is then subtracted from the measured appearance time (ta) to obtain the corrected appearance time (tac).

For a normal curve, the appearance time occurs after 56% of the time from corrected injection signal to time of concentration peak. This fact can be used to estimate the appearance time of the normal dyed blood flow (tn) as the remainder, or $100\% - 56\% = 44\%$ or

QUANTIFICATION OF
RIGHT—TO—LEFT SHUNTS

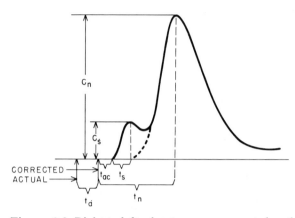

Figure 8.9 Right-to-left shunt: measurement by the forward triangle method of Swan. The area of the initial triangular portion of the curve (proportional to the shunt) can be easily seen to be ½ × time to peak shunt concentration (Ts) × peak shunt concentration (Cs). For the corresponding area of the curve representing systemic blood flow, the time to peak normal concentration is estimated from the observation that the appearance time in the absence of a right-to-left shunt occurs after 56% of the time from the corrected injection signal to the time of concentration peak (Cn). The area of this triangle is therefore, ½ × 0.44 Tn × Cn. Td and Tae are, respectively, the dead-space transit time and the corrected appearance time (see text).

0.44. Assuming (a) that flow is proportional to the area under the dye curve in Fig. 8.9, and (b) that the areas can be approximated by triangles,

(a) shunt flow $= \dfrac{Ts \times Cs}{2}$,

(b) normal (pulmonary) flow $= \dfrac{0.44 \times Tn \times Cn}{2}$,

where Ts, Tn, Cs, and Cn are defined as in equation (8.2).

Since systemic blood flow must be equal to the sum of the right-to-left shunt and pulmonary blood flow,

R-to-L shunt, expressed as percentage of systemic blood flow
$$= \frac{Ts \times Cs \times 100}{Ts \times Cs + 0.44 \times Tn \times Cn} \qquad (8.3)$$

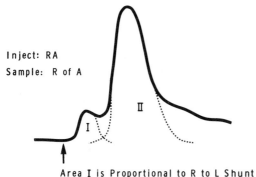

Inject: RA
Sample: R of A

Area **I** is Proportional to R to L Shunt
Area **II** is Proportional to Systemic Blood Flow

Figure 8.10 Schematic of Mook and Zijlstra's method for quantification of right-to-left shunts. The downslope of the initial portion of the dye curve is extrapolated and is subtracted from the dye curve. This process is repeated on the resultant curve and the areas under the respective curves, labeled areas *I* and *II*, are proportional to the shunt and the systemic blood flow.

A slight variation of the method of Mook and Zijlstra (1961) may be used for quantification of right-to-left shunts (Fig. 8.10). In this method, the downslope of the initial part of the dye curve is extrapolated and subtracted from the original curve. This process is repeated on the resulting curve, and the areas under the respective curves are measured. The area under the initial part of the curve is taken as proportional to the right-to-left shunt, whereas the area under the second curve represents systemic blood flow. This method, although slightly more complex, does not require any assumptions other than those of adequate mixing and stationarity which apply to all indicator-dilution techniques. As shown in Table 8.2, there was

Table 8.2 Comparison of formulas for quantification
of right-to-left shunts with calculations based on oximetry

Author and year	n	Slope	Y Intercept	Correlation coefficient(r)
Swan *et al.*, 1953	20	1.22	2.80	0.895
Swan *et al.*, modified	20	1.02	0.41	0.885
Mook and Zijlstra, 1961	8	1.13	2.11	0.977

an extremely high correlation with shunts based on oximetry in the small sample reported by Mook and Ziljstra.

BIDIRECTIONAL SHUNTS

Dye curves in bidirectional shunts combine the characteristics of right-to-left shunts, with early appearance of indicator, and of left-to-right shunts, with early recirculation. Quantification of such bidirectional shunts are fraught with large errors, especially when the shunt in each direction is large, as illustrated in Fig. 8.11.

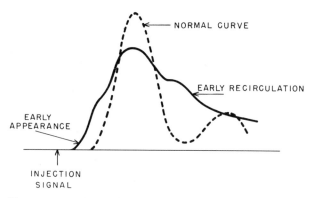

Figure 8.11 Bidirectional shunt with both early appearance and early recirculation.

In the face of a bidirectional shunt, it is tempting to use the dye curves distal to the shunt to calculate the left-to-right shunt. However, when the circulation is fast, as with small patients or high flows, normal systemic recirculation occurs at virtually the same time as the abnormal pulmonary circulation. This results in overestimation of the magnitude of any left-to-right shunt. This is especially true in the small infant with a massive right-to-left shunt.

As in the case of calculations based on oximetry (Schostal *et al.,* 1972), calculations of bidirectional shunts from dye curves may be inaccurate and misleading, especially when the magnitudes of the shunts are approximately equal. It should be noted that calculation of the right-to-left shunt alone is affected by the lowering of the peak concentration of the normal part of the curve (Cn) by the presence of a left-to-right shunt. It is theoretically possible to quantitate bidirectional shunts by using multiple injection and sampling sites, but such experiments are not reported.

This is not to negate the value of indicator-dilution techniques in the presence of bidirectional shunting, but rather to caution against overemphasis and undue reliance upon any figures calculated. Certainly all other pertinent data need to be considered before deciding on the operability of a patient with a bidirectional shunt. Nevertheless, indicator-dilution techniques may be of value in bidirectional shunts

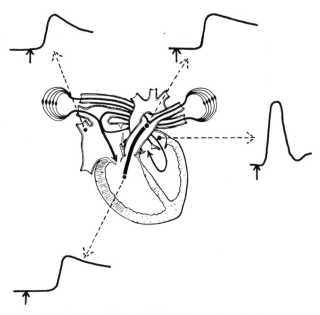

Figure 8.12 Schematic of circulation and systemic artery sampling of indicator-dilution curves in total anomalous pulmonary venous return. The dye curve inscribed after injection into the superior vena cava shows both early appearance of dye and early recirculation, characteristic of a bidirectional shunt. The dye curves from the right ventricle and pulmonary artery are usually similar in contour to the superior vena caval curve but have a longer appearance time; the left atrial dye curve is essentially normal both in appearance time and contour. (Modified from Hetzel, P. S., H. J. C. Swan, and E. H. Wood. 1966. The application of indicator-dilution curves in cardiac catheterization. *In* H. A. Zimmerman (ed.), Intravascular Catheterization, 2nd Ed. Charles C Thomas, Springfield, Ill., p. 539.)

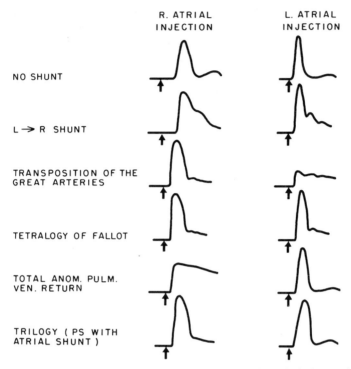

Figure 8.13 Schematic of dye curve pairs obtained in infants with various cardiac anomalies. There is an excellent separation of these patterns, sufficient for a diagnosis, particularly if used in association with the clinical findings. Indicator-dilution techniques allow not only for diagnosis but also for quantification of the shunts which often may be misleading if angiocardiography alone is relied upon.

for purposes of localizing the site of the shunt (Figs. 8.12 and 8.13) or the rough estimates of shunt flows (Fig. 8.14).

SPECIAL TECHNIQUES OF VALUE IN APPLYING INDICATOR-DILUTION TECHNIQUES TO NEONATES AND SMALL ANIMALS

In infants the short circulatory pathways result in rapid pulmonary and systemic circulation times. As has been noted earlier, the transit of blood is so rapid that in the presence of a left-to-right shunt, an indicator injected into the right heart arrives a second time at an arterial sampling site before the inscription of the concentration peak.

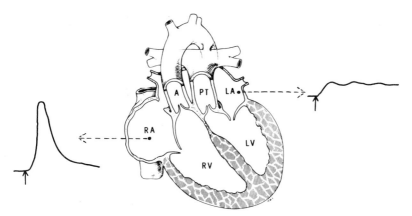

Figure 8.14 Schematic of circulation and systemic arterial sampling of indicator-dilution curves in simple, complete transposition of the great vessels. Note that the dye curves inscribed after injection into the right atrium (*RA*) and left atrium (*LA*) have similar appearance times but markedly different contours. This allows us to conclude that there is only minimal mixing of blood in this patient. As the amount of mixing, at atrial, ventricular, or great vessel levels increases, the dye curves become more alike in contour until, in the cases of single atrium or single ventricle, they are virtually identical.

This then results in the lack of discrimination of left-to-right shunts and also completely negates the usual method of the Stewart-Hamilton calculation of cardiac outputs (see Chapter 5).

Accuracy in the determination of injection times in infants is of considerable importance. In a group of neonates, the average corrected time of appearance for dye injected into the right atrium and sampled in the abdominal aorta was 2.2 seconds (Krovetz and Gessner, 1965). This time interval decreased to 1.0 second for dye injected into the left atrium. Since these times are so brief, delays and errors in recording the time of injection should be minimized. A dye syringe holder with provision for electrical timing of injection (Benson, Krovetz, and Schiebler, 1964) eliminates the need for a manual injection signal. For neonates and small infants, the catheter injection system should also be primed by filling it with the indicator solution, and then an additional measured amount of indicator is injected. By using this combination of dye displacement and electrical timing, reproducibility of appearance times within 0.1 second can be obtained.

Indicator-dilution techniques are extremely valuable for small infants, since the difficulties of oximetric calculations are further compounded by the absence of a truly steady state. It is difficult to obtain a series of blood samples for oximetric calculations without a change in the status of the infant. In contrast, the significant portions of dye curves occur more rapidly, usually in less than 20 seconds. Furthermore, since the Fick principle involves a subtraction to obtain the arterio-venous differences, even relatively small errors in single oximetric determinations of oxygen saturation become greatly magnified when applied to the smaller differences obtained from the formula (Schostal, Krovetz, and Rowe, 1972). This error increases with increasing left-to-right shunts. Oximetry cannot be relied on for left-to-right shunts below approximately 20 to 30% of pulmonary blood flow (Wood, Marshall, and Wood, 1963).

The use of high sampling rates in neonates and small animals can result in significant volume depletion of the vascular bed and marked changes in hemodynamics. To avoid such changes and still have an adequate frequency response of the sampling system, the volume between the artery and the detecting elements of the densitometer (dead space) should be kept as low as possible. The densitometer may be sterilized or wrapped in sterile coverings and brought into the sterile field. By shortening the proximal tubing of the densitometer cuvette and connecting a sterile cuvette directly to the sampling catheter or needle, dead space volumes as low as 0.1 to 0.3 ml are possible. This then allows satisfactory dye-dilution curves with sampling rates as low as 8 to 10 ml/minute.

In the early neonatal period the umbilical vessels are readily available for catheterization. Unfortunately, manipulation of a catheter inserted into the umbilical vein and traversing the ductus venosus is often extremely difficult. Although the catheter will usually enter the right atrium and pass through the foramen ovale into the left atrium, catheter entry of the right or left ventricle has been accomplished fortuitously and infrequently. Fortunately, a great deal of diagnostic and physiologic information can be obtained from a pair of atrial dye curves. Fig. 8.13 shows the results of such a pair of curves obtained from injection into right atrium and left atrium with arterial sampling. (For simplicity, a right-to-left shunt through the ductus arteriosus is assumed absent.) The techniques of indicator dilution provide for simple and rapid detection as well as quantitation of intracardiac or great vessel shunts, particularly where complete cardiac catheterization is either unwarranted or unnecessarily risky.

ACKNOWLEDGMENT

This reseach was supported in part by NHLI Grant No. HL13679–01 and RR–52 from the General Clinical Research Centers Program of the Division of Research Resources, NIH. Dr. Krovetz is the recipient of a Research Career Development Award 5–K3–HE–9761–08.

REFERENCES

Amplatz, K., R. F. Jeffery, F. L. Gobel, Y. Wang, G. E. Gathman, J. H. Moller, and R. V. Lucas. 1969. The Freon test. A new sensitive technique for the detection of small cardiac shunts. Circulation 39:551–556.

Amplatz, K., and J. F. Marvin. 1959. A simple and accurate test for left-to-right shunts. Radiology 72:585–586.

Benson, R. W., L. J. Krovetz, and G. L. Schiebler. 1964. A new type of syringe holder for performance of indicator dilution curves. J. Appl. Physiol. 19:1022–1023.

Braunwald, E., W. W. Pfaff, R. T. L. Long, and A. G. Morrow. 1959. A simplified indicator-dilution technique for the localization of left-to-right circulatory shunts. Circulation 20:875–880.

Braunwald, E., H. L. Tanenbaum, and A. G. Morrow. 1957. Dye-dilution curves from left heart and aorta for localization of left-to-right shunts and detection of valvular insufficiency. Proc. Soc. Exp. Biol. Med. 94:510–512.

Carter, S. A., D. F. Bajec, E. Yannicelli, and E. H. Wood. 1960. Estimation of left-to-right shunts from arterial dilution curves. J. Lab. Clin. Med. 55:77–88.

Clark, L. C., Jr., and L. M. Bargeron, Jr. 1959. Left-to-right shunt detection by an intravascular electrode with hydrogen as an indicator. Science 130:709–710.

Cliffe, P., J. P. Shillingford, and L. McDonald. 1959. Quantitative aspects of indicator-dilution curves. Proc. Roy. Soc. Med. 52:693–695.

Crane, M. G., J. E. Holloway, C. H. Sears, J. A. McEachen, R. H. Selvester, and I. C. Woodward. 1957. Localization of intracardiac shunts by two-site sampling. Circulation 16:870.

Huff, R. L., D. Parrish, and W. Crockett. 1957. A study of circulatory dynamics by means of crystal radiation detectors on the anterior thoracic wall. Circ. Res. 5:395–400.

Krovetz, L. J., and R. W. Benson. 1965. Mixing of dye and blood in the canine aorta. J. Appl. Physiol. 20:922–926.

Krovetz, L. J., and I. H. Gessner. 1965. A new method utilizing indicator-dilution techniques for the estimation of left-to-right shunts in infants. Circulation 32:772–777.

Lucas, R. V. Personal communication.

Marshall, H. W., H. F. Helmholz, Jr., and E. H. Wood. 1962. Physiologic consequences of congenital heart disease. In W. F. Hamilton and P. Dow (ed.), Handbook of Physiology, Vol. I. Amer. Physiol. Soc., Washington, D. C.

McDonald, L. 1959. Indicator-dilution curves in diagnosis of congenital heart disease. Proc. Roy. Soc. Med. 52:689–693.

Mook, G. A., and W. G. Zijlstra. 1961. Quantitative evaluation of intra-cardiac shunts from arterial dye dilution curves. Demonstration of very small shunts. Acta Med. Scand. 170 (Suppl. 6):703–715.

Morrow, A. G., R. J. Sanders, and E. Braunwald. 1958. The nitrous oxide test: an improved method for the detection of left-to-right shunts. Circulation 17:284–291.

Nakamura, T., R. Katori, T. Watanabe, K. Miyazawa, M. Murai, J. Oda, and K. Ishikawa. 1967. Quantitation of left-to-right shunt from a single earpiece dye-dilution curve. J. Appl. Physiol. 22:1156–1160.

Paul, M. H., A. M. Rudolph, and M. D. Rappaport. 1958. Temperature dilution curves for the detection of cardiac shunts. Circulation 18:765.

Phinney, A. O., Jr., P. V. Stoughton, W. P. C. Clason, and C. E. McLean. 1964. The spread to appearance time ratio in the estimation of left-to-right shunts. Amer. Heart J. 68:748–756.

Sanders, R. J. 1958. Use of a radioactive gas (Kr[85]) in the diagnosis of cardiac shunts. Proc. Soc. Exp. Biol. Med. 97:1–4.

Sato, T., H. Onoki, N. Yamauchi, and I. Kano. 1969. A simplified method for calculation of the left-to-right shunt from a single earpiece dye-dilution curve. Tohoku J. Exp. Med. 97:337.

Schostal, S. J., L. J. Krovetz, and R. D. Rowe. 1972. An analysis of errors in conventional cardiac catheterization data. Amer. Heart J. 83:596–603.

Swan, H. J. C. 1954. Diagnostic applications of indicator dilution curves in heart disease. Minnesota Med. 37:123–130.

Swan, H. J. C., and W. Beck. 1960. Ventricular non-mixing as a source of error in the estimation of ventricular volume by the indicator-dilution technique. Circ. Res. 8:989–998.

Swan, H. J. C., and E. H. Wood. 1953. Localization of cardiac defects by dye-dilution curves recorded after injection of T-1824 at multiple sites in the heart and great vessels during cardiac catheterization. Proc. Staff Meet. Mayo Clin. 28:95–100.

Swan, H. J. C., J. Zapata-Diaz, and E. H. Wood. 1953. Dye dilution curves in cyanotic congenital heart disease. Circulation 8:70–81.

Wood, E. H. 1960. Use of indicator-dilution techniques. *In* A. D. Bass and G. K. Moe (eds.), Symposium on Congenital Heart Disease. Amer. Assoc. Adv. Sci., Washington, D. C., p. 209.

Wood, E. H., H. J. C. Swan, and H. W. Marshall. 1958. Technique and diagnostic applications of dilution curves recorded simultaneously from the right side of the heart and from the arterial circulation. Proc. Staff Meet. Mayo Clin. 33:536–553.

Wood, R. C., H. W. Marshall, and E. H. Wood. 1963. Detection and quantitation of intracardiac left-to-right shunts by an oximetric inert gas technique. Circulation 27:351–359.

chapter 9/ Measurement of Valvular Insufficiency

Bruce C. Sinclair-Smith
Vanderbilt University School of Medicine/Nashville, Tennessee

The proportion of stenosis to insufficiency at a diseased cardiac valve assumed therapeutic significance before the general application of open heart surgery (Dexter *et al.,* 1957; Marshall, Woodward, and Wood, 1958; Conn *et al.,* 1957b). Mitral or aortic incompetence of significant hemodynamic degree excluded direct closed heart valvotomies at these sites. Differentiation between stenotic and incompetent lesions required considerable clinical skill while objective tests with high discriminant capacity were constantly pursued. The dispersion of indicator test substances within cardiac chambers having atrio-ventricular dysfunction was one such means of quantitating valvular insufficiency (Meier and Zierler, 1954; Milnor, 1957; Woodward, Burchell, and Wood, 1957; Hoffman and Rowe, 1959; Lacy *et al.,* 1959; McClure *et al.,* 1959; Levinson, Carleton, and Abelmann, 1959; Carleton *et al.,* 1959).

In recent years, a more pragmatic evaluation of valvular dysfunction has followed the widespread and successful use of prosthetic valvular devices. No longer is a precise differentiation between stenosis and insufficiency required, rather an appraisal of ventricular function is emphasized as a major prognostic factor in cardiovascular surgery.

Direct visualization techniques such as selective chamber cineangio-cardiography give visual credence to the presence of valvular dysfunction while providing data from which residual chamber size and ventricular ejection mechanics can be computed. In historical perspective, however, the accurate measurement of regurgitant flows across diseased valves benefited the understanding of the natural history of multivalvular heart disease. The theory of indicator-dilution techniques is still required to interpret the data obtained from the clinical investigation of patients in the cardiac catheterization laboratory or intensive care units and consequently, this chapter is included in the present monograph.

THEORY

An indicator-dilution curve may be distorted from its normal shape (Milnor, 1957; Hoffman and Rowe, 1959; Lacy *et al.*, 1959; McClure *et al.*, 1959) by (1) changes in flow or cardiac output, (2) valvular incompetence between injection and sampling site, (3) uni- or bi-directional intracardiac shunts, (4) a dilutional or "volume" effect from the major mixing chamber component, either the pulmonary blood volume or a giant left atrium, and (5) instrumental factors mainly related to the rate of blood flow through the densitometer system (Lacy, Emanuel, and Newman, 1957; Fox, Sutterer, and Wood, 1957; Milnor and Jose, 1960). With valvular incompetence, peak height concentration of indicator, time to peak concentration and the slope of this wash-out function may all be modified (Fig. 9.1).

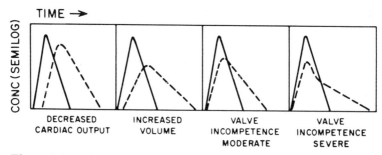

Figure 9.1 Major modifications to systemic arterial dye curves caused by decreased flow, increased volume, and valvular incompetence. The *solid curve* reference indicates the normal curve (modified from Shillingford, 1958, Brit. Heart J. 20:229).

Understanding of these effects may be clarified with a single two-chambered model system comprised of an atrium and a ventricle separated by an incompetent valve structure (McClure *et al.*, 1959) (Fig. 9.2). Stationary flow (with stroke volume V_E) enters and leaves

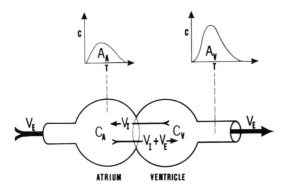

ATRIUM   VENTRICLE

Figure 9.2 Theoretical model of an incompetent atrio-ventricular valve system. V_E, V_I, C_V, *and* C_A represent the stroke volume, regurgitant volume, and ventricular and atrial indicator concentrations after indicator injection into the ventricle. The concentration-time curves of indicator sampled in the atrium and ventricle are shown above with A_A and A_V representing the respective curve areas. At each beat, units of dye moved backwards, $V_I C_V$, and forwards, $(V_I + V_E)C_A$.

the system by a pulsatile ventricular device into which the indicator is injected. The amount of indicator leaving the ventricle by passing forward out of the system with each beat, will be given by the product of the forward stroke volume (V_E) and the concentration of indicator in the ventricle (C_V). The amount of indicator passing backwards into the atrium with each beat will be given by the product of the regurgitant stroke volume (V_I) and the concentration of indicator in the ventricle (C_V). The total amount of indicator leaving the ventricle with each beat is, therefore,

$$V_E C_V + V_I C_V \quad \text{or} \quad V_{(I+E)} C_V.$$

The units of indicator which moved backward into the atrial chamber will establish a new concentration (C_A) which will be, in part, a function of the atrial chamber size. C_A will then be the concentration of indicator moving forward again into the ventricle before the next beat. The flow across the valve will be the sum of the regurgitant and forward stroke volumes ($V_I + V_E$), and the units of indicator moving forward will be the product of this volume and the inflow concentration: $V_{(I+E)}C_A$.

With an increasing number of beats, an equilibration concentration of indicator will be reached between atrial and ventricular chambers, from which point both chambers will then be cleared of indicator in an exponential manner. If observations continue until all the indicator has been removed from the proximal or atrial chamber, the indicator which entered the atrium over this period of time, $V_I C_V t$, is equal to the indicator which left it during the same time, $V_{(I+E)}C_A t$. Since $C_V t$ and $C_A t$ represent the ventricular and atrial curve areas, respectively, the derivation of regurgitant flow in an incompetent system only requires the simultaneous recording of time concentration curves of indicator in both atrial and ventricular chambers after ventricular injection. The above relationship can be expressed as

$$\frac{A_A}{A_V} = \frac{V_I}{V_I + V_E},$$

where A_A and A_V represent the atrial and ventricular curve areas, respectively. Although this relationship is expressed in terms of atrioventricular valvular incompetence, it holds for any valvular incompetent system and may be generally expressed as

$$\frac{A_{PC}}{A_{DC}} = \frac{V_I}{V_I + V_E} \tag{9.1}$$

where PC and DC are proximal and distal chambers, respectively. This ratio is defined as the "regurgitant fraction" and is designated f. Since $f = V_I/(V_I + V_E)$, rearranging for V_I,

$$V_I = \frac{f}{1-f} \times V_E \tag{9.2}$$

for each stroke, or in general terms,

$$\text{regurgitant flow} = \frac{f}{1-f} \times \text{cardiac output.}$$

By substituting (9.1) in (9.2),

$$V_I = \frac{A_{PC}}{A_{DC} - A_{PC}} \times V_E. \tag{9.3}$$

For clinical application, these calculations do not depend upon a knowledge of the over-all size or residual volume of the chambers, nor upon a knowledge of the distribution of indicator in the atrium. The assumption is made that for every cycle at the beginning of ventricular ejection the indicator has been mixed in the ventricle with sufficient uniformity to insure that the concentration in the fluid volume ejected forward V_E is identical with the concentration in the regurgitated or backwardly ejected volume V_I.

With constant infusion techniques, the same principles apply. If indicator is infused at a constant rate until equilibrium is reached, the concentration within distal and proximal chambers will become constant and (9.3) can be written as:

$$V_I = \frac{\text{conc. in proximal chamber}}{\text{conc. in distal chamber} - \text{conc. in proximal chamber}} \times V_E.$$

The regurgitant fraction under these circumstances is the ratio

$$\frac{\text{concentration in proximal chamber}}{\text{concentration in distal chamber}} = f. \tag{9.4}$$

Combined or Double Valvular Insufficiency

When aortic and mitral valvular insufficiency are present in the same patient, indicator injected into the aortic root regurgitates into the left ventricle and then into the left atrium. Therefore, under these special circumstances, we may define: for the mitral valve, $A_{PC} = A_{\text{atrium}}$, $A_{DC} = A_{\text{ventricle}}$, $V_{MI}/(V_{MI} + V_E) = f_m$; and for the aortic valve, $A_{PC} = A_{\text{ventricle}}$, $A_{DC} = A_{\text{aorta}}$, $V_{AI}/(V_{AI} + V_E) = f_a$, where V_{MI} and V_{AI} are the volumes of each stroke regurgitating across the insufficient mitral and aortic valves, respectively, and f_m and f_a are the respective regurgitant fractions. From the following two relationships

$$\frac{A_{\text{atrium}}}{A_{\text{ventricle}}} = \frac{V_{MI}}{V_{MI} + V_E} \quad \text{and} \quad \frac{A_{\text{ventricle}}}{A_{\text{aorta}}} = \frac{V_{AI}}{V_{AI} + V_E},$$

it follows that:

$$\frac{A_{\text{atrium}}}{A_{\text{aorta}}} = \frac{V_{MI}}{V_{MI} + V_E} \times \frac{V_{AI}}{V_{AI} + V_E} = f_m \times f_a.$$

Since $A_{\text{atrium}}/A_{\text{aorta}}$ also represents the curve area ratio of indicator regurgitating serially into the atrium from an aortic injection and is defined as the combined regurgitant fraction (f_c), it follows further that

$$f_c = f_m \times f_a. \tag{9.5}$$

That is, combined regurgitant fraction across both valves is equal to the product of the regurgitant fractions across each valve (the regurgitant product) (Bloomfield, Battersby, and Sinclair-Smith, 1965).

The assumptions that must be fulfilled for the validity of this proposition are that (1) at the beginning of each ventricular systole, the indicator has been mixed in the ventricle in such manner that the ratio of mass of indicator moves from the ventricle to the atrium in volume V_{MI} to the mass moved forward by volume $V_E + V_{AI}$ is proportional to the ratio of those volumes, (2) such relations obtain both for the mass of indicator moved into the ventricle from the aortic root in volume V_{AI} and for the mass leaving the aortic root in volume V_E. If the latter assumption holds, then the area of a curve drawn from a peripheral artery will be identical with the area of the aortic root curve and may be used in place of that area in the above relationships and to calculate forward cardiac output in the usual manner. These assumptions imply that measurement of valvular insufficiency by this indicator-dilution technique is independent of the presence or severity of regurgitation at another valve, and that the same information is present in the estimation of f_c and either f_m or f_a, as is present in the estimation of both f_m and f_a.

Indirect Indices of Valvular Incompetence Derived From Single Curves

The above theoretical analysis applies to incompetence of any cardiac valve so long as certain technical details are observed. Historically, the measurement of valvular incompetence coincided with an increased interest in human mitral valve function, and, therefore, certain indirect indices to assess mitral regurgitation were developed. These earlier indices were derived from systemic arterial curves after pulmonary artery injection. With the advent of left heart and especially transseptal left heart catheterization, left atrial injection eliminated the distorting influence of pulmonary blood volume. However, the indices derived from either time or flow characteristics of the peripheral curve after left atrial injections still contained the distorting influences of large residual chamber volumes and were of limited applied value.

Indices from Single Curves Secondary to Left Heart Injection
The ratio of peak concentration to slope time in arterial dilution curves after left atrial injection has been used as an index of the presence and severity of valvular regurgitation (Woodward, Swan, and Wood, 1957; Sinclair et al., 1960). Under these circumstances,

peak concentration of indicator is relatively insensitive to flow but is reduced by regurgitation, whereas slope time (time taken for the extrapolated semilogarithmic downslope to descend one log cycle) was inversely related to flow and prolonged by regurgitation. The ratio of peak concentration to slope time may be greatly reduced by regurgitation but also by large volume effects from cardiac chambers. Inherent limitations of its general application therefore exist.

Indices from Single Curves Secondary to
Pulmonary Artery Injection
Under these circumstances, variations in cardiac output and calculated central blood volume on the time and concentration components of the curve can be predicted for circumstances where mitral incompetence does or does not exist. Inverse relationships exist both between the time components of the curve (e.g. disappearance time) and the blood flow, as well as between the central blood volume and the concentration components of the curve (e.g. peak height concentration).

Single time or concentration components proved of little value to assess mitral insufficiency. However, the use of the ratio of time components (disappearance time (DT) to build-up time (BT)) to minimize the effects of cardiac output on a time component should theoretically sharpen the discriminatory power of these ratios, since the nonspecific effects of variations in blood flow and blood volume can be partially compensated for in this way. The DT/BT ratio gives a better separation of patients having predominant mitral incompetence than does the single time measurement.

Single ratios with relatively good discriminatory power may be combined with other parameters to improve the separation of patients having predominant stenosis or insufficiency. These combined ratios include (1) the disappearance ratio and the peak concentration and (2) least concentration on downslope of curve to peak recirculation concentration.

Complex Indirect Indices (Korner-Shillingford)
An arterial dilution curve may be considered as a frequency distribution of dye particles, and the effect of regurgitation as a diminished probability of forward movement of a given particle (Korner and Shillingford, 1955; Korner and Shillingford, 1956; Shillingford, 1958; Carleton, Levinson, and Abelmann, 1959). The probability may be assessed as a function of the dispersion of the curve, the dispersion being expressed by either the variance or the reciprocal of the slope

of the descending limb. With valvular incompetence, both the variance and the reciprocal of the slope increased and could be mathematically separated into components related to forward flow, volume, and regurgitant flow. For practical application, this method of assessing mitral regurgitation required the use of complex regression equations and could give absolute flow data inconsistent with clinical circumstances.

A simple empirical method for obtaining an index of regurgitation from dye curves without involved mathematical calculation was described by Shillingford (1958). Using arterial sampling with pulmonary artery injection (to assess total left-sided valvular insufficiency) or right atrial injection (for total right- and left-sided valvular insufficiency), the curves were analyzed for appearance time and spread between upslope and semilogarithmic projection of downslope at one-tenth of the peak concentration. On the basis that the spread of dye particles normally increases with the appearance time, the ratio of spread divided by appearance time was 3.0 or less in the absence of significant regurgitation and 4.0 or more with progressive severity of regurgitation.

The validity of the method, as with the more complex variance method, is reduced by increasing pulmonary vascular or intracardiac volumes.

APPLICATION

General

In the double sampling technique, certain general principles must be observed.

1. Sampling sites should preferably be from chambers on either side of the incompetent valve.

2. Sampling should not take place from a chamber into which injection of indicator has occurred.

3. With ventricular injections, the downstream or distal curve can be inscribed from either the pulmonary artery or aorta depending upon the ventricle under study.

4. Flow tagging (representative dispersion of indicator in laminar velocity profiles) is established once the injectate has passed through an orifice, especially the semilunar valves.

5. Minimal interference with valve leaflets and distortion of papillary muscle contraction should result from catheter placement.

6. Matched densitometers of equal sensitivity should preferably be utilized.

7. Instrumental dead space in the sampling system should be reduced to a minimum with equal withdrawal rates for each densitometer to remove mechanical distortion of curves. Special design pumps are available for simplification.

Mitral Valve

Several reliable techniques of catheter placement are available for mitral valve studies (Carleton *et al.,* 1959).

Combined Transseptal and Retrograde

Left Heart Catheterization

Atrial curves are obtained by a transseptal catheter the tip of which must not lie within the mouth of a pulmonary vein or distort the mitral valve orifice. Ventricular injection is through a retrograde left ventricular catheter and should be made over at least two to three cardiac cycles. Sampling of downstream curves can be from brachial, radial, or femoral arterial sites, or from another catheter placed in the ascending aorta (Fig. 9.3, *A* and *B*).

The transseptal catheter can also be modified for both injection and sampling modalities. Thin-walled nylon tubing can be passed

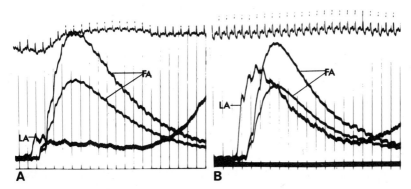

Figure 9.3 Mild (*A*) and severe (*B*) mitral insufficiency. The "forward" dilution curves (*FA*) sampled from the femoral artery have been simultaneously recorded on two sensitivities. The regurgitant dilution curves (*LA*) are sampled from the left atrium after left ventricular injection of indicator. The ratio of *LA* to *FA* curve areas, when corrected for unequal sensitivities of the densitometer-recorder system, gives a measure of severity of the valvular insufficiency.

through the lumen of the transseptal catheter, and the body of the left ventricle can be intubated with pressure monitoring. With the tip of the transseptal catheter in the left atrium, and utilizing an adaptor with side arm orifice and pressure gasket, left ventricular injection and left atrial sampling can be achieved with the use of a single catheter. Minimal distortion of valve function usually occurs from thin-walled tubing "floated" across either atrio-ventricular or semilunar valves.

With suspected combined mitral and aortic insufficiency, evalua-tion of individual and total valvular incompetence can be made. A combined regurgitant fraction may be obtained from injection into the base of the aorta with left atrial sampling. Advancing the trans-septal catheter into the left ventricular cavity allows evaluation of aortic function, and replacing the catheters on either side of the mitral valve permits determination of f_m. A comparison between the derived and directly determined f_c provides some internal check of the experimental procedure. Knowledge of f_c may be of use when the re-gurgitant fraction of only one valve can be determined (usually be-cause of technical difficulties), since f of the other valve can then be calculated from the relationship $f_c = f_a \times f_m$ (Fig. 9.4, *A* and *B*).

Percutaneous Transthoracic Left Atrial Puncture
Percutaneous transthoracic puncture provides a direct approach to the left heart. Surface anatomical landmarks may be established by

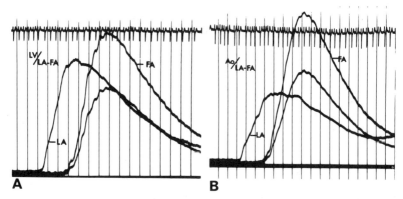

Figure 9.4 Double valvular insufficiency. A, the curves represent direct measurement of mitral insufficiency after left ventricular injection of indicator, and sampling in the left atrium (*LA*) and femoral artery (*FA*, two sensitivities). The corrected ratio of *LA* to *FA* curve areas represents f_m. B, left atrial and femoral artery curves after aortic injection of indi-cator. Corrected ratio of *LA* to *FA* curve areas represent f_c. Aortic valve regurgitant fraction f_a is given by $f_a = f_c/f_m$.

fluoroscopy and thin-walled (18T) needles advanced through the sacro-spinalis muscle into the posterior left atrium. If two needles are inserted, one may be utilized for intubation of the left ventricular chamber with small-bore nylon tubing.

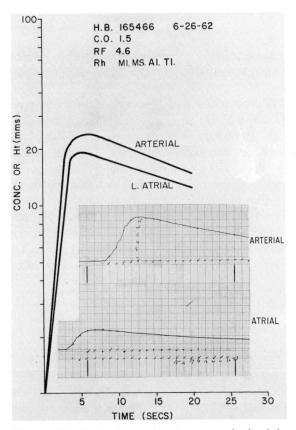

Figure 9.5 Atrial (regurgitant) curve obtained by percutaneous transthoracic left atrial puncture, with equisensitive semilogarithmic replot. Cardiac output is 1.5 liters/minute and regurgitant volume is 4.6 liters/minute.

Minimal instrumental dead space in the sampling system is an advantage of this method, supplying curves with minimal distortion (Fig. 9.5). Evaluation of the aortic valve is technically difficult, and the transthoracic approach is mainly restricted to either investigative or unusual clinical circumstances.

Use of Radioactive Tracers

Radioactive test substances may be substituted for chemical indicators, the elimination of blood sampling and densitometric techniques being a distinct advantage. The accurate collimation of scintillation counters over the left atrium and ventricle introduces additional technical difficulties (Conn *et al.*, 1957a; Heiman *et al.*, 1958).

Aortic Valve

Quantitation of aortic incompetence is possible with central aortic injection and left ventricular and femoral artery sampling, utilizing transseptal and retrograde methods. Reproducibility of results is aided by the following conditions. (1) The tip of the retrograde injecting catheter should be 2 to 3 cm above the plane of the aortic valve, (2) the injection should occupy at least three cardiac cycles to prevent preferential regurgitation of dye in early diastole, (3) the injecting catheter tip should have multiple end holes to prevent a bolus effect in early diastole, and (4) the central and ascending aorta is usually dilated in aortic incompetence and is an adequate mixing chamber (Armelin *et al.*, 1963; Guidry and Wood, 1958).

Certain indirect indices of evaluating aortic incompetence by assessing the retrograde flow of indicator either in the descending aorta or femoral artery provide at the best only limited evaluation of this lesion.

Tricuspid Valve

Because of its ready accessibility to diagnostic techniques, tricuspid valve function has been studied by utilizing the double sampling technique (Bajec *et al.* 1958; Sinclair-Smith, Bloomfield, and Newman, 1965). Unless direct right ventricular puncture is utilized, all methods of right ventricular injection involve passage of catheters across the delicate leaflets of the valve with fears of artifactual production of tricuspid incompetence. Soft polyethylene tubing may be floated into the chamber, but, without a marker, tip position and, therefore, injection site are unknown. Injection proximal to the pulmonary valve during the systolic ejection phase may invalidate the study. Experience suggests that a soft right heart catheter passed from an antecubital vein into the main pulmonary artery and then withdrawn into the midportion of the right ventricular cavity produces no disturbance of tricuspid valve closing. The soft catheter appears molded by the streamline pattern of the blood flow through the valve; so long as no ventricular irritability is produced, satisfactory injection may be obtained. Sampling catheters can be introduced through another forearm vein or percutaneously via the right femoral vein. Representative sampling is more satisfactorily obtained if the tip of the catheter is 2 to 3 cm lateral to the tricuspid orifice, usually in mid-right atrial position.

The isolation of recirculation effects from systemic arterial curves after right ventricular injection may be difficult if those circumstances exist which prolong the downslope of the arterial curve. Small errors in slope may cause large variations in calculation of curve areas and regurgitant fractions. Pulmonary artery downstream sampling is preferential to systemic arterial sampling, and distal-tipped thermistor catheters with thermal signals eliminate recirculation effects. These latter catheters may be floated into the pulmonary artery with minimal interference to right heart function.

PRESENTATION OF DATA

Preferably, curves should be obtained simultaneously from distal and proximal sampling sites with minimal instrumental dead space between the sampling site and the densitometer. Curves should be registered until recirculation effects are seen at the proximal sampling site.

The height of the dye curves on the recording paper is measured at 1-second intervals and tabulated for both sampling sites on semi-logarithmic paper. Correction for differences in sensitivity of the densitometers or of attentuation factors in the recording preamplifiers is then made. Curve areas are calculated by standard methods, the downstream curve permitting determination of cardiac output, whereas the regurgitant fraction and flow are calculated from the upstream and downstream curve ratios according to the preceding formulae (Fig. 9.6).

When moderate to massive valvular incompetence exists, the washout concentration of the arterial curve may be resolved into two exponential components which should be graphed accordingly (Fig. 9.7). Valvular incompetence distorts a normal curve similar to a left-to-right intracardiac shunt. If identifiable, only the second and temporally later exponential curve should be extrapolated to zero base line. Failure to recognize this will lead to overestimation of the regurgitant and forward flows. The inherent disadvantage of the ratio

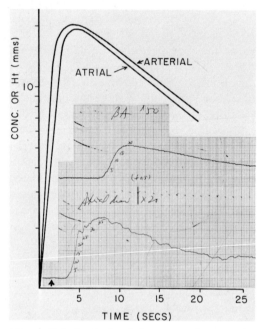

Figure 9.6 Analysis of dye curve data. In this example both atrium and ventricle act as a common mixing chamber with identical washout characteristics.

$f/(1-f)$ is apparent, because, as f approaches unity, the ratio, with its approximation to infinity, suggests regurgitant flows of unphysiological proportion. The use of the regurgitant fraction f as an index of retrograde flow in comparison to forward cardiac output is, therefore, preferred.

CONCLUSION

Accuracy of regurgitant flow measurements across a diseased cardiac valve may be assessed by repeated determinations made during single studies or by an analysis of the reproducibility of multiple studies in a patient population. An independent method of sufficient accuracy is not available for comparison.

Fig. 9.8 shows data from 290 studies in 75 patients with valvular incompetence of predominantly one major cardiac valve (tricuspid 25, mitral 24, aortic 26). Regurgitant fractions f are plotted against the difference between the mean and individual determinations of f for each patient study $(f - f_s)$. For the sake of graphical clarity, individual studies are not plotted, but the envelopes showing $\pm 2\sigma$ of the mean values of f for the three arbitrarily chosen groups are portrayed. Reproducibility of the magnitude of mitral and aortic incompetence is excellent over all ranges of insufficiency. Massive tricuspid insufficiency (regurgitant flows greater than one and a half times forward flow) increased unreliability of the procedure, mainly because of

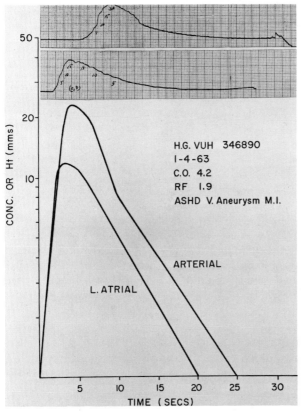

H.G. VUH 346890
1-4-63
C.O. 4.2
RF 1.9
ASHD V. Aneurysm M.I.

ARTERIAL

L. ATRIAL

Figure 9.7 Analysis of curves from left ventricular injection in moderate mitral incompetence. Note the double exponential decay of the arterial curve.

difficulties in analyzing the exponential slope of the systemic arterial curve. Pulmonary artery sampling largely eliminated the source of error.

A major criticism of indicator-dilution techniques in evaluating valvular heart disease has been the effective dispersion of the test substance in the upstream chamber from the valve under study and subsequent nonrepresentative sampling (Sinclair-Smith and Newman, 1969). These criticisms apply especially to the right atrium where superior and inferior caval inflows mix with coronary sinus blood.

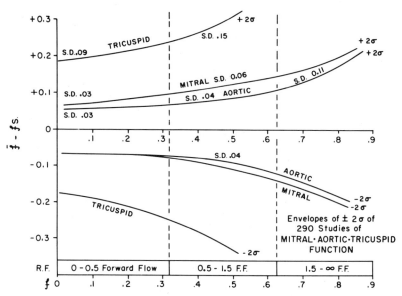

Figure 9.8 Analysis of reproducibility of repeated estimates of valvular incompetence in equal numerical groups of patients with mitral, aortic, and tricuspid insufficiency. The three arbitrarily chosen ranges of f, 0 to 0.33, 0.33 to 0.63, and 0.63 to 0.99, correspond to clinically mild, moderate, and severe disease.

By using thermodilution techniques which have the advantage of multiple and frequent sampling potentialities, the following can be shown.

1. Both right and left ventricles behave as if they have multiple mixing compartments with various rates of clearance. The apical portion of both ventricles, as might be expected, wash out slowly.

For this reason, techniques incorporating simultaneous injection and recording from the same ventricle are subject to criticism.

2. Flow tagging of indicator materials occurs with passage through a single orifice. This orifice may be either the normal pulmonary or aortic valve or an inlet or outlet opening in a diseased atrio-ventricular valve. Downstream curves of equal area are obtained after multiple injections into any portion of a ventricular chamber.

3. Curve areas of equal magnitude are obtained in the right ventricle after either superior or inferior vena caval injections, indicating complete mixing of the thermal signal with one passage across the normal tricuspid valve. The assumption is made that similar mixing exists for retrograde flow across the incompetent tricuspid valve.

4. Careful placement of sampling catheters 2 to 3 cm upstream from an incompetent atrio-ventricular valve will provide curves having equal areas but differing in shape only. Accurate fluoroscopy will prevent withdrawal from the superior or inferior vena cava or the pulmonary vein.

In conclusion, methods exist which possess sufficient accuracy to be of clinical usefulness in the evaluation of regurgitant valvular lesions. The methods are based on sound scientific principles, and, with the correct attention to technical details, they provide quantitative data useful in a better appreciation of the natural history of cardiac disease.

ACKNOWLEDGMENT

This study was supported in part by United States Public Health Service Grant HE-08195.

REFERENCES

Armelin, E., L. Michaels, H. W. Marshall, D. E. Donald, R. J. Cheesman, and E. H. Wood. 1963. Detection and measurement of experimentally produced aortic regurgitation by means of indicator dilution curves recorded from the left ventricle. Circ. Res. 12:269–290.

Bajec, D. F., N. C. Birkhead, S. A. Carter, and E. H. Wood. 1958. Localization and estimation of severity of regurgitant flow at the pulmonary and tricuspid valves. Proc. Staff Meet. Mayo Clin. 33:569–577.

Bloomfield, D. A., E. J. Battersby, and B. C. Sinclair-Smith. 1965. Use of indicator dilution technics in measuring combined aortic and mitral insufficiency. Circ. Res. 18:97–100.

Carleton, R. A., G. E. Levinson, and W. H. Abelmann. 1959. Assessment of mitral regurgitation by indicator dilution: Observations on the principle of Korner and Shillingford. Amer. Heart J. 58:663–674.

Carleton, R. A., G. E. Levinson, G. Katznelson, S. W. Stein, and W. H. Abelmann. 1959. Clinical and physiologic assessment of mitral incompetence. Proc. New Engl. Cardiovasc. Soc. 17:28.

Conn, H. L., Jr., D. F. Heiman, W. S. Blakemore, P. T. Kuo, and S. B. Langfield. 1957a. "Left heart" radio-potassium dilution curves in patients with rheumatic mitral valvular disease. Circ. Res. 15:532–539.

Conn, H. L., Jr., D. F. Heiman, J. C. Wood, B. Jumbala, and W. S. Blakemore. 1957b. Study of mitral regurgitant blood flow in subjects with normal and deformed mitral valves. Clin. Res. Proc. 5:166.

Dexter, L., P. Novak, R. C. Schlant, A. O. Phinney, Jr., and F. W. Haynes. 1957. Mitral insufficiency. Trans. Assoc. Amer. Physicians 70:262–267.

Fox, I. J., W. F. Sutterer, and E. H. Wood. 1957. Dynamic response characteristics of systems for continuous recording of concentration changes in a flowing liquid (for example, indicator dilution curves). J. Appl. Physiol. 11:390–404.

Guidry, L. D., and E. H. Wood. 1958. Application of a method for detecting and estimating severity of aortic regurgitation alone or in association with mitral regurgitation. Proc. Staff Meet. Mayo Clin. 33:596–599.

Heiman, D. F., W. S. Blakemore, H. L. Conn, Jr., B. Jumbala, and H. M. Woske. 1958. Direct estimation of mitral regurgitant blood flow from "left heart" isotope dilution curves. Clin. Res. 6:20.

Hoffman, J. I. E., and G. G. Rowe. 1959. Some factors affecting indicator dilution curves in the presence and absence of valvular incompetence. J. Clin. Invest. 38:138–147.

Korner, P. I., and J. P. Shillingford. 1955. The quantitative estimation of valvular incompetence by dye-dilution curves. Clin. Sci. 14:553–573.

Korner, P. I., and J. P. Shillingford. 1956. Further observations on the estimation of valvular incompetence from dye dilution curves. Clin. Sci. 15:417–431.

Lacy, W. W., R. W. Emanuel, and E. V. Newman. 1957. Effect of the sampling system on the shape of indicator dilution curves. Circ. Res. 5:568–572.

Lacy, W. W., W. H. Goodson, W. G. Wheeler, and E. V. Newman. 1959. Theoretical and practical requirements for the valid measurement by indicator dilution of regurgitant flow across incompetent valves. Circ. Res. 8:454–460.

Levinson, G. E., R. A. Carleton, and W. H. Abelmann. 1959. Assessment of mitral regurgitation by indicator dilution: an analysis of the determinants of the abnormal dilution curve. Amer. Heart J. 58:873–888.

Marshall, H. W., E. Woodward, and W. H. Wood. 1958. Hemodynamic methods for differentiation of mitral stenosis and regurgitation. Amer. J. Cardiol. 2:24–60.

McClure, J. A., W. W. Lacy, P. Latimer, and E. V. Newman. 1959. Indicator dilution in an atrioventricular system with competent or incompetent valves. Circ. Res. 7:794–806.

Meier, P., and K. L. Zierler. 1954. On the theory of the indicator dilution method for measurement of blood flow and volume. J. Appl. Physiol. 6: 731–744.

Milnor, W. R. 1957. Measurement of valvular insufficiency by an indicator dilution method. Clin. Res. Proc. 5:165–166.

Milnor, W. R., and A. D. Jose. 1960. Distortion of indicator dilution curves by sampling systems. J. Appl. Physiol. 15:177–180.

Shillingford, J. P. 1958. Simple method for estimating mitral regurgitation by dye dilution curves. Brit. Heart J. 20:229–232.

Sinclair, J. D., C. P. Newcombe, D. E. Donald, and E. H. Wood. 1960. Experimental analysis of an atrial sampling technic for quantitating mitral regurgitation. Proc. Staff Meet. Mayo Clin. 35:700–709.

Sinclair-Smith, B. C., D. A. Bloomfield, and E. V. Newman. 1965. Venous hypertension, tricuspid valve function and congestive cardiac failure. Trans. Assoc. Amer. Physicians. 78:292–301.

Sinclair-Smith, B. C., and E. V. Newman. 1969. Tricuspid valve disease: an historical and physiological perspective. Trans. Amer. Clin. Climatol. Ass. 81:143–159.

Woodward, E., Jr., H. B. Burchell, and E. H. Wood. 1957. Dilution curves associated with valvular regurgitation. Proc. Staff Meet. Mayo Clin. 32: 518–525.

Woodward, E., Jr., H. J. Swan, and E. H. Wood. 1957. Evaluation of a method for detection of mitral regurgitation from indicator dilution curves recorded from the left atrium. Proc. Staff Meet. Mayo Clin. 32:525–535.

chapter 10/ Measurement of Pulmonary Blood Volume and Pulmonary Extravascular Water Volume

Paul N. Yu
Cardiology Unit/Department of Medicine/
University of Rochester Medical Center/
Rochester, New York

Different methods have been proposed for measuring pulmonary blood volume (PBV) and pulmonary extravascular lung water volume (PEV) in man and in animals. In general, the most frequent approach involves the use of indicator-dilution technique according to the methods of Stewart (1921), Kinsman *et al.* (1929), and Hamilton *et al.* (1930). The indicators which are commonly employed by various investigators include dyes and radioisotopes. It should be emphasized that, whenever indicator-dilution techniques are used to estimate the PBV, only the dynamic or circulating portion of the volume is measured and that, whenever similar techniques are used to estimate PEV, only that portion of extravascular water in the areas close to the patent capillaries is included. In other words, any amount of intravascular blood or extravascular water in the lung which is not in direct contact with the indicator or which is not equally diluted during the first transit through the pulmonary vascular and extravascular beds will not be included in the estimated volume.

PULMONARY BLOOD VOLUME

Methods of Measurement

The first reports of measuring PBV in man by indicator-dilution technique were published by Kunieda (1955) and Fujimoto, Kunieda, and Shiba (1960). The pulmonary artery (PA) was catheterized by means of right heart catheterization, and the left atrium (LA) was catheterized by percutaneous left atrial puncture. They used both single and double injection techniques. The former approach consisted of sequential injections of Evans blue dye into the PA and LA followed by sampling of blood from a systemic artery (BA), whereas the latter consisted of injection of Evans blue dye into the main PA and sampling from the LA.

A few years later, employing the technique similar to that described by Kunieda and associates, Dock, Milnor, and their respective associates reported their results on the estimated PBV in a series of 64 patients. Dock et al. (1959, 1961) made simultaneous injections of Evans blue dye into the PA and radioiodinated human serum albumin (RISA) into the LA, whereas Milnor and associates injected Indocyanine green subsequently into the PA and LA (McGaff et al., 1959; Milnor et al., 1960).

The aforementioned three groups of investigators used the approach of double injection and single sampling. The PBV was estimated according to the conventional formula. Kunieda and co-workers used the median circulation time for calculation, whereas Dock, Milnor, and their respective associates used mean transit time.

PBV has been also estimated by (1) injection of Indocyanine green into the superior vena cava with simultaneous sampling of the dye from PA and LA (Samet et al., 1966; Freitas et al., 1964), and (2) injection of dye into the PA and sampling from LA (Freitas et al., 1964; Levinson et al., 1964). In general, for patients with similar cardiac lesions, the PBV measured by the single injection and double sampling technique is comparable to that estimated by the double injection and single sampling technique. On the other hand, the single injection and single sampling technique consistently overestimated the PBV. The main reason for the overestimation by the latter approach is probably due to absence of a ventricular mixing chamber between the injection and sampling sites, resulting in incomplete mixing of the indicator in the PA as well as in the LA.

In addition to the approach involving arterial sampling, precordial radioisotope dilution curves have been used for the measurement of pulmonary mean transit time and PBV (Lewis *et al.*, 1962; Giuntini *et al.*, 1963; Segre *et al.*, 1965). Most, if not all, of the proposed methods lead a gross overestimation of the pulmonary mean transit time, since the precordial counting invariably includes the passage of the radioisotope through both right and left heart. However, pertinent information concerning PBV derived from the precordial radioisotope technique has been reported by several centers.

At the present time the dilution methods with intracardiac injection and arterial sampling of the dye are most widely used to measure PBV in man. In most of the studies, the indicators employed include Evans blue, Indocyanine green, RISA, and bromsulphthalein. There has been excellent reproducibility of duplicate determinations of PBV reported from various laboratories. After repeated determinations of PBV, no significant changes of intracardiac or intravascular pressures have been observed (Oakley *et al.*, 1962; Yu *et al.*, 1963; Schreiner *et al.*, 1968). If the dye is injected in a site which is proximal to a mixing chamber (either right or left ventricle), there is usually adequate mixing of the dye with the blood. However, if the injection is made very close to the mitral valve orifice, incomplete mixing due to streamline phenomena may occur. The PBV estimated by all of the methods described above includes variable and undetermined amount of left atrial blood volume.

In a separate monograph I have discussed the limitations and potential errors of the indicator-dilution technique for the estimate of PBV (1969). These include the timing and rate of injection, position of the cardiac catheter, mixing problems, sampling system, status of the indicator in the vascular system, and accuracy of recording system.

In our laboratory, cardiac output and PBV are determined by sequential injections of Indocyanine green into the PA and LA and sampling from the BA. Right heart catheterization is performed under local anesthesia and fluoroscopic control. A cardiac catheter is introduced into the pulmonary artery via a right or left antecubital vein, and another catheter is introduced into the left atrium by means of transseptal technique. A no. 18 Longdwell cannula is placed in the brachial artery of the opposite arm for measurement of systemic arterial pressure and blood sampling.

Pulmonary artery, left atrial, and brachial artery pressures are measured through a model 23 DB Statham gauge transducer and recorded on a direct writing recorder. The values of the respective mean pressures are obtained by electronic integration. Zero reference level for the pressure transducer is taken 6.5 cm below the sternal angle. The upper limit of normal mean pulmonary artery pressure (PAm) is 20 mm Hg and that of normal mean left atrial pressure (LAm) is 12 mm Hg.

While the injection of Indocyanine green is made, the arterial blood is continually withdrawn by means of a Harvard pump through the brachial arterial cannula to a Gilford cuvette densitometer at a flow rate of 47 ml/minute. The concentration of dye is inscribed directly on a direct recorder (Fig. 10.1). All blood samples collected

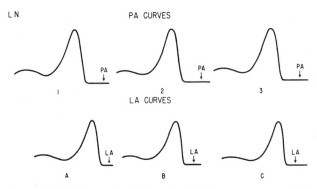

Figure 10.1 Successive indicator-dilution curves recorded from withdrawing systemic arterial blood through a cuvette densitometer after respective injections of indicator to the pulmonary artery (*PA*) and left atrium (*LA*). Note the reproducibility and similarity of these curves.

during the inscription of the curves are returned to the patient under aseptic conditions to minimize blood loss. In addition to the direct recording, all data points are transcribed electronically on an analogue to digital converter and punched by an IBM 526 keypunch. The data are then fed through an XDS Sigma 3 computer for determination of cardiac output and mean transit time of each curve.

The PBV is calculated according to the Stewart-Hamilton formula:

$$PBV = \frac{CI}{60} \times [Tm_{(PA\text{-}BA)} - Tm_{(LA\text{-}BA)}]$$

where PBV = pulmonary blood volume (ml/m^2), CI = mean cardiac index derived from the two sequential curves (ml/m^2), $Tm_{(PA\text{-}BA)}$ =

mean transit time (seconds) from pulmonary artery to brachial artery, $Tm_{(LA-BA)}$ = mean transit time (seconds) from left atrium to brachial artery.

Values of Pulmonary Blood Volume

In the past 12 years, we have measured PBV at rest and under various physiologic and pharmacologic interventions in over 1000 patients with valvular heart disease, coronary artery disease, or cardiomyopathy. In a monograph published previously (Yu, 1969), we reported the values of PBV in 14 "normal" patients and a number of patients with various types of cardiac diseases. Since then six more "normal" patients have been added to the series. The hemodynamic data and pulmonary blood volume in these 20 "normal" subjects are summarized in Table 10.1. The composite values of PBV in various groups of patients with valvular heart disease are also presented.

The mean value of PBV in a series of "normal" subjects is 273 ml/m^2. The magnitude of PBV varies a great deal in cardiac patients of different functional classes according to the criteria adopted by the American Heart Association. PBV is either normal or slightly increased in Class II patients, slightly to markedly increased in Class III patients, and either normal or decreased in Class IV patients. In the vast majority of Class II and Class III patients, particularly in those with valvular heart disease, some degree of pulmonary venous and arterial hypertension is usually recorded. The increased pulmonary distending pressure may cause dilatation of the pulmonary vascular bed resulting in an increase in PBV. In a number of Class IV patients, the pulmonary vascular bed may be compromised by the presence of both functional and organic changes in the pulmonary vessels, hence the PBV may reduce toward the normal level. These functional and organic changes include: medial hypertrophy and endothelial proliferation, thromboembolism, and vasoconstriction secondary to hypoxemia.

Physiologic or Pharmacologic Interventions

Patients with valvular obstruction (mitral stenosis or aortic stenosis) and those with increased left ventricular resistance (aortic stenosis or cardiomyopathy) respond differently to physiologic or pharmacologic interventions from those without valvular lesions, particularly with respect to the changes in PAm and LAm pressures and PBV. The major reason for the difference is due to the fact that any physiologic

Table 10.1 Hemodynamic data and pulmonary blood volume

Subjects	No.	Cardiac index mean	Pulmonary artery mean pressure	Left atrial mean pressure	Pulmonary blood volume
		liters/min/m²	*mm Hg*	*mm Hg*	*ml/m²*
Normal	20	3.40 ± 0.15	13.5 ± 0.7	7.5 ± 0.6	275 ± 9.2
Mitral stenosis Classes I					
and II	45	2.70	23	15	290
Class III	102	2.15	36	21	315
Class IV	12	1.52	46	24	220
Mitral regurgitation Classes I					
and II	26	2.68	19	12	282
Class III	48	2.10	35	20	320
Class IV	5	1.45	40	23	218
Aortic stenosis Classes I					
and II	34	2.90	17	10	272
Class III	56	2.56	21	13	305
Aortic regurgitation Classes I					
and II	26	2.88	16	9	306
Class III	25	2.26	23	15	310

maneuver or pharmacologic agent which induces an increase in cardiac output will cause an appreciably greater rise in the pulmonary vascular pressures in patients with valvular obstruction or increased left ventricular resistance than those without. Thus, a more substantial augmentation of PBV in patients of the former group is expected.

Pulmonary blood volume is altered by physiologic intervention such as change in body posture. During head-up tilt, there was a decrease in cardiac output (CO), PAm and LAm pressures, and PBV. When the legs were raised in supine position, each of the above parameters increased. Exercise in normal and valvular diseased patients caused an increase in PBV. This may be due to expansion of the pulmonary vascular bed secondary to an increase in pulmonary intravascular distending pressure, as a result of recruitment of additional pulmonary blood vessels, or both. Increase in total blood volume, as by rapid fluid infusion, resulted in a substantial increase in PBV, particularly in patients with mitral stenosis. On the other hand, reduction in total blood volume or venous return produced a decline in PBV. Qualitative and quantitative changes in PBV during atrial pacing were similar and occurred for the same reasons as those produced by exercise, but only in patients with mitral stenosis (Table 10.2). In others tachypacing caused a reduction in PBV.

Table 10.2 Effects of exercise, atrial pacing
and isoproterenol in patients with mitral stenosis

Number	Study	Cardiac index mean	Pulmonary artery mean pressure	Left atrial mean pressure	Pulmonary blood volume
		$liters/min/m^2$	$mm\,Hg$	$mm\,Hg$	ml/m^2
I	Control	2.26	26	17	289
	Exercise	3.17	45	30	375
II	Control	2.53	30	20	296
	Atrial pacing	2.69	40	28	368
III	Control	2.20	30	19	315
	Isoproterenol	2.90	36	23	377

Pharmacologic intervention produced varied effects on PBV. Acute digitalization reduced PBV in normal subjects but produced no change in mitral stenotic subjects. Isoproterenol infusion increased PBV in this latter group probably by a direct vasodilating effect (Table 10.2), but reduced PBV in cardiomyopathy where preinfusion elevation of left ventricular diastolic pressure was improved by the drug. Aminophylline and acetylcholine infusions increased PBV by their direct pulmonary vasodilating action. Methoxamine, increasing PBV while producing systemic hypertension and diuretics, reducing PBV together with total blood volume, all illustrate secondary mechanisms for PBV variation. Acute diuresis induced by intravaneous injection of 40 mg of furosemide resulted in a fall in PAm and LAm pressures, CO and PBV. As a general but not invariable rule, increase in PAm, LAm, or total blood volume from any pharmacologic, physiologic, or pathologic cause resulted in an increase in PBV.

PULMONARY EXTRAVASCULAR WATER VOLUME

Methods of Measurement

Utilizing the double radioisotope indicator technique (Chinard, 1951, 1954, 1962), measurement of PEV in man has been reported by workers from many laboratories (Lilienfield et al., 1955; Ramsey et al., 1964; Turino et al., 1968). The upper limit of the normal value is estimated to be 120 ml/m^2 and the mean normal value 90 ml/m^2 (McCredie, 1970; Yu, 1971; Schreiner et al., 1971). These values are corrected by a factor of 0.8, as suggested by Chinard et al. (1969). In some older normal subjects, the PEV may be higher (Bristow and Kirk, 1971). The PEV is frequently increased in patients with valvular heart disease or primary myocardial disease, particularly in those with elevated left atrial or pulmonary wedge pressure. (McCredie, 1967; Korsgren et al., 1969; Sutherland et al., 1971). An increased PEV has been found almost uniformly in association with pulmonary edema (McCredie, 1967; Yu, 1971; Schreiner et al., 1971).

In our laboratory measurement of PEV is carried out in conjunction with the determination of PBV as a special protocol for all patients. After the completion of sequential injections of Indocyanine green to the PA and LA for the determinations of CO and PBV, PEV is estimated by the double radioisotope technique. The indicators are RISA and tritiated water (THO), in a solution of 5 ml or normal saline containing 7 μc of ^{131}I and 80 μc of ^{3}H. Before the injection of the indicators, blood samples are collected for background counts. As soon as the solution of indicators is injected through the cardiac catheter into the PA, arterial blood is sampled either automatically with a special drum containing 30 2-ml heparinized tubes or manually through a small polyethylene tube into a series of 3-ml heparinized tubes. Usually 15 to 20 blood samples are collected, at intervals of 2 seconds, each containing approximately 1.5 ml of blood. Immediately following the collection of samples, the tubes are rotated to mix the blood thoroughly with heparin. Twenty minutes later, two more blood samples are obtained for comparison with the background counts as well as for estimate of plasma volume.

The blood samples are then centrifuged, and an aliquot of 0.5 ml of supernated plasma is carefully pipetted and transferred to a 20-ml capped bottle containing 10 ml of Beckman special scintillation solution. Beta radiation emissions from both ^{131}I and ^{3}H are counted simultaneously through separate windows in a Beckman LS 250 scintillation counter at ambient temperatures. Counting ratios of tritium range from 1,000 to 50,000 counts per minute with background counts of 50 to 200 counts per minute. Counts from standards prepared with known amounts of tritium in whole blood show a linear relationship to concentration up to 50,000 counts per minute. The tritium counts are not affected by the presence of RISA in the blood; the standard error of both ^{131}I and tritium counts is less than 3%.

Time-concentration curves for both RISA and THO are plotted on semilogarithmic paper after subtracting background counts from the actual counts (Fig. 10.2). The values of the points on each curve

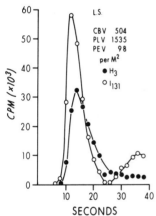

Figure 10.2 Simultaneous radioisotope dilution curves obtained after injections of radioiodinated serum albumin (^{131}I) and tritiated water (^{3}H) into the pulmonary artery and recovered from a systemic artery. The concentrations of the radioactivity of the two radioisotopes were replotted. The amount of central blood volume (CBV), plasma volume (PLV), and pulmonary extravascular water volume (PEV) are expressed in milliliters per m[2] BSA. The curves were obtained from a patient with normal hemodynamics. Note the normal values of all the volumes.

are punched onto data cards, and the CO and the mean transit time are derived from both RISA and THO curves using the aforementioned computer program. The values of mean transit time determined by Indocyanine green and RISA curves in 31 patients were compared. The average mean transit time from RISA curve was about 0.5 seconds longer than that from Indocyanine green curve. Although there was a slight systematic difference, it was considered of no practical importance. The PEV is calculated according to the following formula:

$$PEV = \frac{CI}{60} [Tm_{THO \ (PA-BA)} - Tm_{RISA \ (PA-BA)}],$$

where PEV = pulmonary extravascular water volume (ml/m^2), CI = cardiac index (ml/m^2/minute), $Tm_{THO\ (PA\text{-}BA)}$ = mean transit time (seconds) of tritiated water from pulmonary artery to brachial artery, and $Tm_{RISA\ (PA\text{-}BA)}$ = mean transit time (seconds) of radioiodinated serum albumin from pulmonary artery to brachial artery.

The values obtained in each determination are adjusted by a factor of 0.8, as suggested by Chinard *et al.* (1969). To provide a comparison between patients of different body sizes, values of PEV are expressed in terms of m^2 of body surface area.

To demonstrate adequate mixing of the two radioisotope indicators with injection into the PA, repeat determinations have been made in 11 patients by injecting a second dose of the indicator into the right atrium. As shown in Fig. 10.3, the difference between mean transit

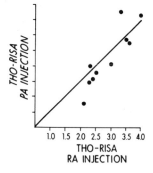

Figure 10.3 Difference between the mean transit time (*MTT*) of tritiated water (*THO*) and that of radioiodinated serum albumin (*RISA*) obtained during respective injections of indicators to the pulmonary artery and to the right atrium.

time of THO and that of RISA in the first determination is not significantly different from that in the second determination.

Measurements of PEV have been carried out in 12 patients with valvular heart disease to document the reproducibility of the technique. The results of the first and second determinations are less than 10% in eight patients and less than 20% in the remaining four patients (Fig. 10.4). We use 120 ml/m^2 as the upper limit of "normal" value.

Pulmonary Extravascular Water Volume in Cardiac Patients at Rest

Our initial data are derived from results obtained from 124 patients. Five patients (Schreiner *et al.,* 1971) were considered "normal," who were referred for study because of the presence of a systolic murmur. Forty-eight patients had documented coronary artery disease (CAD),

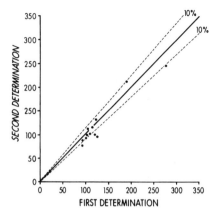

Figure 10.4 Duplicate determinations of the pulmonary extravascular water volume (ml/m^2) made in 12 patients with cardiac disease. In eight patients the difference between the first and second determinations was less than 10%. In the remaining four patients the difference was 20%.

25 had aortic valve disease (AVD), and 46 had predominant mitral stenosis (MS). One-hundred-and-fourteen patients were in either Class II or Class III, and five patients were in Class IV. The classification of the functional capacity is according to the criteria adopted by the American Heart Association. However, it is evident that in patients with MS and AVD the functional disability is usually due to exertional dyspnea with associated left atrial hypertension, whereas in patients with CAD the disability is due to angina pectoris. In the latter group, when the patients are free from pain, the LAm pressure is usually normal.

In 63 patients with normal LAm pressure there was no correlation between the PEV and cardiac index. On the other hand, as shown in Fig. 10.5, LAm pressure was significantly correlated with the PEV in all 124 patients ($r = 0.50$, $p < 0.01$).

When all the patients were grouped according to etiology and functional capacity, the PEV was similar among the "normal" patients, Class II or III patients with CAD, Class II or III patients with AVD, and Class II patients with MS. However, the PEV was significantly elevated in 5 patients of Class IV with either CAD or AVD and in 33 patients with MS of Class III (Fig. 10.6). In addition to an absolute increase in PEV in these 38 patients, the ratio of PEV to PBV is also increased in each group with an average value of 0.51 (Figs. 10.7 and

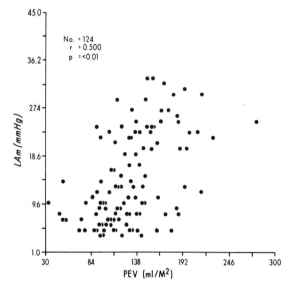

Figure 10.5 Pulmonary extravascular water volume (*PEV*) plotted against mean left atrial pressure (*LAm*) in 124 patients with various types of cardiac disease. A significant positive correlation between the two parameters was observed.

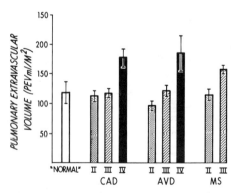

Figure 10.6 Pulmonary extravascular water volume (mean ± SEM) measured in four groups of patients: no cardiovascular disease and normal hemodynamics (*"normal"*), coronary artery disease (*CAD*), aortic valve disease (*AVD*), and mitral stenosis (*MS*). The functional classes of these patients are indicated, respectively, by *II, III,* and *IV*.

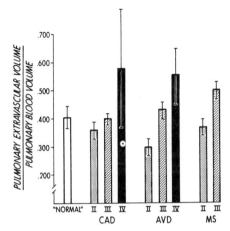

Figure 10.7 Ratio between pulmonary extravascular volume and pulmonary blood volume in patients with no cardiovascular disease and normal hemodynamics ("normal"), coronary artery disease (CAD), aortic valve disease (AVD), and mitral stenosis (MS). The ratio was significantly increased in Class IV patients with CAD or AVD and in Class III patients with MS.

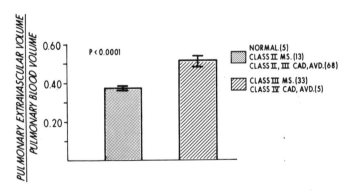

Figure 10.8 Ratio between pulmonary extravascular volume and pulmonary blood volume in two major groups of patients. The first group consists of 5 patients with normal hemodynamics, 13 Class II patients with mitral stenosis (MS), and 68 Class II and Class III patients with coronary artery disease (CAD) or aortic valve disease (AVD). The second group consists of 33 Class III patients with MS and 5 patients with CAD or AVD. There is a significant difference between the mean values of the two groups.

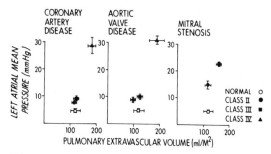

Figure 10.9 Pulmonary extravascular water volume (PEV) plotted against mean left atrial pressure (LAm) in three groups of patients. The values of the "normal" patients are represented by the hollow circles at the bottom. In Class II and Class III patients with coronary artery disease (CAD) or aortic valve disease (AVD) and in Class II patients with mitral stenosis (*MS*), LAm was either slightly or moderately elevated but PEV was within normal limits. In Class IV patients with CAD or AVD and in Class IV patients with MS both LAm and PEV were markedly elevated.

10.8). When this value compares to a ratio of 0.37 for the remaining 86 patients, the difference is highly significant.

In Figs. 10.9 and 10.10 are plotted the values (mean ± SEM) of average LAm and PEV in all the patients studied. "Normal" patients

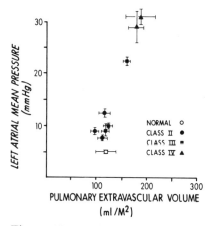

Figure 10.10 Composite diagram of pulmonary extravascular water— water plotted against left atrial mean pressure in all patients studied as shown in Fig. 10.7.

and patients with CAD or AVD of Classes II and III had normal LAm pressures and PEV which averaged 115 ml/m². Patients with MS of Class III had slightly elevated LAm pressure but similar average PEV. Those with MS of Class III had an average LAm pressure of 23 mm Hg and PEV which averaged 160 ml/m². The five patients of Class IV with either CAD or AVD had a marked elevation of both LAm pressure and PEV, with average values of 30 mm Hg and 190 ml/m², respectively. There is a linear relationship between these variables. Regression equations for each group of patients, considered separately, did not significantly differ from one another.

Evaluation of the degree of pulmonary congestion based upon the changes in the roentgenograms of the chest was made by a radiologist who had no knowledge of either the clinical status or the hemodynamic findings of the patients. According to the X-ray findings, the patients were grouped in four classes: 0 = normal, 1+ = pulmonary venous congestion, 2+ = presence of interstitial fluid, and 3+ = frank pulmonary edema. The average values of PEV in each group are shown in Fig. 10.11. Patients with normal pulmonary vasculature had an average PEV of 116 ml/m² and normal LAm pressure. This group of patients consisted almost entirely of patients with CAD. In those

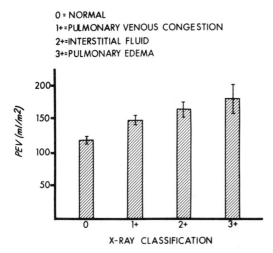

Figure 10.11 Pulmonary extravascular water volume (PEV) in four groups of patients classified according to the changes in the roentgenograms. There was a gradual increase in PEV from patients with pulmonary venous congestion to those with frank pulmonary edema.

patients with radiologic changes consistent with pulmonary venous congestion, the PEV averaged 146 ml/m² and LAm pressure of 20 mm Hg. Patients with radiologic evidence of interstitial fluid had an average PEV of 160 ml/m² and an average LAm pressure of 24 mm Hg. The majority of this group of patients had MS. In patients with frank pulmonary edema demonstrated in the chest roentgenograms, the average PEV was 178 ml/m² and the average LAm pressure was 28 mm Hg (Fig. 10.12). Thus, the roentgenograms of the chest were

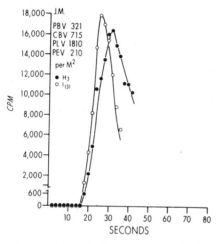

Figure 10.12 Similar curves as obtained in Fig. 10.2 from a patient with chronic congestive heart failure and pulmonary edema. Pulmonary blood volume (*PBV*), central blood volume (*CBV*), and plasma volume (*PLV*) were slightly elevated, but the pulmonary extravascular volume (*PEV*) was markedly increased. All the volumes are expressed in milliliters per m² BSA.

reliable in separating normal from degrees of pulmonary vascular congestion and edema in terms of both PEV and LAm pressure. However, they are not specific enough to predict magnitudes of PEV and left atrial hypertension when varying degrees of pulmonary congestion are present. For instance, in patients with moderately severe or severe mitral valve obstruction, considerable overlapping of the magnitude of PEV and LAm pressure is observed.

To compare healthy subjects with patients with cardiac disease, it must be assumed that the chronic vascular changes associated with valvular heart disease do not affect the technique by impairing the diffusion of THO and prolonging its transit time. A barrier to THO diffusion through the capillary bed would result in shorter transit time and reduce the calculated PEV. Since there is no evidence to show that significant amount of THO is lost, it appears that prolongation of transit time by trapping or loss of the indicator in the extravascular space is not a significant problem. Furthermore, the error of transit time associated with slow indicator passages by the Stewart-Hamilton method should be eliminated or minimized by the subtraction of the similarly prejudiced time for simultaneously injected intravascular indicators.

Most authors found a good correlation between PEV and the LAm or Pw pressure. Although LAm or Pw pressure may not accurately represent pulmonary capillary pressure because of the uncertain level of the alveolar pressure, in the supine position, LAm pressure should give an approximate reflection of pulmonary capillary pressure.

Pulmonary capillary pressure, as reflected by LAm pressure, could affect the pulmonary extravascular volume in at least three ways. It increases perfused capillary surface areas by dilating previously perfused capillaries and recruiting new capillary beds. Capillary surface area is thought to be a direct factor in capillary fluid balance. Capillary hydrostatic pressure is also essential to the Starling theory of intra- and extravascular fluid exchange. Increased pulmonary capillary pressure leads to fluid transudation from the pulmonary capillaries to the interstitial space and a new balance is established at a higher extravascular volume.

O'Connor and associates measured the PEV in patients before and after cardiac surgery in patients with MS (1972). They found a good correlation before surgery between the PEV and LAm pressure. After successful cardiac surgery the LAm pressure was reduced, as was the PEV; and again PEV correlated well with LAm pressure.

Animal studies have shown that the threshold of pressure for the formation of acute pulmonary edema is approximately 24 mm Hg of LAm pressure (Guyton and Lindsay, 1959; Levine et al., 1965). In man, acute elevation of LAm pressure to similar level is also thought to produce pulmonary edema. However, many patients with prolonged elevation of pulmonary venous pressure, as in MS, are somehow protected from pulmonary edema at LAm pressure far in excess of 20 mm Hg. The site of this protection is unknown, but two possibilities could exist: (1) alteration in factors which control the fluid balance across the capillary membrane, i.e. a barrier to transudation, and (2) a barrier to intra-alveolar edema existing at the level of the alveolar cell. Given this mechanism, the response to increased hydrostatic pressure would be increased pressure and volume in the interstitial space. In our study we have observed greatly augmented accumulation of pulmonary extravascular water in many patients with prolonged elevation of LAm pressure, and yet no signs of alveolar edema. Thus, in this connection the second mechanism seems to be more likely. However, it is not possible to exclude the first mechanism.

There are two possible ways that the fluid in the pulmonary capillaries may move into the extravascular spaces in the lung. (1) To reach the interstitial space, the fluid passes only through the wall of the pulmonary capillaries, which is relatively a thin barrier; and (2) to reach the alveolar sacs the fluid has to penetrate the alveolar epithelium, which is probably less permeable to water than that of the capillary endothelium.

It is a common finding radiologically that in the early stage of pulmonary edema the excessive fluid is located in the interstitial space. The interstitial space in the alveolar wall is connected to that surrounding the large pulmonary vessels and airways. In the perivascular and peribronchial spaces, many lymphatics are present and have the function to carry the excessive edema fluid toward the hilum. Thus, interstitial edema is not infrequently associated with prominant so-called perivascular and peribronchial cuffing. Clinically this condition is difficult to recognize in man. When pulmonary rales are heard, the pulmonary edema is already in the late stage as a result of entrance of excessive fluid into the alveolar spaces.

Pulmonary Extravascular Lung Water Volume in Cardiac Patients during Physiologic or Pharmacologic Interventions

Exercise produced a nonuniform response, with increase or no change in PEV. These findings are in agreement with those of Korsgren and associates (1969). Moderate rate (130/minute) atrial pacing in patients with coronary artery disease did not appreciably effect PEV, except in those who developed angina pectoris when both PBV and PEV were increased. More prolonged pacing caused increases in PEV in most patients who demonstrated elevation in LAm and PBV, but all parameters returned to control levels on discontinuance of pacing. After acute diuresis in patients with chronic left atrial hypertension, the change in PEV was inconstant. It would appear that the excessive fluid in the lungs may be firmly bound with the collagen in the interstitial tissues and released rather slowly after diuresis.

Pulmonary Extravascular Lung Water Volume Measurement in Animals

Utilizing the double indicator technique, many investigators have measured PEV in animals at rest and during hemodynamic interventions (Ramsey and associates, 1964; Levine and co-workers, 1965, 1967, 1970; Goresky et al., 1969; Pearce and associates, 1965). In the normal dog, the PEV is about 3 to 4 ml/kg. This volume is approximately 50 to 60% of the total lung water determined by drying. The reason for this discrepancy is not clear. It is likely that there are nonperfused portions of the lungs where tissue water is not accessible to the diffusible indicator and that all the pulmonary water volume may not be adjacent to permeable branches of the pulmonary arterial system. Whether the same finding applies to human subjects remains to be determined. Nevertheless, despite the ability of the diffusible technique to quantitate changes in PEV, measured PEV is probably not the same as the total noncellular extravascular fluid volume.

ACKNOWLEDGMENTS

I express my thanks and appreciation to my colleagues, Drs. Bernard F. Schreiner, Gerald W. Murphy, Pravin M. Shah, Morrison Hodges, and David H. Kramer for their contributions and assistance in the preparation of this manuscript. The secretarial help of Mrs. Judy Lukensmeyer is gratefully acknowledged.

This research was supported in part by grants HL 03966 and HL 05500 and a MIRU contract No. PH 43-68-1331 from the National Heart and Lung Institute, National Institutes of Health, Bethesda, Maryland.

REFERENCES

Bristow, G. K., and B. W. Kirk. 1971. Venous admixture and lung water in healthy subjects over 50 years of age. J. Appl. Physiol. 30:552–557.

Chinard, F. P. 1951. Capillary permeability. Bull. Johns Hopkins Hosp. 88:489–492.

Chinard, F. P., and T. Enns. 1954. Transcapillary pulmonary exchange of water in the dog. Amer. J. Physiol. 178:197–202.

Chinard, F. P., T. Enns, and M. F. Nolan. 1962. Pulmonary extravascular water volumes from transit time and slope data. J. Appl. Physiol. 17:179–183.

Chinard, F. P., W. Perl, and R. M. Effros. 1969. Lung water: physiological and clinical significance. Trans. Amer. Clin. Climat. Assoc. 81:85–97.

Dock, D. S., W. L. Kraus, L. B. McQuire, J. W. Hyland, F. W. Haynes, and L. Dexter. 1961. The pulmonary blood volume in man. J. Clin. Invest. 40:317–328.

Dock, D. S., W. L. Kraus, E. Woodward, L. Dexter, and F. Haynes. 1959. Observations on pulmonary blood volume in man. Fed. Proc. 18:37.

Freitas, F. M., E. Z. de Faraco, N. Nedel, D. F. Azevedo, and J. Zuduchliver. 1964. Determination of pulmonary blood volume by single intravenous injection of one indicator in patients with normal and high pulmonary vascular pressures. Circulation 30:370–380.

Fujimoto, K., R. Kunieda, and T. Shiba. 1960. Lung blood volume in acquired heart disease. Jap. Heart J. 1:442–455.

Giuntini, C., M. L. Lewis, A. Sales Luis, and R. M. Harvey. 1963. A study of the pulmonary blood volume in man by quantitative radiocardiography. J. Clin. Invest. 42:1589–1605.

Goresky, C. A., R. F. P. Cronin, and B. E. Wangel. 1969. Indicator dilution measurements of extravascular water volumes in the lungs. J. Clin. Invest. 48:487–501.

Guyton, A. C., and A. W. Lindsay. 1959. Effect of elevated left atrial pressure and decreased plasma protein concentration in the development of pulmonary edema. Circ. Res. 7:649–657.

Hamilton, W. F., J. W. Moore, J. M. Kinsman, and R. G. Spurling. 1930. Blood flow and intrathoracic blood volume as determined by the injection method and checked by direct measurement in perfusion experiments. Amer. J. Physiol. 93:654–655.

Kinsman, J. M., J. W. Moore, and W. F. Hamilton. 1929. Studies on circulation. I. Injection method, physical and mathematical consideration. Amer. J. Physiol. 89:322–330.

Korsgren, M., R. Luepker, B. Liander, and E. Varnauskas. 1969. Pulmonary intra and extra vascular fluid volume changes with exercise. Cardiovasc. Res. 3:1–6.

Kunieda, R. 1955. Evaluation of pulmonary blood volume in mitral valve disease by T-1824 method. Resp. Circ. 3:510.

Levine, O. R., R. B. Mellins, and A. P. Fishman. 1965. Quantitative assessment of pulmonary edema. Circ. Res. 17:414–426.

Levine, O. R., R. B. Mellins, and R. M. Senior. 1970. Extravascular lung water and distribution of pulmonary blood flow in the dog. J. Appl. Physiol. 28:166–171.

Levine, O. R., R. B. Mellins, R. M. Senior, and A. P. Fishman. 1967. The application of Starling's law of capillary exchange to the lungs. J. Clin. Invest. 46:934–944.

Levinson, G. E., M. J. Frank, and H. K. Hellems. 1964. The pulmonary vascular volume in man. Measurement from atrial dilution curves. Amer. Heart J. 67:734–741.

Lewis, M. L., C. Giuntini, L. Donato, R. M. Harvey, and A. Cournand. 1962. Quantitative radiography. III. Results and validation of theory and method. Circulation 26:189–199.

Lilienfield, L. S., E. D. Freis, E. A. Partenope, and H. J. Morowitz. 1955. Transcapillary migration of heavy water and thiocyanate ion in the pulmonary circulation of normal subjects and patients with congestive heart failure. J. Clin. Invest. 34:1–8.

McCredie, M. 1967. Measurement of pulmonary edema in valvular heart disease. Circulation 36:381–386.

McCredie, R. M. 1970. Pulmonary oedema in lung disease. Brit. Heart J. 32:66–70.

McGaff, C. J., A. D. Jose, and W. R. Milnor. 1959. Pulmonary, left heart and arterial volume in valvular heart disease. Clin. Res. 7:230–231.

Milnor, W. R., A. D. Jose, and C. J. McGaff. 1960. Pulmonary vascular volume, resistance and compliance in man. Circulation 22:130–137.

Oakley, C., G. Glick, M. N. Luria, B. F. Schreiner, and P. N. Yu. 1962. Some regulatory mechanisms of the human pulmonary vascular bed. Circulation 26:917–930.

O'Connor, N. E. 1972. Discussion on the paper entitled, "Extravascular lung water and gas exchange in cardiac patients," by M. Mariani, C. Giuntini, A. Barsotti, F. Fazio, A. Bulbarini, and L. Odoguardi. In C. Giuntini (ed.), Central Hemodynamics and Gas Exchange, Minerva Medica, Torino, Italy, p. 364.

Pearce, M. L., J. Yamashita, and J. Beazell. 1965. Measurement of pulmonary edema. Circ. Res. 16:482–488.

Ramsey, L. H., W. Puckett, A. Jose, and W. W. Lacy. 1964. Pericapillary gas and water distribution volumes of the lung calculated from multiple indicator-dilution curves. Circ. Res. 15:275–286.

Samet, P., W. H. Bernstein, A. Lopez, and S. Levine. 1966. Methodology of true pulmonary blood volume determination. Circulation 33:847–853.

Schreiner, B. F., G. W. Murphy, D. H. James, and P. N. Yu. 1968. Effect of isoproterenol on the pulmonary blood volume in patients with valvular heart disease and primary myocardial disease. Circulation 37:220–231.

Schreiner, B. F., G. W. Murphy, P. M. Shah, D. H. Kramer, and P. N. Yu. 1971. Pulmonary extravascular volume in cardiac patients. Circulation 43–44 (suppl. II):39.

Segre, G., G. L. Turco, and F. Ghemi. 1965. Determinations of pulmonary weighting functions, of the mean pulmonary transit time, and of the pulmonary blood volume in man by means of radiocardiograms. Cardiologia 46:295–311.

Stewart, G. N. 1921. The pulmonary circulation time, the quantity of blood in the lungs and the output of the heart. Amer. J. Physiol. 58:20–44.

Sutherland, P. W., J. F. Cade, and M. C. F. Pain. 1971. Pulmonary extravascular fluid volume and hypoxaemia in myocardial infarction. Aust. N. Z. J. Med. 2:141–145.

Turino, G. M., N. H. Edelman, R. M. Senior, E. C. Richards, and A. P. Fishman. 1968. Extravascular lung water in cor pulmonale. Bull. Physio-Pathologie Respiratoire 4:47–64.

Van deWater, J., J. Shek, N. O'Connor, I. Miller, and E. Milne. 1970. Pulmonary extravascular water volume: measurement and significance in critically ill patients. J. Trauma 10:440–449.

Yu, P. N. 1969. Pulmonary Blood Volume in Health and Disease. Lea and Febiger, Philadelphia.

Yu, P. N. 1971. Lung water in congestive heart failure. Mod. Conc. Cardiovasc. Dis. 40:27–32.

Yu, P. N., G. Glick, B. F. Schreiner, and G. W. Murphy. 1963. Effects of acute hypoxia on the pulmonary vascular bed of patients with acquired heart disease. Circulation 27:541–553.

chapter 11/ Measurement of Ventricular Volumes by Thermodilution and Other Indicator-Dilution Methods

Elliot Rapaport
Cardiopulmonary Laboratory/San Francisco General Hospital/
and Department of Medicine and Cardiovascular Research Institute/
University of California/San Francisco, California

The measurement of ventricular volume aids materially in the evaluation of ventricular performance. Although ventricular function curves traditionally have been constructed over the years by relating stroke work or stroke output to ventricular filling pressure, serious potential errors exist with their use. First, improved myocardial contractility is generally assumed to take place after a given intervention if a Frank-Starling curve shifts to the left such that a greater degree of work or output is achieved with the same or less filling pressure. However, if the intervention itself changes the diastolic compliance or pressure-volume characteristics of the ventricle, an increase or decrease in end-diastolic pressure need not necessarily reflect a respective increase or decrease in ventricular end-diastolic volume (or fiber length) but may represent a change in ventricular compliance. Second, Starling's law of the heart states that the energy of myocardial contraction is a function of its end-diastolic fiber length. The utilization of stroke output or even stroke work as an index of the energy of myocardial contraction neglects the fact that more than one-half of the myocardial oxygen consumption is utilized during isovolumetric contraction or before the actual expulsion of blood and the performance of any physical work.

Clearly, mean systolic force would be superior to stroke work in constructing a ventricular function curve. The measurement of ventricular volume permits the estimation of the force developed by the ventricular wall.

Finally, perhaps the greatest concern in using the traditional Frank-Starling curve to measure ventricular performance is the observation that intrapericardial pressure becomes rapidly positive as right or left ventricular end-diastolic pressure is increased experimentally. Over the range of normal to pulmonary edema levels of filling pressure for the left ventricle, the actual physiologic distending pressure or transmural distending pressure (the difference between the left ventricular end-diastolic and intrapericardial end-diastolic pressures at the same hydrostatic level) changes only 5 or 6 mm Hg (Holt, 1967). Thus accurate construction of a ventricular function curve requires the plotting of an index of the energy of the myocardial contraction against the transmural distending pressure rather than the absolute left ventricular end-diastolic pressure. Since the distending pressure varies only slightly over a wide range of absolute pressure changes, relatively small changes in end-diastolic pressure are hazardous to use in judging changes in ventricular performance. Thus, one must take into consideration intrapericardial pressure obtained from regression equations constructed previously from the pressure volume relationships of the ventricle (Holt et al., 1960) or one must have a catheter not only within the left ventricle but also within the pericardial space with its tip at the same hydrostatic level as the intraventricular catheter. Under these circumstances, the difference between the pressure inside and outside of the ventricle can be measured accurately and used as an index of changes in the transmural distending pressure and, therefore, changes in volume of the ventricle. It becomes clear that direct measurement of ventricular volume is necessary if ventricular performance is to be judged accurately in terms of changes in the shape of the Frank-Starling curves.

The measurement of ventricular volume may also assist in the evaluation of myocardial contractility by another method. Recently, the ejection fraction or the ratio of the stroke volume to the end-diastolic volume has been used as an index of myocardial contractility. Although the ejection fraction is influenced by changes in heart rate, preload, and afterload Krayenbühl et al., 1968a), it is sensitive to changes of mycardial contractility as well and therefore becomes reduced below normal limits in the presence of compromised myocardial function.

Knowledge of ventricular volume also permits one to look at the relationship between the velocity of fiber shortening and myocardial force. If one measures the ventricular volume and then continuously measures the volume ejected from the ventricle throughout systole (i.e. by an implanted electromagnetic flowmeter in the experimental animal), one can in fact plot the velocity of circumferential fiber shortening against myocardial force and thus construct a force-velocity curve for the period of ejection. In man, one may be limited to expressing merely the average velocity of fiber shortening for the period of ejection or the velocity at a given stress and observing whether this value falls in a distinctly abnormal area when compared with normal control values.

A number of techniques have been employed in past years to measure changes in ventricular volume by observing a particular dimensional change of the ventricle itself. Hewitt et al. (1971) have sutured gauges to sections of the myocardium and observed the change in length occurring under that section of the myocardium. Erickson et al. (1971) have implanted sonarcardiometer crystals and observed dimensional changes of the diameter being observed. These techniques are obviously limited to experimental animals. Harrison et al. (1963) followed by cineradiography the movement of a segment of the left ventricle postoperatively in cardiac patients in whom radiopaque markers had been sutured in place at the time of surgery. More recently, considerable interest has developed in the use of ultrasound as a method of observing changes in ventricular dimensions in man. By focusing on the plane of the left ventricle at a level of the chordae just below the mitral valve, one can measure the diameter during systole and diastole between the lateral wall of the ventricle and the septum near its equatorial center. This technique may also be used to estimate ventricular volume in accordance with the method described by Feigenbaum et al. (1969). Although this method gives an estimate of the ventricular volume in ventricles with normal wall mobility, its use in patients with heart disease, particularly chronic coronary artery disease and/or acute myocardial infarction where portions of the wall may be moving dyskinetically or even paradoxically, the estimation of volume based upon changes in a single diameter is susceptible to serious error.

Investigators have generally used one of the following methods for estimating ventricular volume in the intact animal or man: (1) isotope or contrast angiocardiography and (2) indicator-dilution techniques. The rest of this chapter is devoted to a discussion of the use of an indicator-dilution technique to measure ventricular volume, particularly using negative heat as the indicator or the so-called thermodilution method.

PRINCIPLE OF THE METHOD

A mathematical model of a two-chamber pulsatile heart simulating an atrium and ventricle reveals that when indicator is injected into either the atrium or ventricle, assuming complete mixing, its loss into the vessel leading from the ventricle is exponential. The slope of the exponential washout into the artery outside the heart is governed by the chamber with the largest volume. If the indicator is injected into the atrium, the washout slope of indicator into the artery from the ventricle will be governed by the residual fraction of the ventricle only if end-diastolic ventricular volume is greater than maximal atrial volume. On the other hand, if the indicator is injected into the ventricle, the washout slope into the aorta (or pulmonary artery) will also be exponential and will always be dependent on the residual fraction of the ventricle provided valvular regurgitation of the atrioventricular valve system is not present.

In practice, it is customary for the indicator to be injected rapidly directly into the ventricle. If the detector sampling the time-concentration curve outside the ventricle has a sufficiently rapid response time, one will see a series of decremental steps with each heart beat as correspondingly less concentrated indicator is washed out of the heart with each succeeding beat. The ratio of the concentration of one beat to that of the previous beat will be constant, reflecting the exponential washout. Knowledge of this ratio and the stroke volume itself permits calculation of the absolute ventricular volume by the formula, $EDV = SV/(1 - K)$, where EDV = end-diastolic volume, SV = stroke volume, $K = (Cn + 1)/Cn$, and $Cn + 1$ and Cn = the concentrations of indicator in the aorta at end-systole after beats n and n_1, respectively. It should be pointed out that the presence of atrio-ventricular valve regurgitation results in a downslope that is not monoexponential (Polissar and Rapaport, 1961). It reflects an initial fast washout slope, determined primarily by the ventricular volume, that then becomes more gradual as indicator stored within the atrium is washed out more slowly at a rate influenced by the atrial volume and the degree of regurgitation.

A second and mathematically different approach consists of injecting the total amount of indicator (i) in one diastole directly into the ventricle using a power-driven device. If one then assumes that the indicator mixes evenly with the residual volume within the ventricle as well as with the entering stroke volume from the atrium, the ultimate volume of dilution will be the end-diastolic volume of the ventricle. Consequently, a detector sampling the time-concentration curve just outside the ventricle during systole of the following beat would record a concentration representative of indicator diluted by the end-diastolic volume of the previous beat. Consequently, $EDV = i/C_1$.

These mathematical principles were first applied in an attempt to measure ventricular volume in man by Bing and associates (1951). They attempted to measure right ventricular volume from a pulmonary arterial dye-dilution curve obtained after right ventricular injection of Evans blue dye. However, because of the poor response time of the original system, these initial attempts were unsuccessful since the method requires the recording of an accurate time-concentration curve immediately outside of the ventricle. This has tended to discourage the use of dyes such as Evans blue dye and Indocyanine green, since blood must be withdrawn at rapid flow rates through a catheter connected to a densitometer to avoid catheter distortion of the concentration curve. Nevertheless, if one uses such rapid flow rate sampling through a catheter, one can obtain indicator-dilution curves that show square wave steps of decreasing concentration as the ventricle washes out indicator with each beat, allowing accurate estimates of ventricular volume provided the densitometer has an appropriately fast response time (Levinson et al., 1967). More recently, fiberoptic catheters have been used to overcome the necessity of withdrawing blood rapidly (Hugenholtz et al., 1968; Bussmann, et al., 1971). These catheters permit the concentration curve to be recorded from the base of the aorta through the fiberoptic system at the end of the catheter without the necessity of sampling blood. For many years, Holt et al. (1956, 1962) used changes in the electrical conductivity of blood after injection of hypertonic saline to construct washout curves from both ventricles in a variety of species. Using a conductivity cell mounted near the tip of the catheter, the sampling rate of blood drawn by the cell to obtain an undistorted time conductivity curve is minimized.

Radioisotope angiography has also been used to accomplish ventricular volume measurements by the indicator-dilution principle. The washout curve of appropriately tagged albumin particles or other substances introduced into the ventricle is obtained by rapid external counting techniques by using a gamma camera (Folse and Braunwald, 1962; Ishii and MacIntyre, 1971).

However, the technique that has probably received the greatest attention is the thermodilution method. The method consists of introducing negative heat (usually cold saline or blood) into the ventricle (or atrium) and sampling the resultant time-temperature curve from the aorta or pulmonary artery, permitting the measurement of either left or right ventricular volume in the experimental animal or in man. The technique has the advantage that the thermistor-tipped catheter directly senses changes in aortic or pulmonary artery temperature obviating drawing blood through the catheter. Additionally, the use of cold saline or blood as the injectate has the advantage that the indicator is not a foreign substance or a dye, and one need not worry about its accumulation in the circulation after repeated injections. Thus, one can perform unlimited ventricular volume measurements.

TECHNIQUE OF THERMODILUTION MEASUREMENT

Detection of the time-temperature curve outside the ventricle is generally accomplished by the use of a thermistor bead although, infrequently, resistance thermometers and thermocouples have also been used. The thermistor bead is mounted in the end of a cardiac catheter which is passed retrogradely to the base of the aorta or into the pulmonary artery. If chronic implantation is desired to measure left ventricular volume in the awake animals, the bead is mounted in the end of the bevel of a no. 25 needle specially mounted on a brass hub and inserted through the wall of the aorta, with the hub lying on the outside of the aorta next to an electromagnetic flow probe (Keroes and Rapaport, 1972). The bead will then lie in the middle of the aortic stream just above the aortic valve.

Thermistor beads have the characteristic of changing their electrical resistance in response to changes in temperature, although this is not a linear function over a wide range of temperature. Over the small range of temperature changes that occur in the aorta in response to the injection of 4 to 5 ml of cold saline or blood into the ventricle, the change is essentially linear when the resistance of the bead is wired as one arm of a Wheatstone bridge. The excitation voltage of the bridge should be less than 0.3 volts to insure that the bead does not heat, serving as a velocity detector for flowing blood instead of recording blood temperature accurately. Critical to the accuracy of the method is that the bead must have an extremely fast response time or the slope will be distorted. To accomplish this, beads of extremely small size are used; they are mounted in such a way as to minimize the amount of cement substance holding them in place within the catheter tip or needle bevel.

Essentially two methods have been used for calculation of end-diastolic volume by the single injection indicator-dilution technique. In the first, the bolus of cold injectate is injected into the ventricle during one diastole, and the end-diastolic volume is calculated from the formula i/C_1, where i = the amount of indicator or heat injected into the ventricle and C_1 is the indicator concentration (or temperature change) of the first beat just outside of the ventricle. In the second method, the bolus is rapidly injected but not necessarily within one diastole. The end-diastolic volume is equal to the stroke volume divided by $1 - K$, where K is the ratio of the indicator concentration or temperature change of one beat to that of the beat before averaged for the series of washout beats. Ordinarily, this calculation is begun with the second beat and is accomplished by the sixth beat since the curve becomes more difficult to measure as it approaches the original base line. This difficulty reflects not only a greater arithmetical error as the deviation from the base line becomes small, but also the fact that there will be a proportionately larger effect from some storage of cold within the walls of the ventricle (reflecting transfer of cold temperature from blood to ventricular walls after the injection), which is being released back into the warmer blood entering the ventricle. Studies using both methods have suggested either that the washout slope method is more likely to be the correct method for estimating ventricular volume (Krayenbühl et al., 1968b, 1969a) or that there is no significant difference between the two methods (Hugenholtz and associates, 1968). The technique utilizing the ratio of the amount injected to its concentration at the end of the first beat assumes complete mixing within the first beat and therefore is dependent on the manner in which the injection is carried out as well as on the assurance that the injection was completed within one beat. In the case of the second method in which the downslope stroke is used, the method does not depend on the manner of mixing the initial injectate with the ventricular volume, but is dependent solely on the adequacy with which the ventricular residual volume mixes with the entering stroke volume from the left atrium by the time the ventricular discharge reaches the thermistor bead in the aorta. Additionally, the temperature of beat C_1 is likely to be somewhat artifactually lowered due to the immediate transfer of cold from ventricular injectate to ventricular wall. This gradient is maximum for the first beat and tends to diminish with subsequent beats. Therefore, the artifactual effect is likely to be greatest in the first beat. There is probably not too significant an error involved in view of the general agreement of simultaneous volume determinations by these two methods, albeit the single beat dilutional method tends to give an average 15% higher result using thermodilution, although there appears to be no significant difference between the two techniques when dye is used (Hugenholtz et al., 1968).

Knowledge of stroke volume from an independent technique such as the direct Fick method greatly simplifies calculation of ventricular volume by thermodilution using the washout slope method. Under these circumstances, it is unnecessary to know the temperature of the injectate as it emerges from the injection catheter, the volume of the injectate, the temperature of the aortic blood, or the specific heat capacity and specific gravity of both blood and the injectate, all of which are required if stroke volume is to be obtained from the same thermodilution curve or if one calculates end-diastolic volume by the first beat dilutional method. Instead, one need only measure the uncalibrated deflection of each decremental plateau from the base line after each beat. The average of the ratios of each deflection from the base line to that of the beat before for the series of washout beats is K and is equal to the residual fraction of the ratio of the end-systolic to end-diastolic volume of the ventricle. The ratio of the stroke volume to $1 - K$ gives one the absolute value for the end-diastolic volume.

Fig. 11.1 illustrates a typical thermodilution curve obtained from a thermistor bead implanted in the ascending aorta of a dog. The bead, imbedded in the tip of a no. 25 needle attached to a hub welded to a brass plate, was positioned so that the needle punctured the wall and the bevel with the exposed bead lay in the middle of the aortic stream. The brass base was anchored to the outside wall alongside an electromagnetic flowmeter probe to determine stroke volume independently. The dog was then allowed to recover from surgery. This experimental preparation allows ventricular volume to be determined under a variety of experimental situations. The thermodilution curve shown in Fig. 11.1 was obtained while the dog was standing quietly on a treadmill after acute anemia was produced through the exchange of 40 ml of blood per kg with equal amounts of low molecular dextran.

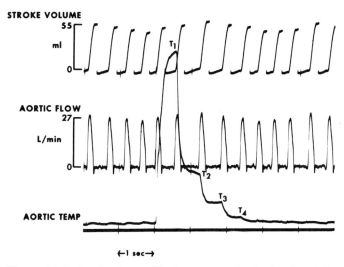

Figure 11.1 Aortic thermodilution curve obtained, after left ventricular injection of iced saline, from a thermistor bead implanted transaortically in a dog 1 week earlier. Acute isovolemic anemia was produced before the study by exchange with dextran. Aortic flow and stroke volume were recorded by an electromagnetic flowmeter probe around the ascending aorta. Ejection fraction $(K) = 0.59$, stroke volume $= 54$ ml/m², heart rate $= 124$ beats/min, and end-diastolic volume $= 132$ ml/m².

CONSTANT INFUSION TECHNIQUE

Just as a continuous infusion of indicator resulting in a time-concentration curve approaching an asymptote can be used to calculate cardiac output in place of the single bolus injection method, so can a similar constant infusion technique be utilized to calculate ventricular volume. Shaffer (1964) originally derived the mathematical formulation and demonstrated its applicability in the dog. Subsequently, the technique has also been utilized by Lee, McClelland, and Zaragoza (1970) to study right ventricular volumes in patients with and without right ventricular failure. These workers used Indocyanine green as the indicator and sampled blood rapidly through a catheter-densitometer system. From knowledge of the plateau concentration, the number of beats required to reach it, and the stroke volume, the end-diastolic volume may be calculated by the following formula:

$$\text{ESV} = \frac{\text{SV}}{C_{\max}} \left[nC_{\max} - \sum_{i=1}^{n} C(i) \right],$$

where n is the number of heart beats required to reach C_{max}, and $C(i)$ represents the concentration after each cardiac cycle. End-diastolic volume is then calculated as the sum of the end-systolic and stroke volumes. Thermodilution is particularly suited for this technique because the plateau concentration achieved is unlikely to be distorted materially by recirculation of indicator, owing to the fact that the large heat sink created by the peripheral capillary bed regulates the returning temperature to the heart such as to minimize the effects of returning indicator. Additionally, rapid sampling of blood from the catheter is unnecessary since the detector is at the end of the catheter itself. Nevertheless, the constant infusion technique has no real advantage over the recording of a series of single step function thermodilution curves, since, theoretically, whatever errors are introduced in the determination by the single injection method, such as incomplete mixing, are equally applicable to the constant infusion method.

MIXING ERRORS

The validity of the thermodilution and other indicator-dilution methods has been questioned primarily on the adequacy of ventricular mixing of indicator after its injection. Although the initial extent of dispersion of indicator after a bolus injection is important in the calculation of ventricular volume when using the first beat dilution method, it is of no consequence in the calculation of ventricular volume by the usual formulation in which stroke volume is divided by $1 - K$. This latter method does not use the concentration of indicator after the first beat. Instead, one averages the temperature (T) ratios, $(Tn + 1)/Tn$ to obtain K, generally starting with the second beat and progressing until the fifth or sixth beat is reached. Thus, the adequacy of mechanical mixing of indicator in the left ventricle for calculation of K depends on the extent to which the stroke volume entering from the atrium normally mixes with the residual volume of the ventricle with each beat and is totally independent of the manner in which the initial injection was carried out.

Those who question the adequacy of such mixing generally cite the experimental work of Swan and Beck (1960) to support the contention that the entering left atrial stroke volume does not adequately mix normally with the end-systolic ventricular volume. Swan and Beck demonstrated that, after the introduction of angiocardiographic contrast medium into the left ventricle, there was a tendency for accumulation to take place in the more dependent part of the ventricle, thus implying incomplete mixing. Those who cite these experiments neglect the authors' own comments that a possible explanation for this tendency of nonmixing may be directly attributable to the high specific gravity of the contrast medium and that it therefore does not necessarily hold true that indicator particles would behave similarly. Swan and Beck also reported that the shape of the indicator-dilution curve obtained just above the aorta occasionally showed an overshoot early in systole with only a late plateau of the curve, suggesting incomplete mixing of the left ventricular contents as they were being expelled into the aorta. The overshoot was most frequent when the catheter was placed immediately above the aortic valve. However, the overshoot evidence of left ventricular nonmixing was seen in only 16% of their curves. In our experience with thermodilution, such overshoots rarely occurred; decremental plateaus, such as those illustrated in Fig. 11.1, were either obtained or the plateau sloped downward or was overdamped, due to a poor time constant in some cases, particularly when heart rates were very fast. Even if one accepts that one out of six curves may reveal evidence of nonmixing, the final end plateau values are not necessarily at a different concentration than the average concentration within the ventricle, and, therefore, the curve need not give a specious value for the calculation of K. In any case, the thermodilution technique permits one to record 10 to 12 or more curves in rapid succession. If one or two of the curves are grossly distorted by such overshoots, they can readily be eliminated from the calculations, and one can average the remaining curves where the contour is acceptable. Finally, it should be noted that Swan and Beck compared curves obtained from two catheters placed in different locations in the pulmonary artery after right ventricular injection and in the ascending aorta after left ventricular injection. After right ventricular injection, only 2 of the 17 determinations showed disagreements greater than 5% between the two curves. In the case of the left ventricle, 6 of the 16 determinations did show greater than 5% variability, but in only 1 was it great. This analysis of Swan and Beck's work clearly shows that those who cite it as evidence of lack of validity of the indicator-dilution method for calculating ventricular volume are over-interpreting the data. Also, work reported in more recent years from Swan's laboratory showed excellent agreement between angiocardiographic and thermodilution volume calculations from the left ventricle of the dog (Swan *et al.,* 1968).

Nevertheless, other experimental evidence suggests that ventricular mixing may normally be incomplete. Irisawa and associates (1960), using electrical conductivity cells, found different concentrations at different sites within the ventricle for one or two beats after injection or indicator into the ventricle. However, the concentrations were essentially the same in the majority of their recordings after three beats. Furthermore, Holt (1966) emphasized that the method is dependent on the measurement of indicator concentration in aortic blood and not in blood within the ventricle. Theoretically, the method does not require complete mixing of blood within the ventricle, but only that the average of the ratios of concentrations (or temperature changes) measured from aortic blood is the same as the average of similar ratios obtained at the end of each diastole from the ventricle. The mixing of injected thermal indicator is not only dependent on the adequacy of the mechanical mixing of the entering atrial stroke volume with the residual ventricular volume, but also on heat conduction or transfer of kinetic energy from 1 molecule of blood to another. Thus, theoretically, one would expect that equilibration of temperature would take place faster and more thoroughly than would mixing by diffusion of large ions or dye particles within the ventricle. Rapaport (1966), using two thermistor beads located in different areas of the ventricle, showed good agreement of temperatures by the second beat after injection. Freis and Heath (1964) also showed that the temperature equalized immediately, sensed by two different thermistor probes at the aortic root, after the onset of systole when cold saline was injected suddenly into the area.

LOSS OF INDICATOR

A potential error arises from the possibility that the volume of distribution of injected cold saline or blood may be larger than the ventricular volume due to diffusion of heat into the ventricular walls. Such exchange would be influenced by the differences in the thermal conductivity between blood and the ventricular endocardium, the rate of myocardial blood flow, the actual temperature gradient, the surface area of the ventricle in contact with blood, the actual value of the residual fraction (cold exchange would be enhanced at lower values of K), and the heart rate. Theoretical considerations and experimental data suggest that the extravascular volume of distribution is negligible during the first one-half dozen beats after injection from which ventricular volume is calculated. Simultaneous determinations of ventricular volumes by thermodilution and aortic dye-dilution curves after the injection of cold dye into the ventricle have shown close agreement between the two methods (Weigand and Jacob, 1965). Wilcken (1965) also demonstrated close agreement between the residual fractions calculated from a thermodilution curve and a simultaneous dye curve obtained from a densitometer connected to a catheter, the tip of which allowed blood to be sampled from the base of the aorta after injection of cold Indocyanine green into the left ventricle. Recently, Bussmann, Krayenbühl, and Rutishauser (1971) carried out simultaneous measurements of dye concentration using a fiberoptic catheter and the temperature curves from a thermistor catheter in dogs. Not until the ratio of the concentration of the sixth beat compared to the fifth beat was reached was the value by thermodilution significantly higher than by the fiberoptic technique. Thus, if one begins with the second beat and terminates his calculation of K by the sixth beat, there is no significant difference in calculations by using the two methods. The authors concluded that the error for K introduced by cold exchange between blood and myocardium is not more than 2 to 4% for a K of about 0.65. Furthermore, they concluded that this error is even lower when only the second to the fifth beat is analyzed. It would, therefore, seem reasonable to assume that the determination of end-diastolic volume by the thermodilution method is as valid as by the dye-dilution method using a fiberoptic catheter provided washout concentrations between the second and the sixth beat are used for the calculation of K.

COMPARISON WITH ANGIOCARDIOGRAPHIC TECHNIQUES

Early studies using the indicator-dilution method, particularly thermodilution, in the dog, tended to give values for ejection fractions that were significantly lower and ventricular volumes that were appreciably higher than the values obtained from angiocardiographic estimates (Holt, 1956; Hallermann et al., 1963; Carleton et al., 1966; Sanmarco et al., 1966; Bartle and Sanmarco, 1966b; Rapaport et al., 1962). These early washout curves resulted in ejection fractions that were consistently less than 50% and frequently were as low as 25 to 35%. Furthermore, early studies comparing end-diastolic volume obtained by dye or thermodilution techniques with sequential or simultaneous left ventricular volume measured by angiocardiography in the same dog seemed to confirm that higher volumes were obtained by the indicator-dilution methods (Bartle and Sanmarco, 1966a). Most observers assumed that the angiocardiographic method was the more valid estimate, since ventricular volumes measured at autopsy as well as in heart models with known volumes gave values that agreed well with those calculated from angiographic films taken at the same time. However, thermodilution also accurately estimates volumes of models and of the left ventricle in the few dogs in which the actual volume was measured after sacrifice of the animal (Salgado and Galletti, 1964; Rolett et al., 1964). I believe that there are several reasons why apparent discrepancies between the two techniques were described.

First, most of the differences were observed in early studies, whereas more recent studies seem to show no apparent or very little difference between volumes calculated by the two methods (Hugenholtz et al., 1968; Swan et al., 1968; Fleming and Hamer, 1968; Frank et al., 1971). In large part, this may reflect improvement in the instrumentation being used in thermodilution studies, particularly in the use of very rapid response thermistor beads exposed more directly to flowing blood. This results in shorter time constants with resulting clear-cut decremental step plateaus from which ratios of temperature changes can be calculuated more accurately. This may also explain why Carleton, Bowyer, and Graettinger (1966) obtained left ventricular volumes by the indicator-dilution method in six dogs that were consistently higher than volumes calculated by angiocardiography. They used asorbate concentration curves detected by a platinum electrode to inscribe the indicator-dilution curve. Unfortunately, the time constant of the system (0.3 second) was between two and four times the time constant usually obtained with thermodilution. Clearly, the curves would be overdamped when used in the dog, with fast heart rates leading to spurious values by the dye-dilution method.

Secondly, many of the initial indicator-dilution studies in dogs were performed in animals under barbiturate anesthesia, which is now known to produce dilatation of the ventricle and impaired emptying. In addition, some of the dogs in which thermodilution measurements were obtained were open chest preparations, which also tended to reduce the ejection fraction markedly. Frequently, in the early studies, particularly those involving simultaneous comparisons or sequential comparisons of angiocardiographic and dye or thermal volumes, only a single indicator-dilution curve was obtained rather than averaging multiple curves, which is necessary if one is to obtain a high degree of reliability for any given estimate (Rolette *et al.*, 1964).

Failure to insure comparable stroke volumes and heart rates was another source of possible error. If one is to compare the validity of the end-diastolic volume obtained by two different methods, he must establish that stroke volume and heart rate are, in fact, comparable between the two series of experiments he is comparing. Otherwise, there may be actual differences in end-diastolic volumes, reflecting biologic rather than methodologic differences. Finally, indicator-dilution measurements of left ventricular volume frequently were carried out with the dog supine, after which the dog was rotated into the right anterior oblique position to perform the single plane angiographic measurement. Hallermann *et al.* (1963) suggest that such rotation may produce a significant hemodynamic effect since the heart rate and stroke volumes listed vary substantially during the period between sequential studies.

In examining possible errors found in the angiocardiographic methods, it has been pointed out that stroke volume increases during the filming period after injection of contrast material, due to the effects of the agent itself; one may actually get different end-diastolic volumes if one uses an early film rather than a late film in making angiocardiographic measurements of the left ventricle. Additionally, the area-length method of volume calculation (Dodge *et al.*, 1960) now appears to be a more standard technique than the Arvidsson or spatial vector method used earlier (1961), a method that consistently gives abnormally low volume values. Also, it is not infrequent to stop the respirator in the dog and/or arrest respiration and use positive intrathoracic pressure through breath holding in man during the shooting of angiocardiographic films. These procedures are likely to reduce ventricular volume and thereby lower the resultant angiographic estimation in relationship to the values obtained by the thermal or dye methods. Finally, the exact delineation of the left ventricular border and the volume occupied by papillary muscles and the trabeculae carnae are imprecise and, therefore, a potential source of error in measuring volume by the angiocardiographic method. For example, as much as 30% of the left ventricular end-diastolic volume may be occupied by intracavitary musculature (Gribbe *et al.*, 1959).

Despite these sources of possible error, the agreement between angiocardiographic estimates of left ventricular end-diastolic volume and those obtained by indicator-dilution methods is surprisingly close in normal people. Angiocardiographic estimates in subjects without hemodynamic evidence of cardiovascular impairment range from as low as 70 ml/m² to as high as 105 ml/m² (Kennedy et al., 1966; Hood et al., 1968; Hermann and Bartle, 1968; Krayenbühl et al., 1969b, Miller, 1969; Mahler et al., 1970). Values obtained by the indicator-dilution method, including thermodilution, are more or less comparable, ranging from 82 to 107 ml/m² (Levinson et al., 1967; Krayenbühl et al., 1969a; Wilcken, 1965; Paley et al., 1971; Bristow et al., 1964; Ueda et al., 1965; Gorlin et al., 1964). Table 11.1 summarizes the values reported in the literature with the indicator-dilution method for left ventricular end-diastolic volume in subjects who clearly had no hemodynamically significant abnormalities. It is not surprising to note that the largest end-diastolic volumes are in those studies in which the largest stroke volumes were observed since, often within individual studies, a linear correlation was observed between stroke volume and left ventricular end-diastolic volume (Levinson et al., 1967; Krayenbühl et al., 1969a; Lüthy et al., 1968).

Surprisingly little data have been obtained in establishing normal values for right ventricular end-diastolic volume either by the angiocardiographic or the indicator-dilution methods. Angiocardiographic determination of right ventricular volume has been difficult to accomplish because of the geometric shape of the right ventricular cavity. Nevertheless, several groups have reported values based on different geometric assumptions (Arcilla et al., 1971; Goerke and Carlsson, 1967; Carlsson et al., 1971; Reedy and Chapman, 1963). Right ventricular volume in subjects without heart disease has also been estimated using indicator-dilution methods. Freis, Rivara, and Gilmore (1960) obtained an average value of 79.1 ml/m² in six subjects without heart disease. By using a scintillation camera with precordial counting after technetium⁹⁹ᵐ injection, Ishii and MacIntyre (1971) reported an average right ventricular end-diastolic volume of 88.3 ml/m² (left ventricular end-diastolic volume averaged 81.1 ml/m²) in subjects without hemodynamic abnormalities. Rapaport et al. (1965), using thermodilution, found an average right ventricular end-diastolic volume of 103 ml/m² in 15 subjects without heart disease. Finally, Lee, McClelland, and Zaragoza (1970), using the constant-infusion, indicator-dilution method, obtained average values of 150

Table 11.1 Left ventricular end-diastolic volume in man measured by indicator-dilution methods

Reference	Method	No.	LVEDV[a] ml/m^2	SV ml/m^2	HR $beats/min$	Comments
Bristow et al., 1964	Thermodilution	5	99	37	83	Miscellaneous post-operative surgical patients with normal LV function at rest
Gorlin et al., 1964	Thermodilution	7	103	46	73	Normal subjects
Ueda et al., 1965	External monitoring of radioisotope dilution	3	95	42	82	Normal subjects
Wilcken, 1965	Dye dilution	4	104	57	93	Four subjects (1 normal) without pulmonary hypertension or mitral stenosis
Levinson et al., 1967	Dye dilution	11	82	45	83	Normal subjects
Krayenbühl et al., 1969a	Thermodilution	11	107	55	74	Subjects without LV abnormalties
Paley et al., 1971	Thermodilution	5	97	40	67	Normal subjects

[a]Abbreviations: LV, left ventricular; EDV, end-diastolic volume; SV, stroke volume; and HR, heart rate.

ml/m². It would appear both from indicator-dilution methods and from angiographic studies that right ventricular volume is consistently larger than left ventricular volume in the normal subject.

In summary, indicator-dilution methods including thermodilution afford a reasonable and reproducible estimation of ventricular volume when careful attention is paid to technical details and adequate instrumentation.

ACKNOWLEDGMENT

This research was supported in part by United States Public Health Service Grant HL-06285.

REFERENCES

Arcilla, R. A., P. Tsai, O. Thilenius, and K. Rænniger. 1971. Angiographic method for volume estimation of right and left ventricles. Chest 60:446–454.

Arvidsson, H. 1961. Angiocardiographic determination of left ventricular volume. Acta Radiol. 56:321–339.

Bartle, S. H., and M. E. Sanmarco. 1966a. Comparison of angiocardiographic and thermal washout techniques for left ventricular volume measurement. Amer. J. Cardiol. 18:235–252.

Bartle, S. H., and M. E. Sanmarco. 1966b. Measurement of left ventricular volume by biplane angiocardiography and indicator-washout techniques: a comparison in the canine heart. Circ. Res. 19:295–306.

Bing, R. J., R. Heimbecker, and W. Falholt. 1951. An estimation of the residual volume of blood in the right ventricle of normal and diseased human hearts in vivo. Amer. Heart J. 42:483–502.

Bristow, J. D., R. L. Crislip, C. Farrehi, W. E. Harris, R. P. Lewis, D. W. Sutherland, and H. E. Griswold. 1964. Left ventricular volume measurements in man by thermodilution. J. Clin. Invest. 43:1015–1024.

Bussmann, W. D., H. P. Krayenbühl, and W. Rutishauser. 1971. Simultaneous determination of the stroke volume and the left ventricular residual fraction with the fiberoptic- and thermodilution method. Cardiovasc. Res. 5:136–140.

Carleton, R. A., A. F. Bowyer, and J. S. Graettinger. 1966. Overestimation of left ventricular volume by the indicator dilution technique. Circ. Res. 18:248–256.

Carlsson, E., R. J. Keene, P. Lee, and R. J. Goerke. 1971. Angiocardiographic stroke volume correlation of the two cardiac ventricles in man. Invest. Radiol. 6:44–51.

Dodge, H. T., H. Sandler, D. W. Ballew, and J. D. Lord, Jr. 1960. The use of biplane angiocardiography for the measurement of left ventricular volume in man. Amer. Heart J. 60:762–776.

Erickson, H. H., V. S. Bishop, M. B. Kardon, and L. D. Horwitz. 1971. Left ventricular internal diameter and cardiac function during exercise. J. Appl. Physiol. 30:473–478.

Feigenbaum, H., S. B. Wolfe, R. L. Popp, C. L. Haine, and H. T. Dodge. 1969. Correlation of ultrasound with angiocardiography in measuring left ventricular diastolic volume. Amer. J. Cardiol. 23:111.

Fleming, J., and J. Hamer. 1968. Left ventricular volume in aortic stenosis measured by an angiocardiographic and a thermodilution method. Brit. Heart J. 30:475–482.

Folse, R., and E. Braunwald. 1962. Determination of fraction of left ventricular volume ejected per beat and of ventricular end-diastolic and residual volumes: experimental and clinical observations with a precordial dilution technique. Circulation 25:674–685.

Frank, M. J., P. E. Cundey, Jr., T. L. Crews, and W. J. Lewis III. 1971. Comparison of left ventricular volumes by single-plane cineangiography and by indicator dilution. J. Lab. Clin. Med. 77:580–593.

Freis, E. D., G. L. Rivara, and B. L. Gilmore. 1960. Estimation of residual and end-diastolic volumes of the right ventricle of men without heart disease, using the dye-dilution method. Amer. Heart J. 60:898–906.

Freis, E. D., and W. C. Heath. 1964. Hydrodynamics of aortic blood flow. Circ. Res. 14:105–116.

Goerke, R. J., and E. Carlsson. 1967. Calculation of right and left cardiac ventricular volumes. Method using standard computer equipment and biplane angiocardiograms. Invest. Radiol. 2:360–367.

Gorlin, R., E. L. Rolett, P. M. Yurchak, and W. C. Elliott. 1964. Left ventricular volume in man measured by thermodilution. J. Clin. Invest. 43:1203–1221.

Gribbe, P., L. Hirvonen, J. Lind, and C. Wegelius. 1959. Cineangiocardiographic recordings of the cyclic changes in volume of the left ventricle. Cardiologia 34:348–366.

Hallermann, F. J., G. C. Rastelli, and H. J. C. Swan. 1963. Comparison of left ventricular volumes by dye dilution and angiographic methods in the dog. Amer. J. Physiol. 204:446–450.

Harrison, D. C., A. Goldblatt, and E. Braunwald. 1963. Studies on cardiac dimensions in intact, unanesthetized man. I. A description of techniques and their validation. Circ. Res. 13:448–467.

Hermann, H. J., and S. H. Bartle. 1968. Left ventricular volumes by angiocardiography: comparison of methods and simplification of techniques. Cardiovasc. Res. 4:404–414.

Hewitt, R. L., M. L. Meistrell, and T. Drapanas. 1971. Continuous determination of left ventricular volume by measurement of a single external dimension. J. Appl. Physiol. 30:569–574.

Holt, J. P. 1956. Estimation of the residual volume of the ventricle of the dog's heart by two indicator dilution techniques. Circ. Res. 4:187–195.

Holt, J. P. 1966. Indicator-dilution methods: indicators, injection, sampling and mixing problems in measurement of ventricular volume. Amer. J. Cardiol. 18:208–225.

Holt, J. P. 1967. Ventricular end-diastolic volume and transmural pressure. Cardiologia 50:281–290.

Holt, J. P., E. A. Rhode, and H. Kines. 1960. Pericardial and ventricular pressure. Circ. Res. 8:1171–1181.

Holt, J. P., E. A. Rhode, S. A. Peoples, and H. Kines. 1962. Left ventricular function in mammals of greatly different size. Circ. Res. 10:798–806.

Hood, W. P., Jr., C. E. Rackley, and E. L. Rolett. 1968. Wall stress in the normal and hypertrophied human left ventricle. Amer. J. Cardiol. 22:550–558.

Hugenholtz, P. G., H. R. Wagner, and H. Sandler. 1968. The in vivo determination of left ventricular volume. Comparison of the fiberoptic-indicator dilution and the angiocardiographic methods. Circulation 37:489–508.

Irisawa, H., M. F. Wilson, and R. F. Rushmer. 1960. Left ventricle as a mixing chamber. Circ. Res. 8:183–187.

Ishii, Y., and W. J. MacIntyre. 1971. Measurement of heart chamber volumes by analysis of dilution curves simultaneously recorded by scintillation camera. Circulation 44:37–46.

Kennedy, J. W., W. A. Baxley, M. M. Figley, H. T. Dodge, and J. R. Blackmon. 1966. Quantitative angiocardiography. I. The normal left ventricle in man. Circulation 34:272–278.

Keroes, J., and E. Rapaport. 1972. Ventricular volume measurement in the awake dog using implanted thermistor beads. J. Appl. Physiol. 32:404–408.

Krayenbühl, H. P., W. D. Bussmann, M. Turina, and E. Lüthy. 1968a.

Is the ejection fraction an index of myocardial contractility? Cardiologia 53:1–10.

Krayenbühl, H. P., G. Noseda, G. De Sepibus, and G. Fricke. 1968*b*. Zur Problematik der Kammervolumenmessung mit Hilfe der Thermodilutions-Methode. Verh. Deut. Ges. Kreislaufforsch. 34:149–154.

Krayenbühl, H. P., W. Rutishauser, P. Wirz, G. Noseda, and E. Lüthy. 1969*a*. Das enddiastolische Volumen der linken Kammer beim Menschen, bestimmt mit der Thermodilutions Methode. Arch. Kreislaufforsch. 58:1–35.

Krayenbühl, H. P., H. J. Ssimon, W. Rutishauser, P. Wirz, and B. O. Preter. 1969*b*. Die enddiastolische Wandspannung bei linksventrikularer Volumenbelastung. Verh. Deut. Ges. Kreislaufforsch. 35:352–358.

Lee, S. J. K., A. R. McClelland, and A. J. Zaragoza. 1970. The right ventricular function during exercise in patients with and without right ventricular failure. Acta Cardiol. (Brux.) 25:313–325.

Levinson, G. E., M. J. Frank, M. Nadimi, and M. Braunstein. 1967. Studies of cardiopulmonary blood volume. Measurement of left ventricular volume by dye dilution. Circulation 35:1038–1048.

Lüthy, E., W. D. Bussmann, G. Fricke, M. Turina, and H. J. Ssimon. 1968. Die Bedeutung des enddiastolischen Volumens der linken Kammer bei der Myokardinsuffizienze. Vehr. Deut. Ges. Kreislaufforsch. 34:160–163.

Mahler, F., H. P. Krayenbühl, W. Rutishauser, P. Wirz, and R. Miotti. 1970. Cineangiokardiographische Bestimmung des enddiastolischen Volumens und der Muskelmasse des linken Ventrikels. Verh. Deut. Ges. Kreislaufforsch. 36:265–273.

Miller, G. A. H. 1969. Angiographic measurement of left atrial and ventricular volume. Brit. J. Radiol. 42:556.

Paley, H. W., I. G. McDonald, J. Blumenthal, and J. Mailhot. 1971. The effects of posture and isoproterenol on the velocity of left ventricular contraction in man. The reciprocal relationship between left ventricular volume and myocardial wall force during ejection on mean rate of circumferential shortening. J. Clin. Invest. 50:2283–2294.

Polissar, M. J., and E. Rapaport. 1961. Some theoretical aspects of quantification of mitral valve regurgitation by the indicator-dilution method. Sufficient and insufficient experiments. Circ. Res. 9:639–663.

Rapaport, E. 1966. Usefulness and limitations of thermal washout techniques in ventricular volume measurement. Amer. J. Cardiol. 18:226–234.

Rapaport, E., B. D. Wiegand, and J. D. Bristow. 1962. Estimation of left ventricular residual volume in the dog by thermodilution method. Circ. Res. 11:803–810.

Rapaport, E., M. Wong, R. E. Ferguson, P. Bernstein, and B. D. Wiegand. 1965. Right ventricular volumes in patients with and without heart failure. Circulation 31:531–541.

Reedy, T., and C. B. Chapman. 1963. Measurement of right ventricular volume by cineangiofluorography. Amer. Heart J. 66:221–225.

Rolett, E. L., H. Sherman, and R. Gorlin. 1964. Measurement of left ventricular volume by thermodilution: an appraisal of technical errors. J. Appl. Physiol. 19:1164–1174.

Salgado, C. R., and P. M. Galletti. 1964. In vitro evaluation of the thermodilution technique for the measurement of end-diastolic ventricular volume. Fed. Proc. 23:302.

Sanmarco, M. E., K. Fronek, C. M. Philips, and J. C. Davila. 1966. Continuous measurement of left ventricular volume in the dog. II. Comparison of washout and radiographic techniques with the external dimension method. Amer. J. Cardiol. 18:584–593.

Shaffer, A. B. 1964. Estimation of ventricular volumes by a constant infusion indicator dilution technique. Circ. Res. 15:168–178.

Swan, H. J. C., and W. Beck. 1960. Ventricular nonmixing as a source of error in the estimation of ventricular volume by the indicator-dilution technique. Circ. Res. 8:989–998.

Swan, H., V. Ganz, J. C. Wallace, and K. Tamura. 1968. Left ventricular end-diastolic volume (EDV) by angiographic and thermal methods in a single diastole. Circulation 38 (suppl. 6): 193.

Ueda, H., Y. Sugishita, A. Nakanishi, I. Ito, H. Yasuda, M. Sugiura, Y. Takabatake, K. Ueda, T. Koide, and K. Ozeki. 1965. Clinical studies on the cardiac performance by means of transseptal left heart catheterization. II. Left ventricular function in high output heart diseases, especially in hyperthyroidism. Jap. Heart J. 6:396–406.

Weigand, K. H., and R. Jacob. 1965. Zur Frage der Restvolumenbestimmung des linken Ventrikels im natürlichen Kreislauf. Arch. Kreislaufforsch. 46:97–114.

Wilcken, D. E. L. 1965. The measurement of the end-diastolic and end-systolic, or residual, volumes of the left ventricle in man, using a dye-dilution method. Clin. Sci. 28:131–146.

chapter 12/ Measurement of Hepatic-Splanchnic Blood Flow in Man by Dye Techniques

Loring B. Rowell
Departments of Physiology and Biophysics and of Medicine/
University of Washington School of Medicine/Seattle, Washington

Measurement of hepatic-splanchnic blood flow under local steady state conditions is a well established application of the Fick principle. This chapter makes no attempt to reiterate the many details of the measurement and specific methodology, the subject of voluminous literature reviewed by Bradley (1962) and Grayson and Mendel (1965), but focuses upon some basic problems in applying the method and interpreting the data.

Methods commonly applied to measure hepatic blood flow in man also measure total splanchnic blood flow (SBF), since all organs in the splanchnic system are in series with their venous blood draining into several hepatic veins (with exception of the pathological case where portal-caval shunting of blood is significant) (Caesar *et al.,* 1961). The technique as originally applied in 1945 by Bradley and colleagues has been used in laboratories over the world with a consistent spread of values for SBF averaging 1.5 liters/minute or 867 ml/M^2 of body surface area per minute by a variety of different methods. Thus, SBF normally is about 25% of resting cardiac output in man, making it the largest regional circulation. Reflex changes in SBF and splanchnic vascular resistance are important to the organism for two reasons.

First, since only about 20% of the available oxygen is removed from blood in its transit through the splanchnic system, decrements in SBF can, in effect, redistribute substantial quantities of oxygen for use in other tissues when necessary (Rowell *et al.,* 1964). Second, the region serves as a major means of reflexly maintaining blood pressure or restoring it to its normal level when this regulation is challenged by some stress (Rowell *et al.,* 1972). In short, the splanchnic region plays a major role in correcting disparities between cardiac output and regional oxygen supply and between cardiac output and peripheral vascular resistance. The techniques described herein allow some quantification of these functions.

MEASUREMENT OF SPLANCHNIC BLOOD FLOW— GENERAL PRINCIPLES

Blood flow through the splanchnic organs can be measured by a simple application of the law of conservation of mass called the Fick principle. This principle states that, under conditions of constant flow, the volume of blood ($\dot{Q}$) flowing through an organ during a time interval can be measured by determining the amount of indicator (R) removed during that time by the organ and the difference in the concentration of indicator going into the organ (C_i) and that coming out (C_o), and may be stated:

$$\dot{Q} = \frac{R}{C_i - C_o}.$$
(12.1)

Requirements for the valid application of this principle are the following.

1. Representative samples of blood must be obtained containing true average concentrations of indicator flowing both in and out of the organ.

2. The indicator must be removed exclusively by the organ, and its total exchange must be measurable. It should neither be created nor destroyed elsewhere in the system.

3. If the blood flow is not constant, corrections must be made for differences between delivery of indicator to and from the organ and the actual removal rate of indicator by the organ.

Thus, three major considerations in the measurement of splanchnic blood flow are the sites and conditions for blood sampling, the properties of the indicator, and the constancy of the blood flow.

Sites and Conditions for Blood Sampling

The measurement of SBF requires catheterization and blood sampling from an hepatic vein and either a peripheral vein or artery for determination of the fraction of dye extracted from blood by hepatic (and splanchnic) tissues. Once dye is fully mixed, its concentration in arterial and peripheral venous blood becomes equal (Weigand et al., 1960). Thus, either arterial or peripheral venous blood samples can be used along with hepatic venous samples to calculate efficiency of dye extraction. A third catheter must be placed in a superficial vein for infusion of dye.

A major problem in the measurement of SBF is that hepatic venous blood does not exist as a pooled volume representing truly mixed contents of all hepatic lobes. Indicator concentration is usually measured in only one of several hepatic veins and, as this may not represent a true average, yields what is commonly called "estimated" SBF. Selection of the site and mode of sampling which will avoid significantly atypical outflow dye concentrations and, hence, erroneous flow measurements are crucial factors and will be treated in some detail.

The best site for hepatic venous sampling has been discussed by Hultman (1966). It is essential that the largest hepatic vein, which is located on the right side and drains this region of the liver, be used whenever possible. The location of this vein as it appears with others during fluoroscopy is shown in unwedged position in Fig. 12.1. There

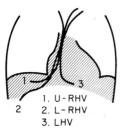

1. U-RHV
2. L-RHV
3. LHV

Figure 12.1 Simulated fluoroscopic view of hepatic venous catheter positions. Site 2 (*L-RHV*) represents the position of the major right hepatic vein, the preferred and most frequently used site. Sites *1* and *3* are easily reached during catheterization, but should be avoided because of their smaller size and the ease with which catheter wedging and sampling difficulties occur. Sites *1* and *2* are those from which data were derived in Figure 12.2. (From Rowell, L. B. 1971. The liver as an energy source in man during exercise. *In:* B. Pernow and B. Saltin (eds.), Muscle Metabolism During Exercise, pp. 127–141. Plenum Press, New York.

are two advantages in using this site. First, this vein drains a major fraction of the liver, minimizing effects of variation in venous contents from various portions of the organ. Second, its large size permits deep penetration of the catheter into the organ, which in turn reduces the chance of wedging the catheter. Furthermore, this location negates the possibility of contaminating hepatic venous blood with refluxed inferior vena caval blood which contains dye at arterial concentration. Such contamination causes a decrease in measured hepatic dye extraction, resulting in a false increase in calculated SBF. Variation in hepatic venous-inferior vena caval pressure gradients and instantaneous blood flow with respiration poses another important problem. Bradley *et al.* (1945) noted that reflux of caval blood into hepatic veins could occur during heavy breathing. Brauer (1963) reported regurgitation of caval blood into hepatic veins during inspiration in some animals. The constancy of hepatic venous concentration of oxygen, dye, and other substances released into the right hepatic vein, plus the high efficiency of dye extraction—even during severe exercise (Rowell *et al.*, 1966, 1968, 1972), indicates that the sampling problems observed in animals by Brauer and by Sapirstein and Reininger (1956) are essentially eliminated by the use of this vein. This conclusion is borne out by the similarity of values obtained using the dye clearance and extraction methods and those employing indicators (e.g. chromphosphate, radiogold, and ^{131}I-albumin) which are almost completely cleared in their first passage through the liver and therefore require no measurement of their hepatic venous concentration (Hultman, 1966). Direct measurements of hepatic arterial and portal venous flows using flowmeters on dogs also agree with simultaneous measurements using the dye method (Drapanas *et al.*, 1960).

There are further potential inaccuracies to be avoided. Withdrawal of blood from a wedged catheter raises concentration of both dye and oxygen in the hepatic venous sample (Brauer, 1963). Portal flow is deflected by a wedged catheter to adjacent regions, as evidenced by detection of substances injected into the hepatic artery and failure to detect all but minute quantities of substances injected into the portal route (Bradley, 1962; Brauer, 1963). Blood may also be rapidly aspirated from the high pressure arterial supply so that "venous" concentrations are further reduced. However, with attention to these details, it would appear that hepatic venous sampling errors are not, in general, very large.

In man, the greatest sampling difficulty occurs when one of the smaller right or left hepatic veins are used (Hultman, 1966; Rowell *et al.,* 1964, 1966, 1968), particularly during exercise when sampling from them becomes more difficult. The catheter becomes wedged either because of its movement deeper into the vein or, more likely, because the vein constricts around it (Bradley, 1962). A sign that the catheter rests in too small an hepatic vein, although blood withdrawal may be easy at the time, is production of right upper quadrant discomfort during a sudden 2- to 3-ml injection of saline into the vein.

Despite these problems, a slow and gentle withdrawal of blood from smaller hepatic veins even during upright exercise can sometimes provide suitable estimates of the changes in SBF and regional metabolism (Rowell, 1971*a*). For example, simultaneous measurement of hepatic venous dye concentration from a smaller upper and the large right hepatic veins (Fig. 12.2) in a man during heavy upright exercise (walking on a treadmill at an intensity which required 2.3 liters of O_2/minute) yielded quite similar values of SBF. The values were 704 ml/minute from the large hepatic vein and 863 ml/minute from the small hepatic vein where dye concentration was, as one would predict from the preceding discussion, slightly higher. The calculated decrements in SBF from this subject's resting value of 1.7 liter/minute were 50 and 58% from the two sites. Note the rise in hepatic venous dye concentration from the smaller of the two veins (*U-RHV* in Fig. 12.2) toward the end of the study when sampling from this vein became difficult.

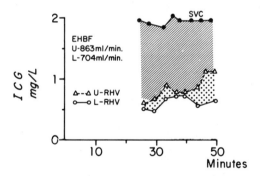

Figure 12.2 Simultaneous measurement of dye concentration from a small right hepatic vein (*open triangles*), the largest right hepatic vein (*open circles*), and a systemic artery (*closed circles*) during heavy upright exercise. Average SBF determined at the two venous sites were similar but a rise in *U-RHV* dye concentration due to progressive sampling difficulty between 43 to 50 minutes raised the average SBF value determined from this site. (Modified from Rowell, as in Fig. 12.1. With permission.)

Fewer sampling problems are encountered when a no. 7 or 8 NIH type dacron catheter is used. The closed end and side holes of the catheter reduce the chance of withdrawing from a wedged position. The withdrawal rate is most important. Depending upon the ease of withdrawal, the collection of each 3-ml sample over a 10- to 20-second period is optimal. Since dead space of the catheter is 1 ml or more, the flush withdrawal preceding the sample must also be made at an equally slow rate.

Indicators for Splanchnic Blood Flow Measurement

Sulfobromphthalein sodium (bromsulphalein) and Indocyanine green are the two indicators commonly used for determining hepatic blood flow in man.

The properties of bromsulphalein and the mechanism and kinetics of its removal from blood have been reviewed (Bradley, 1962; Grayson and Mendel, 1965; Brauer, 1963; Fauvert, 1959). The Fick principle requires that the indicator for organ blood flow be removed exclusively by the organ; since a small extra-hepatic removal does occur, measurements of SBF using this indicator are systematically overestimated when compared to measurements using Indocyanine green. However, these overestimates are not quantitatively very significant (Caesar *et al.*, 1961; Hultman, 1966; Winkler *et al.*, 1965; Leevy *et al.*, 1962).

Indocyanine green has proved to be an ideal indicator for measuring hepatic blood flow (Fox and Wood, 1960). It is cleared from plasma exclusively by the liver and excreted in unconjugated form into bile without significant enterohepatic circulation (Hunton *et al.*, 1960; Cherrick *et al.*, 1960). Quantitative recovery of 97% infused dye from the bile duct has been made in dogs (Wheeler *et al.*, 1958). In studies where five different indicators were simultaneously employed, Winkler *et al.* (1965) found Indocyanine green to have the highest extraction efficiency and to provide the most reliable measurements of SBF.

Important physical properties and the kinetics and mechanisms of removal of this dye from plasma, where it is tightly bound to the albumin fraction, have been described by Cherrick *et al.*, (1960). At high infusion rates, extraction mechanisms of hepatic parenchymal cells becomes saturated so that, as C_A rises to high levels, extraction efficiency is dependent upon infusion rate (Hunton *et al.*, 1961), but very high doses must be used in man to cause such saturation (Cherrick et al., 1960). Details pertaining to the analysis of indicator concentration are deferred to the end of this chapter.

In summary, the adaption of the general principle for the specific measurement of SBF can be stated as:

$$\dot{Q} = \frac{R}{C_A - C_{HV}} = \frac{I}{C_A - C_{HV}} \qquad (12.2)$$

where $\dot{Q}$ is the rate of blood flow through the splanchnic organs, R is the dye removal rate which is constant and equal to the dye infusion rate (I), and C_A and C_{HV} are arterial and hepatic venous dye concentrations, both of which are constant.

MEASUREMENT OF SPLANCHNIC BLOOD FLOW— SINGLE INJECTION METHOD

Since the introduction of Indocyanine green, a single injection technique for measurement of SBF has been commonly used. This technique has been discussed by several investigators and its advantages are simplicity and speed. A determination of SBF by single injection requires only 12 to 14 minutes, whereas 15 to 20 additional minutes are required in the constant infusion method for C_A to reach a steady state after infusion begins. Repeated determinations are possible with less risk of saturating hepatic extraction mechanisms for the dye. Also it is not necessary to construct standard curves for determination of dye concentration as the following calculations will reveal. Values for SBF by single injection methods agree closely with those measured using constant infusion (Caesar et al., 1961; Wiegand et al., 1960; Rowell et al., 1966; Rowell, 1971a; Reemtsma et al., 1960). Rowell et al. (1964) found SBF measured by single injection of Indocyanine green to average 1.61 liters/minute with an extraction ratio of 0.77 in seven normal young men at rest, supine. In 12 normal young men, equivalent measurements of SBF by constant infusion of Indocyanine green yield an average value of 1.63 liters/minute with an extraction ratio of 0.85 (Rowell et al., 1970, 1971b, 1972). In one experiment, SBF was determined during the first 20 minutes of prolonged heavy exercise by single injection and during the final 30 minutes by constant infusion of the dye. Flows so determined were 487 and 426 ml/minute, respectively (Rowell, 1966, 1971a).

As shown in Fig. 12.3, plasma Indocyanine green concentrations are plotted against the midpoint of withdrawal time on a semilog scale

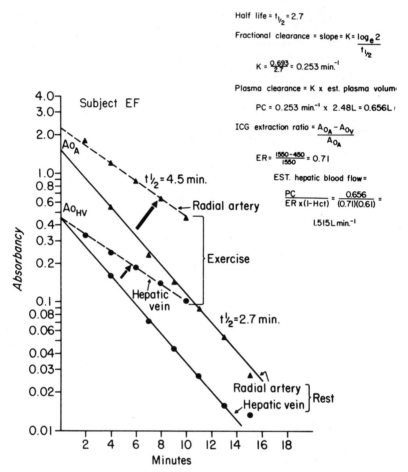

Figure 12.3 Illustration of the single injection method using Indocyanine green (*ICG*) to determine SBF during rest and heavy exercise in a young male subject. Note the increase in dye extraction efficiency (increased distance between arterial and hepatic venous slopes) during exercise. Note also the very high arterial dye concentrations early in exercise and the low hepatic venous values at the end of rest. The measured percentage decrease in SBF was 49%; the percentage decrease calculated from the percentage change in K or $t_{1/2}$ was 40%, the difference being due to increased extraction. (From Rowell, L. B., J. R. Blackmon, and R. A. Bruce. 1964. Indocyanine green clearance and estimated hepatic blood flow during mild to maximal exercise in upright man. J. Clin. Invest. 43: 1677–1690.)

from which the half-time $(t_{1/2})$ of dye clearance and the disappearance rate constant or fractional clearance rate (K) are derived. The constant K is calculated from the equation:

$$K = \frac{\log_e C_1 - \log_e C_2}{t_2 - t_1}. \qquad (12.3)$$

If $t_2 - t_1$ represents the time to clear 50% of the dye concentration at $t = 0$, then

$$K = \frac{\log_e 2}{t_{1/2}} = \frac{0.693}{t_{1/2}}, \qquad (12.4)$$

where K represents the fraction of the total quantity of injected dye that disappears from the plasma in 1 minute (or alternatively the fraction of the total plasma volume cleared in 1 minute). To calculate SBF from K, it is necessary to measure or estimate total plasma volume and the percentage of dye extracated from blood in its transit through the organ. Then SBF is given as:

$$\dot{Q} = \frac{K \cdot PV}{(A_a - A_{hv}/A_a)(1 - Hct)} = \frac{K \cdot PV}{ER(1 - Hct)}. \qquad (12.5)$$

PV is plasma volume, $(A_a - A_{hv}/A_a)$ is the extraction ratio (ER) as it is calculated from the ratio of the differences in arterial (A_a) and hepatic venous (A_{hv}) absorbancy of dye. $(1 - Hct)$, the fraction of whole blood which is plasma, converts the measurement from splanchnic plasma flow to splanchnic blood flow.

The single injection technique has a number of disadvantages. The most obvious is the need to measure plasma volume. At rest, the absolute value of SBF has an error which is proportional to the error in measuring or estimating plasma volume. In normal young men with normal lean body mass, plasma volume averages 45.6 ± 5.3 ml/kg of body weight with the coefficient of variation, 11.5% (Gregersen, 1961). Thus, one potential error in measuring SBF is of this magnitude. Supported by the finding that the volume distribution of Indocyanine green is equal to that of ^{131}I-labeled albumin and Evans blue dye, some have estimated plasma volume from C_A at zero time by extrapolation of the slope of C_A (Wiegand et al., 1960; Cherrick et al., 1960). If this is done, the quantity of injected dye must be determined by sterile weighing as the weight specified on the bottles (Cardio-Green®, Hynson, Wescott and Dunning, Baltimore, Md.) is often low by several milligrams.

The injected dose must be sufficient to give plasma dye concentrations within the most accurate optical density range of the spectrophotometer. To achieve this the conventional dose injected is 0.5 mg per kg of body weight. Nevertheless, there is always a compromise between obtaining values of C_{HV} which are too low and those for C_A which are too high for greatest spectrophotometric accuracy (Fig. 12.3). Initial C_A values often have to be read on an attenuated and therefore less precise scale of the spectrophotometer. On occasions where high C_A values can be anticipated, sufficient blank plasma can be withdrawn to allow quantitative dilution of the more concentrated plasma samples. The low values of C_{HV} toward the end of resting measurements and the very high C_A values at the beginning of clearance measurements when SBF is reduced, as by exercise (Fig. 12.3), can be important limitations of the single injection method.

Despite these limitations, changes in SBF can be reliably followed by using this technique. If applied during exercise, corrections must be made for reductions in plasma volume. Again, if the injected dose of dye is weighed, changes in plasma volume can be calculated from changes in C_A at zero time. When the time between repeated measurements of SBF is not sufficient to allow complete clearance of dye from plasma, correction must be made for background C_A at the time of injection. This can be complicated by the gradual deceleration of dye clearance which some have observed to begin 10 to 20 minutes after injection (Leevy et al., 1962; Cherrick et al., 1960). Since the decrements in plasma volume during various levels of exercise are predictable and range from 5 to 15%, changes in plasma volume can also be estimated. To offset initial rapid changes in albumin concentration, plasma volume and plasma optical density, it is necessary to wait approximately 5 minutes after the start of exercise to draw preinjection plasma blank samples. Earlier injection also produces an initial non-linearity of the dye disappearance slope.

The possibility of extra-hepatic removal of Indocyanine green secondary to exercise or other stresses suddenly reducing plasma volume, is another circumstance which would overestimate SBF. The albumin-bound dye may pass through capillary walls during exercise, but any leaked albumin appears to be rapidly returned to the circulation because of accelerated lymph flow during exercise, thus precluding any net loss of albumin-dye complex from blood. Ekelund (1967) found the loss to be less than 2%. The fact that Indocyanine green disappearance follows first order kinetics after initial alterations in plasma volume argues against significant extra-hepatic circulation or changing SBF with time. It was further noted that no measurable quantity of dye appeared in urine even after several bouts of moderate to severe exercise (Rowell *et al.*, 1964).

Finally, uncertainty in estimating the transit time of blood through hepatic sinusoids results in an error of unknown significance in the calculation of the extraction ratio. To allow for the transit time, a delay of 10 to 15 seconds after the mid-point in volume of arterial sampling is commonly made before hepatic venous samples are drawn. When average splanchnic transit time is increased, as by exercise, the direction of error is toward decreased extraction efficiency and increased SBF due to early hepatic venous sampling, i.e. C_{HV} is probably too high with respect to the paired value of C_A.

Despite these problems which tend to reduce the extraction ratio, this ratio tends to increase when SBF is reduced by exercise or other stresses (Rowell *et al.*, 1970, 1971*b*, 1972; Bradley *et al.*, 1952).

Since hepatic extraction ratio for Indocyanine green does increase slightly with exercise (8.2%), percentage changes in K or $t_{1/2}$ can be used to calculate percentage changes in SBF. Using this approach it was found, as expected, that the percentage increase in K or $t_{1/2}$ of dye clearance underestimated the measured decrease in SBF by 8.2% due to the increase in the extraction ratio. Clearly, this approach will not work in cases of hepatic dysfunction where the extraction ratio may fall during exertion or other stresses. Nor will it work when SBF is increased; the extraction ratio for bromsulphalein falls when SBF increases above the normal (Bradley, 1946).

In summary, the single injection technique should be regarded as providing a good approximation of SBF (as measured by the more reliable constant infusion technique). Its major advantage is that it allows more rapid and repeated measurements. Although both techniques are subject to the same hepatic venous sampling errors, these errors are exaggerated in the single injection technique when C_{HV} falls to very low values. The extraction ratio for the dye is similar for both techniques but a potential error, peculiar to the single injection method, arises from lack of knowledge of the transit time for dye through the liver. From sources of error described above, one major point is clear: these errors tend to underestimate any decrease in SBF.

MEASUREMENT OF SPLANCHNIC BLOOD FLOW—
CONSTANT INFUSION METHOD

This technique, with its problems and limitations, has been reviewed by Bradley, 1962; Grayson and Mendel, 1965; Hultman, 1966; Winkler et al., 1965; and Brauer, 1963. Little can be added to Bradley's thorough and critical treatment of the method. Its application to resting subjects under steady state conditions is straight-forward. The usual application is to situations which allow an extended equilibration time and a prolonged period of flow measurement. After a priming dose of approximately 10 to 12 mg of Indocyanine green, infusion of dye at the conventional rate of 0.5 mg/minute is made from a calibrated syringe driven at known speed by an infusion pump. In a man with normal blood volume of 5 to 6 liters and SBF of 1.5 liters, C_A approaches its equilibrium value exponentially with a 95% response time of 15 to 18 min (Rowell et al., 1972).

Accuracy of I in the numerator in equation (12.2) depends upon the accuracy with which dry dye is weighed, diluted, and infused. The analytical accuracy of determining C_A as mentioned later is very good. In the constant infusion technique, the major limitation in accuracy of determining SBF under steady state conditions is the measurement of C_{HV}. With this technique the limitation is anatomical, i.e. because of the several venous sampling sites and their associated problems, rather than analytical. During constant infusion of bromsulphalein, Bradley (1945) found variations in C_{HV} between sites to be as large as 20%. However, Winkler et al.'s data (1965) and data from this laboratory indicate that such large variation is not the usual case.

The major problems with this technique arise when SBF changes during the course of the measurement. Then the dye infusion rate (I) no longer equals the dye removal rate (R) in equation (12.2). Bradley et al. (1960) proposed that R could still be calculated from the rate of change of $C_A(\dot{C}_A)$. That is

$$\dot{Q} = \frac{I - [(C_{A_2} - C_{A_1})/\Delta t] \cdot BV}{C_A - C_{HV}}. \tag{12.6}$$

During any sampling interval, R equals I minus the amount of dye ($\Delta C_A \cdot BV$) required to change C_{A_1} to C_{A_2} over the sampling interval, Δt. For greatest accuracy, the values C_A and $C_A - C_{HV}$ are interpolated to the midpoint of the sampling interval. The calculation used by Bradley is an instantaneous equation which can be expressed functionally as

$$\dot{Q} = (I - BV \cdot \dot{C}_A)/(C_A - C_{HV}). \tag{12.7}$$

This method has essentially two major sources of error. The major determinant of its accuracy for calculating SBF at any instant is the accuracy of measuring $\dot{C}_A$. With an analytical error of 0.006 mg/liter (see below), $C_A - C_{HV}$ is in error by less than 3%. The error from the removal rate term $(C_{A_2} - C_{A_1})/\Delta t$ yields a $10/\Delta t\%$ error. Thus, the total error from this source is 8% for a 2-minute sampling interval and 5% for a 5-minute sampling interval. This requires, of course, that the midpoint of arterial sampling be carefully timed.

The second source of error is an effect of discontinuous sampling. These are errors in tracking true values of C_A. To determine whether a significant change in C_A between samples could be missed, the deviation of C_A from a straight line between two samples was ascertained for a given change in the magnitude of $\dot{Q}$ or SBF over a given time interval. This was done for fast and slow changes of SBF to a new level (Fig. 12.4, *bottom graph*). If C_A were to track SBF closely, large errors in calculated SBF could result if SBF changed rapidly over the measurement interval. Taking a rigorous case,[1] a square wave change in SBF (Fig. 12.4), the solution for equation (12.6) is

$$C_A = (C_{A_0} - C_{HV_0} - I/\dot{Q})e^{-(\dot{Q}/BV)t} + I/\dot{Q} + C_{HV}, \tag{12.8}$$

where C_{A_0} and C_{HV_0} represent C_A and C_{HV} at time t zero or $t = $ zero, and $\dot{Q}$ is the final value of SBF. C_A approaches the equilibrium value exponentially with a time constant of 4 to 6 minutes (time required for C_A to reach within $1/e$ of its final value) in a man with normal blood volume and SBF. Thus, a sampling interval of 2 minutes or more is appropriate to the dynamic properties of the system and the dynamics of the expected change in SBF. If Δt is less than 2 minutes, the potential measurement error for C_A and for $\dot{C}_A$ becomes proportionally much greater as Δt gets smaller. On the other hand, Δt should not exceed the duration of any expected change in SBF; although the accuracy of determining the final value increases with increased Δt, the time course of the change would be missed.

[1]Actually a more rigorous but unlikely case would be a rapid pulselike change in SBF during a sampling interval. Such a change would approximately double the theoretical errors discussed above.

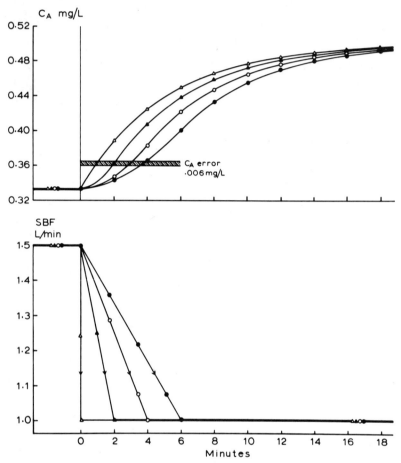

Figure 12.4 Computer solutions to equations (12.6) and (12.7) for changes in Indocyanine green concentration of arterial whole blood (C_A) (*upper graph*) where SBF has been changed in 0.5 liter/minute ramps of different rates (*lower graph*) and in a square wave. Calculations were based on a dye infusion rate of 0.5 mg/minute, a control SBF of 1.5 liter/minute, a blood volume of 5 liters, and a constant hepatic venous concentration (C_{HV}). Width of *shaded bar* in the *upper graph* for C_A shows the analytical error of dye measurement.

For other forms of $\dot{Q}(t)$, equation (12.7) was solved for C_A using a computer language for digital simulation. Values so generated were also used to compute SBF using Bradley's method of interpolation between discrete sample points. Flows used to generate C_A values (Fig. 12.4) and the flows calculated from these values agreed within 1 to 2% for SBF changing between the limits of 2.0 to 0.5 liter/minute and sampling interval of 2 minutes. Thus, the maximum theoretical error in computing flows by this method is less than 2%. The total error in SBF from both tracking and measurement errors over any 2 to 5 minutes Δt is about 10% in the most rigorous case tested, a large square wave change in SBF. Analysis of ramplike changes in SBF yields lower errors. An error of $x\%$ in estimating blood volume yields a maximum error of $x\%$ in estimating a change in SBF during Δt when Δt is very small (less than 2 minutes). This error falls rapidly when Δt is increased so that over a period of 10 to 15 minutes it has a negligible effect on the calculated change in SBF. For example, if blood volume were to change 10% during a 10-minute period of measurement, the error in estimating SBF would be less than 2%. These errors serve mainly to dampen or exaggerate variations in SBF during Δt when Δt is small.

Variations in splanchnic blood volume appear not to have been considered by Bradley as a source of error. Since the dye techniques described herein measure splanchnic outflow rather than inflow, rapid changes in splanchnic blood volume, which comprises 20 to 25% of blood volume, could for short intervals change calculated outflow while inflow actually remained constant. For instance, splanchnic blood volume would be expected to decrease during splanchnic vasoconstriction mainly because of passive recoil of veins at reduced distending pressure. Vasodilatation would produce the opposite effect. Splanchnic blood volume can be altered by gravitational effects due to sudden changes in posture and may be further reduced by reflex venoconstriction during some stresses, but this is not well established in man. Inflow is clearly the desired measurement when change in total splanchnic vascular resistance is to be derived.

Modifications have been made in Bradley's equation which assume that the dye is distributed at C_A only outside of the SBF, i.e. dye is distributed at C_{HV} throughout splanchnic blood volume or in 0.2 blood volume (Castenfors et al., 1961). But since all Indocyanine green (and most of bromsulphalein) is extracted by the liver, the dye will be at C_A not C_{HV}, in the extra-hepatic splanchnic blood volume. Because of this, variations in extra-hepatic splanchnic blood volume (which is the major portion of total splanchnic blood volume) will have no effect upon calculated SBF, i.e. splanchnic outflow. However, within the liver some undetermined fraction of its venous volume does contain dye at C_{HV}. Were the volume at C_{HV} large, major errors in calculated changes in SBF could conceivably attend sudden shifts in visceral volume. Taking 1600 g as the normal human hepatic mass along with published data (Greenway and Stark, 1971) for blood content per 100 g of liver for various classes of large and small hepatic vessels, hepatic blood volume is estimated at approximately 400 ml. Assuming that 50% of this volume has been exposed to the dye-removing parenchymal cells at any given instant, the volume of blood at C_{HV} should be about 200 ml.

To calculate SBF when changes in hepatic volume are anticipated, the correct expression becomes:

$$\dot{Q}_{out} = \frac{I - \dot{C}_A(BV - V_{HV}) + \dot{C}_{HV}(V_{HV}) - C_{HV}(dV_{hv}/dt)}{C_A - C_{HV}}, \quad (12.9)$$

where $\dot{Q}_{out}$ is SBF out of the liver and V_{HV} is the volume of blood at $\dot{Q}_{HV}$. The terms C_{HV}, dV_{hv}/dt, V_{HV}, and $\dot{C}_{HV}$ are very small compared to others in this equation. Based upon a sudden reduction in V_{HV} of 50%, equation (12.7) yields less than a 5% error in the calculation of a change in SBF over a 1-minute interval. As Δt increases, this source of error becomes negligible. Thus, when Δt is 2 to 5 minutes or when changes in SBF are to be followed over longer periods, equation (12.7) yields appropriate values. As indicated by the simulated results in Fig. 12.4, the dynamic properties of the system allow one, within the small analytical errors for C_A and $C_H - C_{HV}$, to reliably follow rapid changes in SBF using 2-minute or longer sampling intervals. Particular care is obviously needed in handling and analysis of blood at C_A; minor variations in dilution with heparin or turbidity, for example, could cause rather large, spurious changes in SBF.

Castenfors et al. (1961) took another approach to demonstrate the reliability of measuring SBF under dynamic conditions. C_A for bromsulphalein was altered experimentally by step changes in dye infusion rate during periods of constant SBF. Despite an overestimate of the distribution of dye at C_{HV}, the error in estimating SBF while C_A was rapidly rising was only 3.6% (the Δt of 5 minutes minimized any effect of changing splanchnic blood volume).

This treatment of dynamic errors does not, of course, consider absolute errors in SBF which have been discussed and arise primarily from the technique of hepatic venous sampling. These are base line errors which do not influence estimates of change in SBF.

ANALYSIS OF INDOCYANINE GREEN

Indocyanine green has numerous and important advantages over bromsulphalein as an indicator for SBF measurement. It is both exclusively and more efficiently extracted by the liver. In its dry form (as provided with 1 to 3% water), it is stable for long periods, but its optical density slowly decreases over a period of days once it is mixed with its solvent (Fox and Wood, 1960). However, in this laboratory, no significant changes in absorbancy were noted in standard concentrations mixed for periods of 2 to 3 hours, and this has been true also at high ambient temperatures (43.3°C). The dye is also relatively easily analyzed in blood samples. However, measurement of Indocyanine green by continuous flow densitometers is unsatisfactory for SBF analysis. The electrical drift in densitometers tested for this purpose is too great to permit accurate estimation of C_A for longer than a few minutes. Drift is a particularly serious problem at low hepatic venous concentrations where densitometer gain settings must be very high for adequate quantification. It is unfortunate that the desirable spectral properties of this dye cannot be utilized in this way.

Specific details for the measurement of Indocyanine green in plasma samples have been discussed in the literature. Basically, the measurement requires that plasma cells be well separated by centrifugation. This should be done as soon as possible after withdrawal to minimize turbidity (Nielsen, 1965), as well as hemolysis. Two-fold separation reduces chances of turbidity from residual white blood cells and other nonspecific causes.

Although the practice in this laboratory has always been to measure dye concentrations immediately following a procedure, the separated plasma samples along with the standard dilutions can be refrigerated so that dye concentrations can be determined one or more days after the experiment. No significant change in optical density of samples so treated has been noted for as long as 4 days. Since any small change in optical density equally affects all samples and standards, no change in calculated SBF results. This is at odds with one report of a progressive fall of 2% per hour in the optical density of Indocyanine green in plasma samples (Caesar *et al.*, 1961). However, if the samples are frozen, they frequently become turbid after thawing, and readings cannot be made even after repeated centrifugation.

Spectrophotometric determinations should be made as quickly as possible on each sample as prolonged exposure of the dye to heat and light in the spectrophotometer may slightly reduce its optical density. Spectrophotometric measurement of the dye in plasma is impossible without special procedures if the subject has ingested lipid within a few hours preceding the study. Although lipemic serum can be cleared by ultracentrifugation (1 million g minutes), it is technically difficult to draw off clear serum below suspended chylomicrons without remixing some of the particles.

Hemolysis is less important at 805 nm than at other wave lengths, but it can still significantly raise the optical density. Procedures have been described which allow correction for dissolved hemoglobin, but they are time consuming (Warren, 1958). The number of samples taken should be sufficient to permit discarding hemolyzed samples without loss of crucial data.

Blood samples are generally mixed with heparin to prevent clotting. If the heparin contains a preservative such as sodium bisulfite, which is a reducing agent, the dye is rapidly discolored to the extent that blood flow measurements are invalidated (Cobb and Barnes, 1965). Addition of heparin can also be an important source of variation in both turbidity and the estimation of dye concentration. Nielsen (1965) pointed out that since heparin acts as a "clearing" factor *in vitro,* heparin concentration must be uniform in all samples. Since heparin is generally mixed with blood by filling the dead space of the syringe, it is important that the same amount of blood be taken into the same size syringes for each sample, particularly during periods when SBF is changing. To determine the analytical error in measuring C_A and C_{HV}, an experiment was simulated by repeatedly withdrawing 3-ml blood samples into heparinized 5-cc syringes from a reservoir of mixed dye and recently drawn whole blood. The analytical error was found to be 0.006 mg of Indocyanine green/liter (Rowell et al., 1972). Winkler and Tygstrup (1960) found the standard deviation of 57 duplicate determinations to be ±0.003 mg of dye/100 ml of plasma. If less than 3 ml of blood are taken into the syringes, the error increases due to the increased fraction of heparin diluting the sample. Three milliliters of blood in a 5-ml syringe represents the best compromise between conserving the subject's blood on one hand and the best analytical accuracy on the other, with ease of separation and reading in the spectrophotometer included as factors. The significance of this error is important during periods of changing SBF when C_A values must be used at each sampling time to compute flow changes over the interval Δt as in (12.6).

Changes in hepatic venous oxygen content have been used to indirectly estimate changes in SBF (Bishop *et al.*, 1957). The method, which simply estimates the change in SBF from the change in arterial-hepatic venous oxygen difference, rests upon the assumption that splanchnic oxygen uptake is constant. This approach has worked well when subsequently compared with dye techniques while SBF was decreased by brief periods of mild (Wade *et al.*, 1956) to severe exercise (Rowell *et al.*, 1964) and splanchnic oxygen uptake did not change. However, during prolonged heavy exercise wherein splanchnic oxygen uptake may rise 1.5 times, such an approach would overestimate any decrease in SBF by this factor (Rowell *et al.*, 1966, 1968).

REFERENCES

Bishop, J. M., K. W. Donald, S. H. Taylor, and P. N. Wormald. 1957. Changes in arterial-hepatic venous oxygen content difference during and after supine leg exercise. J. Physiol. (London) 137:309–317.

Bradley, S. E. 1946. Liver function as studied by hepatic vein catheterization. *In* Transactions of the Fifth Conference on Liver Injury. Josia Macy, Jr. Foundation, New York, pp. 38–42.

Bradley, S. E. 1960. Estimation of hepatic blood flow. *In* H. D. Bruner (ed.), Methods in Medical Research, Vol. 8, pp. 275–283. Year Book Publishers, Chicago.

Bradley, S. E. 1962. The hepatic circulation. *In* Handbook of Physiology, Circulation, Vol. 3, pp. 1387–1438. Amer. Physiol. Soc., Washington, D.C.

Bradley, S. E., F. J. Ingelfinger, and G. P. Bradley. 1952. Determinants of hepatic haemodynamics. *In* Visceral Circulation, pp. 219–232. Ciba Foundation, J & A Churchill, London.

Bradley, S. E., F. J. Ingelfinger, G. P. Bradley, and J. J. Curry. 1945. The estimation of hepatic blood flow in man. J. Clin. Invest. 24:890–897.

Brauer, R. W. 1963. Liver circulation and function. Physiol. Rev. 43:115–213.

Caesar, J., S. Shaldon, L. Chiandussi, L. Guevara, and S. Sherlock. 1961. The use of Indocyanine green in the measurement of hepatic blood flow and as a test of hepatic function. Clin. Sci. 21:43–57.

Castenfors, H., J. Eliasch, and E. Hultman. 1961. Estimation of the splanchnic blood flow in man by the bromsulphalein (BSP) method; the effect of continuous variation of the peripheral plasma concentration of BSP. Scand. J. Clin. Lab. Invest. 13:489–502.

Cherrick, G. R., S. W. Stein, C. M. Leevy, and C. S. Davidson. 1960. Indocyanine green: observations on its physical properties, plasma decay and hepatic extraction. J. Clin. Invest. 39:592–600.

Cobb, L. A., and S. C. Barnes. 1965. Effects of reducing agents on indocyanine green dye. Amer. Heart J. 70:145–146.

Drapanas, R., D. N. Kluge, and W. G. Schenk. 1960. Measurement of hepatic blood flow by bromsulphthalein and by the electromagnetic flowmeter. Surgery 48:1017–1021.

Ekelund, L.-G. 1967. Circulatory and respiratory adaptation during prolonged exercise. Acta Physiol. Scand. 70:(Suppl. 292).

Fauvert, R. E. 1959. The concept of hepatic clearance. Gastroenterology 37:603–616.

Fox, I. J., and E. H. Wood. 1960. Indocyanine green: physical and physiological properties. Proc. Staff Meet. Mayo Clin. 35:732–744.

Grayson, J., and D. Mendel. 1965. Physiology of the Splanchnic Circulation. The Williams & Wilkins Co., Baltimore.

Gregersen, M. I. 1961. Blood volume, hemorrhage and shock. In P. Bard, (ed.) Medical Physiology, 11th Ed. C. V. Mosby, St. Louis, p. 280.

Greenway, C. V., and R. D. Stark. 1971. Hepatic vascular bed. Physiol. Rev. 51:23–65.

Hultman, E. 1966. Blood circulation in the liver under physiological and pathological conditions. Scand. J. Clin. Lab. Invest. 18 (Suppl. 92): 27–41.

Hunton, D. B., J. L. Bollman, and H. N. Hoffman. 1960. Studies of hepatic function with Indocyanine green. Gastroenterology 39:713–724.

Hunton, D. B., J. L. Bollman, and H. N. Hoffman II. 1961. The plasma removal of indocyanine green and sulfobromothalein: effect of dosage and blocking agents. J. Clin. Invest. 40:1648–1655.

Leevy, C. M., C. L. Mendenhall, W. Lesko, and M. M. Howard. 1962. Estimation of hepatic blood flow with Indocyanine green. J. Clin. Invest. 41:1169–1179.

Nielsen, N. C. 1965. On the causes of variations in turbidity in heparinized human plasma. Scand. J. Clin. Lab. Invest. 17:349–356.

Reemtsma, K., G. C. Hottinger, A. C. DeGraff, Jr., and O. Creech, Jr. 1960. The estimation of hepatic blood flow using Indocyanine green. Surg. Gynecol. Obst. 110:353–356.

Rowell, L. B. 1971a. The liver as an energy source in man during exercise. *In* B. Pernow and B. Saltin (eds.), Muscle Metabolism During Exercise. Plenum Press, New York, pp. 127–141.

Rowell, L. B., J. R. Blackmon, and R. A. Bruce. 1964. Indocyanine green clearance and estimated hepatic blood flow during mild to maximal exercise in upright man. J. Clin. Invest. 43:1677–1690.

Rowell, L. B., G. L. Brengelmann, J. R. Blackmon, and J. A. Murray. 1970. Redistribution of blood flow during sustained high skin temperature in resting man. J. Appl. Physiol. 28:415–420.

Rowell, L. B., G. L. Brengelmann, J. R. Blackmon, R. D. Twiss, and F. Kusumi. 1968. Splanchnic blood flow and metabolism in heat-stressed man. J. Appl. Physiol. 24:475–484.

Rowell, L. B., J.-M. R. Detry, J. R. Blackmon, and C. Wyss. 1972. The importance of the splanchnic vascular bed in human blood pressure regulation. J. Appl. Physiol. 32:213–220.

Rowell, L. B., J.-M. R. Detry, G. R. Profant, and C. Wyss. 1971*b*. Splanchnic vasoconstriction in hyperthermic man—role of falling blood pressure. J. Appl. Physiol. 31:864–869.

Rowell, L. B., K. K. Kraning II, T. O. Evans, J. W. Kennedy, J. R. Blackmon, and F. Kusumi. 1966. Splanchnic removal of lactate and pyruvate during prolonged exercise in man. J. Appl. Physiol. 21:1773–1783.

Sapirstein, L. A., and E. J. Reininger. 1956. Catheter induced error in hepatic venous sampling. Circ. Res. 4:493–498.

Wade, O. L., B. Combes, A. W. Childs, H. O. Wheeler, A. Cournand, and S. E. Bradley. 1956. The effect of exercise on the splanchnic blood flow and splanchnic blood volume in normal man. Clin. Sci. 15:457–463.

Warren, J. V. 1958. Dye method for determining cardiac output. *In* J. V. Warren (ed.), Methods in Medical Research. Year Book Publishers, Chicago, pp. 62–68.

Wheeler, H. O., W. I. Cranston, and J. I. Meltzer. 1958. Hepatic uptake and biliary excretion of Indocyanine green in the dog. Proc. Exp. Biol. Med. 99:11–14.

Wiegand, B. D., S. G. Ketterer, and E. Rapaport. 1960. The use of Indocyanine green for the evaluation of hepatic function and blood flow in man. Amer. J. Dig. Dis. 5:427–436.

Winkler, K., J. A. Larsen, T. Munkner, and N. Tygstrup. 1965. Determination of the hepatic blood flow in man by simultaneous use of five test substances measured in two parts of the liver. Scand. J. Clin. Lab. Invest. 17:423–432.

Winkler, K., and N. Tygstrup. 1960. Determination of hepatic blood flow in man by Cardio Green. Scand. J. Lab. Clin. Invest. 12:353–356.

chapter 13/ Measurement of Blood Volume and Effusion Volume

Dennis A. Bloomfield/A. M. Safiqul Khan
Division of Cardiology/Maimonides Medical Center/
Brooklyn, New York

MEASUREMENT OF BLOOD VOLUME

General Principles

Blood volume measurements were originally attempted by noting the amount drained from opened arteries in animals and decapitated criminals. The first indirect measurement is credited to Valentin (1838), who suggested that if a known volume of distilled water was infused into the circulation, the blood volume could be estimated from the resulting dilution of total blood solids. This proposal has limited usefulness by virtue of the large infusion volume necessary and, as a consequence, the possible significant changes in parameters to be measured. However, a restatement of this dilution principle is used in all current methods of blood volume measurement. The basic proposal is that in any fluid space V, in which a quantity of indicator m is dissolved or distributed, the concentration, c, is given by:

$$c = \frac{m}{V}.$$ (13.1)

If both m and c are known, the equation may be solved for V. The limitations and conditions previously stated for indicator-dilution theory apply equally to the valid measurement of intravascular blood volume, as detailed in Chapter 2. Two basic requirements, however, need special definition. First, the value taken for c must be equal to the mean concentration for the entire vascular system. Second, the amount m at the time the concentration is obtained must be measureable since it may not be equal to the amount originally injected.

Calculation of blood volume is further unique amongst indicator-dilution methods, since there is no technically acceptable indicator which is partitioned proportionately between the cells and the plasma. That is, there is no single indicator which labels blood as such. The indicators which reflect red cell volume are carbon monoxide (historically) and radioactive iron, chromium, and phosphorus. The indicators which reflect the apparent circulating plasma volume are protein-bound dyes such as Evans blue (T-1824) and radioactive tracers such as iodine-131. It follows, therefore, that a mean intravascular hematocrit or plasmacrit value is necessary to calculate a circulating blood volume with any indicator. In order not to exclude the approximately 2% contribution of the white cells and platelets to the total blood volume, plasmacrit values must be read at the top of the packed cellular column and the hematocrit readings at the top of the buffy coat of centrifuged samples. Since the special considerations mentioned above influence very substantial variations in calculated blood volume, they will be treated in some detail.

Circulatory Mixing and Loss of Indicator

The theoretical requirements and practical achievement of mixing have been discussed in relation to cardiac output measurement and dye curve morphology (Chapter 4). These considerations were confined to only a part of the first circulation of the indicator. The circulatory path of interest in cardiac output measurements may not include a capillary bed and almost never includes the systemic capillaries. A different set of criteria is required to validate mixing throughout the entire vascular system. A bolus injection of indicator becomes gradually mixed in the circulation by processes of flow fractionation into circuits with different transit times and recombination with blood having different activity at circuit confluences. After the first circulation of indicator, attenuated waves may usually be observed up to the third or fourth recirculation, but their disappearance from the central circulation does not assure generalized complete mixing. This cannot occur until the peripheral circuits, as a group, are returning indicator to the central circulation at the same rate as they receive it. Before that time, there is a net loss of indicator from the central to the peripheral circulation, and this cannot be distinguished from extravascular loss of impermanent indicator. All plasma labels, for example, continue to leave the central circulation after mixing until they finally disappear completely. If a liberal amount of time is allowed for mixing, a significant proportion of the indicator may be lost from the intravascular space, and correction of the value of m at the time of measurement of c is required. Most plasma labels disappear at a nearly exponential rate after mixing. It is usual, therefore, to extrapolate the logarithm of the indicator concentration back through the mixing period to an intercept at the zero time ordinate (Gibson and Evans, 1937). This method of correcting for the unobserved loss during the mixing period can be only an approximation, because, as the amount of indicator cleared is concentration-dependent, an excessive loss must occur during the mixing period when the concentration is high.

Realistic criteria for uniform mixing are difficult to establish. Direct sampling of venous outflow of individual circuits is hard to interpret because of the time lapse in transit through these circuits. Studies of tissue concentration of indicator demonstrate wide time variations in establishing stability between tissue and blood concentrations (Friedman, 1960). Consequently, to attain a uniform concentration of indicator throughout the cardiovascular system is not a practical consideration, and it is sufficient that the concentration in the sampled blood represents the mean for the system. This requirement is considered satisfied by finding a constant concentration in the central circulation over a long period (when using a "permanent" indicator) or a constant rate of concentration decline (when using a "temporary" indicator).

The observed time required for mixing of labeled cells varies with both the indicator and the circulatory state of the patient. Values ranging from 7.5 to 15 minutes have been cited (Berson and Yalow, 1952; Nylin, 1947). This figure is prolonged with splenomegaly, low cardiac output, and shock. A similar range (4 to 15 minutes) is required for mixing of labeled plasma (Pritchard et al., 1955). No constant difference in mixing time has been observed when measured by various plasma indicators (Noble and Gregersen, 1946).

Some inconsistency appears, however, in reports comparing the mixing times of cell and plasma indicators. Although no regular difference was found even in congestive heart failure (Hedlung, 1953; Moir *et al.*, 1956; Schreiber *et al.*, 1954), shorter plasma mixing times were found in the experiments of Lawson (1962). These irregularities could be explained by sequestration of cells or plasma and present a source of inaccuracy in the method. Immobilization of cells in areas of the circulatory system inaccessable to indicators has been shown in experimental shock by Gibson *et al.* (1947). It may be assumed the plasma is sequestered along with cells whenever blood stops circulating in abnormal states, but estimation of its volume has been of uncertain reliability. In fact, it has not been possible to demonstrate plasma sequestration in most studies because of active mechanisms to replenish plasma volume (Lawson and Rehm, 1945; Huggins *et al.*, 1957*a*).

Distribution of Cells and Plasma in the Circulatory System

All procedures for blood volume measurement, whether they assess indicator concentration in whole blood or after blood has been separated into cell and plasma fractions, use hematocrit directly or indirectly. However, as the ratio of cells:plasma is not the same in all parts of the circulation (the spleen is cell-rich; the smaller vessels are cell-poor), indicator concentration varies throughout the system when compared to concentration in drawn blood samples, even after complete mixing has occurred. Nevertheless, blood drawn from any definable vascular site has essentially the same hematocrit as blood drawn from any other vascular site, and the value represents the hematocrit of the central circulation. It has been suggested that this central hematocrit, although not representing mean hematocrit, bears a constant relationship to mean hematocrit, so that the latter may be derived from specimens of drawn blood (Chaplin *et al.*, 1953; Reeve *et al.*, 1953*a* and *b*).

The ability to measure an accurate central hematocrit poses a preliminary problem. Although centrifugation readily produces cell-free plasma, it is impossible to drive plasma completely out of the packed cell column in this way. The percentage of remaining plasma has been variously considered at between 4% (Gregersen and Schiro, 1938) and 8.5% (Chapin and Ross, 1942). However, a constant correction cannot be applied to the centrifugal hematocrit, as the value is dependent on centrifugal force and duration, height of blood in the centrifuge tube and the actual hematocrit itself.

When cell and plasma labels are injected simultaneously and the virtual distribution volumes are calculated for each, the ratio of cells to plasma may be computed for the whole circulatory system and the mean circulatory hematocrit H_m is calculated as:

$$H_m = \frac{CV}{CV + PV}, \tag{13.2}$$

where CV and PV are the distribution volumes of the cell indicator and the plasma indicator, respectively. Many studies relating H_m to the central hematocrit H_c have been performed (Berson and Yalow, 1952; Chaplin and associates, 1953; Gray and Frank, 1953; Verel, 1954). The average ratio H_m/H_c in normal adults is reported to range between 0.89 and 0.94. It is reduced to 0.87 in the newborn (Mollison et al., 1950) and to 0.81 in early pregnancy (Caton et al., 1951). If the ratio could be assumed constant, it would be possible to use a single indicator for either cells or plasma, to estimate the unlabeled compartment from this relationship. If, for example, a plasma indicator were used and PV were measured, the CV_h', the calculated cell volume (CV_h) corrected for differences between the mean and central hematocrits is given by:

$$CV_h' = PV \times \frac{H_c'}{1 - H_c'}, \tag{13.3}$$

where H_c' is H_m estimated from the observed value for H_c using the predicted relationship between H_m and H_c. The range of the H_m/H_c ratio in man and the large and unexplained differences reported in experimental animals (Huggins et al., 1957b; Reeve et al., 1953a) present a further source of error in calculations of blood volume.

Local Circulatory Hematocrits and Excess Plasma

A variety of methods have shown that the ratio of cellular indicator concentration to plasma indicator concentration in individual vascular areas is different from the ratio in the central circulation. Within a short time after injection, plasma indicators expand their virtual distribution space in most tissues out of proportion to the expansion of the cell indicators. Although it is believed that firmly bound cellular indicators remain intravascular, no such assumption can be made with plasma indicators. As no methods distinguish intra- from extravascular plasma indicators, no anatomical or quantitative definition of the extra plasma space can be made for local circuits (Lawson, 1962). The volume of cell-free, or "excess" plasma space can be readily calculated, however, for the total vascular system. This excess plasma (reflected in the error in calculating CV_h and in the difference between

H_m and H_c) is the difference in plasma volume measured from the distribution of the plasma label and the volume which would be required to suspend the total cell volume in plasma at the hematocrit of the central circulation. It is considered to be about 11% of the total vascular space. These qualifications in the quantitation and description of the plasma volume and the differences in distribution of individual plasma indicators are reviewed by Lawson (1962).

MEASUREMENT OF EFFUSION VOLUME

Theory

The volume of a fluid confined within a serous cavity can be measured in the same manner as previously described for total blood volume. The principle that the volume is given by the ratio of the amount of indicator injected and its subsequent concentration in the fluid was established in Chapter 2 and has been stated as

$$V = \frac{m}{C},$$

where V is the volume, m is the amount of indicator, and C is the concentration of indicator. When applied to effusion collections, the requirements, limitations, and assumptions defined for intravascular indicator dilution still hold, but two of the conditions assume special significance.

Mixing

Complete mixing of the indicator in the native fluid and the subsequent obtaining of a representatively mixed sample of fluid is more difficult to achieve in the absence of a flowing stream, branching and reuniting of flow paths, and areas of turbulence. Other methods of achieving adequate mixing are required.

Confinement of Indicator within the Volume

For validity, the dilution principle requires no loss or gain of indicator from the system under measurement. Under conditions where the effusion has accumulated because of pathological or physiological change in the state of the confining walls, through which fluid may pass in either direction, special precautions are required to determine that no significant shift occurs during the period of measurement.

Method

Quantitative assessment of pericardial effusion volume has been carried out by Khan *et al.* (1973), using Indocyanine green, in nine patients. The technique was as follows: A needle was introduced into the pericardial sac under local anaesthesia, and an initial 100 cc of fluid was removed for calibration purposes. A known volume (1 to 2 cc) of Indocyanine green was injected through the needle, and a 5-minute period was allowed to elapse while mixing took place. This was aided by repeatedly aspirating and reinfusing about 30 cc of the effusion fluid before a sample was drawn for concentration measurement. Calibration was carried out by an adaption of the method described in Chapter 3 and illustrated here in Fig. 13.1. Alternately, direct spectrophotometric analysis of concentration can be used.

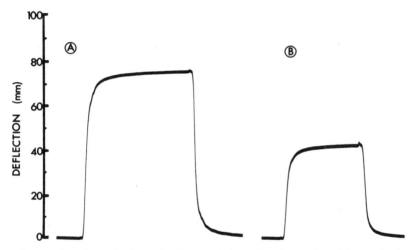

Figure 13.1 Quantitation of volume. *A*, base line is produced by undyed pericardial fluid drawn through the densitometer. The deflection in millimeters (73 mm) is produced by drawing through the densitometer a concentration of Indocyanine green equivalent to the injected volume diluted in 500 ml of plasma. *B*, base line is again produced by undyed pericardial fluid. The deflection (43mm) is produced by pericardial effusion sample containing Indocyanine green. The effusion volume in this case is calculated as $73/43 \times 500 = 848$ cc.

The technique was verified by Kahn and associates in experimental animals and clinical subjects. Where effusion was produced in animals by acutely distending the pericardium with infused plasma, the calculated volume was consistently less than the known infused volume by a range of 4 to 8 cc, representing 7 to 10%. Since difficulty was experienced in sealing the entrance site of the infusion catheter into the pericardial sac to avoid leakage, and total aspiration of the fluid consistently produced less than had been infused, the smaller calculated volumes were thought to result from fluid loss out of the pericardial space before uniform mixing of the indicator.

The adequacy of mixing was determined by obtaining effusion samples at 1-minute intervals for 20 minutes after the injection of indicator. The concentration reached a stable level, indicating uniform mixing after 5 minutes (Fig. 13.2). That there was no exchange of

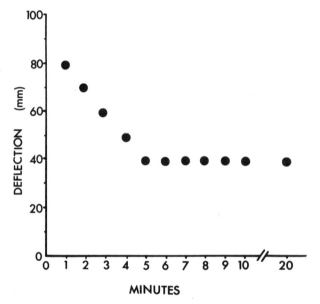

Figure 13.2 Mixing of indicator in pericardial fluid. The concentration of indicator is plotted on the ordinate; time in minutes is plotted on the abscissa. During the initial period, there is progressive fall in the concentration of the indicator until complete mixing has occurred at the end of 5 minutes. There is no further fall in the concentration after this period up to 20 minutes, indicating no loss of indicator from the pericardial space.

tagged effusion fluid with the general body pool during the time course of the measurement was indicated by the constant indicator concentration that was maintained after mixing had occurred.

Pleural effusion was measured by the same technique by Khan and co-workers using radiologic assessment to determine complete emptying, and close correlation was obtained with the calculated volumes.

Ascitic volume determinations have been performed by Mankin and Lovel (1948) using vital red and Evans blue indicators; by Eisenmenger and co-workers (1950) using Evans blue; by Warter, Mandel, and Metais (1950) with polyvinyl-pyrrolidone; by Baker and co-workers (1952) using bromsulfalein, ^{131}I, and para-aminohippurate; and Khan et al. (1973) using Indocyanine green. Due to anatomical considerations, the condition of mixing in the intraperitoneal space is slower and more difficult to achieve than in the pericardial or pleural spaces. Mankin and his associates allowed 30 to 60 minutes, and Baker and co-workers required as long as 120 minutes before mixing of indicator and effusion fluid was completed and indicator concentration reached an equilibrium. A further difficulty in utilizing this technique of volume measurement lies in satisfying the second assumption for validity. Ascitic fluid is not a stagnant pool but is in relative dynamic equilibrium with plasma water and protein. Many factors combine to effect clearance of an indicator from the fluid and produce a decrement in concentration which is linear with time. Extrapolation back to "zero time" will yield the indicator concentration occurring at instantaneous equilibrium before loss from any route. The volume computation for distribution of an indicator in ascitic fluid utilized the extrapolated concentration.

All indicators used gave similar values when the above assumptions were satisfied. The calculated volumes were greater than the aspirated volumes by about 400 cc, or 5%, which most likely reflects the inability to completely drain the ascitic fluid by aspiration methods.

Indirect Methods

Indirect methods, such as angiocardiography with radiopaque contrast (Burch and Phillips, 1962), radioisotope scanning (Wagner et al., 1960), negative contrast angiocardiography after injection of carbon dioxide (Durant, 1964), and ultrasound (Goldberg et al., 1967) have been used to establish the diagnosis of pericardial effusion and to distinguish it from cardiac enlargement.

However, although these methods are reasonably accurate for qualitative assessment, quantitation of the volume of fluid cannot be obtained. Even after pericardiocentesis, incomplete removal of fluid may cause difficulty in distinguishing relative contribution to the radiologic cardiac silhouette by an enlarged heart and by the residual effusion. Further difficulty in this assessment is attendant on the peculiarity of fluid distribution about the heart (Bryk et al., 1969).

Volume of pericardial fluid has been semiquantitatively assessed by correlation of X-ray and ^{131}I-cholografin and iodinated human serum ^{131}I-albumin heart scan by Sklaroff, Charkes, and Morse (1964). These investigators measured the pericardial contents in 23 patients undergoing open heart surgery and in 11 patients at autopsy or by pericardiocentesis. Isotopic photoscans of the heart were made, and these were superimposed on the chest films of the patients. The ratio of the maximum transverse cardiac diameters on scan and roentgenogram was taken as an index for diagnosis and quantitation of pericardial fluid. In patients with less than 100 cc of pericardial fluid, this ratio of internal to external transverse cardiac diameter was greater than 0.8; it was less than 0.8 in cases of effusion of 200 cc or more. This technique appears to be fairly accurate for diagnosis of pericardial effusion and semiquantitation of volume. Correlation of the diameter ratio with volume of effusion was not possible since the full pericardial contents were not always measured. Moreover, cardiac dilatation and/or hypertrophy decreases the sensitivity of the technique.

The ability to predict this volume has been found to be of considerable value in the management of pericardial effusion. The most significant aid is the knowledge of volume remaining when aspiration becomes difficult or ceases. The operator will be guided as to desist or persist in his efforts to aspirate further fluid by comparing the volume aspirated to that calculated to be present.

The method is also valuable in the consideration of loculation. When the size of the cardiac chambers can be determined, and the remaining cardiac silhouette suggests an effusion volume out of proportion to the calculated volume, it can be presumed that loculation is present. Further, in patients with repeated pericardial effusions, certain radiologic over-all heart sizes are associated with relatively constant figures for fluid volume. Khan et al. (1973) described effusion volume measurements on two occasions in one patient with uremic pericarditis when the radiologic heart size appeared similar. Values of 1550 and 1700 cc were obtained. Consequently, loculation can be diagnosed when, on a repeated examination with the same heart size, the calculated effusion volume is significantly lower. Patients with repeated pericardial effusions can be followed with assurance that loculation is not occurring, when, at reproducible radiologic heart sizes, similar estimations of fluid volume are obtained.

Despite all precautions during pericardiocentesis, the needle may inadvertently be inserted into a cardiac chamber. In cases of hemorrhagic effusion, when the hematocrit difference between the circulating blood and the effusion may be as small as 3%, it may be difficult for the operator to know whether the aspirated material is from the heart or from the pericardial space. If indicator is injected into a cardiac cavity and sampled after 5 minutes, virtually no indicator can be recovered. Recovery of measurable concentrations of indicator after the 5-minute mixing confirms the catheter position in the pericardial space.

The technique is simple and safe, and can be applied to bloody as well as clear effusion fluids as long as the same fluid is used for blank and standard samples. The method can be applied to either pericardial or pleural fluid collections, and the measurement is of clinical value in diagnosis and therapy.

REFERENCES

Baker, L., R. C. Puestow, S. Kruger, and J. H. Last. 1952. Estimation of ascitic fluid volume. J. Lab. Clin. Med. 39:30–35.

Berson, S. A., and R. S. Yalow. 1952. The use of K^{42} or P^{32} labeled erythrocytes and I^{131} tagged human serum albumin in simultaneous blood volume determinations. J. Clin. Invest. 31:572–580.

Bryk, D., I. G. Kroop, and J. Bundow. 1969. The effect of heart size, cardiac tamponade, phase of cardiac cycle on the distribution of pericardial fluid. Radiology 93:273–278.

Burch, G. E., and J. H. Phillips. 1962. Methods in the diagnostic differentiation of myocardial dilatation from pericardial effusion. Amer. Heart J. 64:266–281.

Caton, W. L., C. C. Roby, D. E. Reid, R. Caswell, C. J. Maletskos, R. G. Fluharty, and J. G. Gibson. 1951. The circulating red cell volume and body hematocrit in normal pregnancy and the puerperium. Amer. J. Obst. Gynec. 61:1207–1217.

Chapin, M. A., and J. F. Ross. 1942. The determination of the true cell volume of dye dilution, by protein dilution, and with radioactive iron. The error of the centrifuge hematocrit. Amer. J. Physiol. 137:447–455.

Chaplin, H., P. L. Mollison, and H. Vetter. 1953. The body/venous hematocrit ratio: its constancy over a wide hematocrit range. J. Clin. Invest. 32:1309–1316.

Durant, T. M. 1964. Negative (gas) contrast angiocardiography. Amer. Heart J. 61:1–4.

Eisenmenger, W. J., S. A. Blondheim, A. M. Bongiovanni, and H. G. Kunkel. 1950. Electrolyte studies on patients with cirrhosis of the liver. J. Clin. Invest. 29:1491–1499.

Friedman, J. J. 1960. Distribution of red blood cells between tissues of mouse. Proc. Soc. Exp. Biol. Med. 103:80–83.

Gibson, J. G., and W. A. Evans. 1937. Clinical studies of the blood volume: application of a method employing the azodye Evans blue and the spectrophotometer. J. Clin. Invest. 16:301–316.

Gibson, J. G., A. M. Seligman, W. C. Peacock, J. Fine, J. C. Aub, and R. D. Evans. 1947. The circulating red cell and plasma volume and the distribution of blood in large and small vessels in experimental shock in dogs, measured by radioactive isotopes of iron and iodine. J. Clin. Invest. 26:126–144.

Goldberg, B. B., B. J. Ostrum, and H. J. Isard. 1967. Ultrasonic determination of pericardial effusion. J. A. M. A. 202:103–106.

Gray, S. J., and H. Frank. 1953. Simultaneous determination of red cell mass and plasma volume in man with radioactive sodium chromate and chromic chloride. J. Clin. Invest. 32:1000–1004.

Gregersen, M. I., and H. Schiro. 1938. The behavior of the dye T-1824 with respect to its absorption by red blood cells and its fate in blood undergoing coagulation. Amer. J. Physiol. 121:284–292.

Hedlung, S. 1953. Studies on erythropoiesis and red cell volume in congestive heart failure. Acta Med. Scand. Suppl. 284.

Huggins, R. A., E. L. Smith, S. Deavers, and R. C. Overton. 1957a. Changes in cell and plasma volume produced by hemorrhage and reinfusion. Amer. J. Physiol. 189:249–252.

Huggins, R. A., E. L. Smith, and S. Deavers. 1957b. Distribution of red cells and plasma in the dog. Amer. J. Physiol. 191:163–166.

Khan, A. M. S. I., M. Nejat, and D. A. Bloomfield. 1973. The measurement of pericardial effusion volume. Chest, 63:762–766.

Lawson, H. C. 1962. The volume of blood—a critical examination of methods for its management. In W. F. Hamilton and P. Dow (eds.), Handbook of Physiology, Vol. I, pp. 23–49. Amer. Physiol. Soc., Washington, D. C.

Lawson, H. C., and W. S. Rehm. 1945. The effect of hemorrhage and replacement on the apparent volume of plasma and cells. Amer. J. Physiol. 144:199–205.

Mankin, H., and A. Lovel. 1948. Osmotic factors influencing the formation of ascites in patients with liver cirrhosis. J. Clin. Invest. 27:145–153.

Moir, T. W., W. H. Pritchard, and A. B. Ford. 1956. The early disappearance of I^{131} serum albumin from the circulation of edematous subjects and its implications in the clinical determination of blood volume. J. Lab. Clin. Med. 47:503–512.

Mollison, P. L., N. Veall, and M. Cutbush. 1950. Red cell and plasma volume in newborn infants. Arch. Dis. Childhood 25:242–253.

Noble, R. P., and M. I. Gregersen. 1946. Mixing time and disappearance rate of T-1824 in normal subjects and in patients in shock. J. Clin. Invest. 25:158–171.

Nylin, G. 1947. The effect of heavy muscular work on the volume of circulating red corpuscles in man. Amer. J. Physiol. 149:180–184.

Pritchard, W. H., T. W. Moir, and W. J. MacIntyre. 1955. Measurement of the early disappearance of iodinated (I^{131}) serum albumin from circulating blood by a continuous recording method. Circ. Res. 3:19–23.

Reeve, E. B., M. I. Gregersen, T. H. Allen, and H. Sear. 1953a. Distribution of cells and plasma in the normal and splenectomized dog and its influence on blood volume estimates with P^{32} and T-1824. Amer. J. Physiol. 175:195–203.

Reeve, E. B., M. I. Gregersen, T. H. Allen, H. Sear, and W. W. Walcott. 1953b. Effects of alteration in blood volume and venous hematocrit in splenectomized dogs on estimates of total blood volume with P^{32} and T-1824. Amer. J. Physiol. 175:204–210.

Schreiber, S. S., A. Bauman, R. S. Yalow, and S. A. Berson. 1954. Blood volume alterations in congestive heart failure. J. Clin. Invest. 33:578–586.

Sklaroff, D. M., D. N. Charkes, and D. Morse. 1964. Measurement of pericardial fluid correlated with the I-cholografin and IHSA heart scan. J. Nucl. Med. 5:101–111.

Valentin, 1838. Cited by Erlanger, J. 1941. In Blood volume and its regulation. Physiol. Rev. 1:177.

Verel, D. 1954. Observations on the distribution of plasma and red cells in disease. Clin. Sci. 13:51–59.

Wagner, H. N., Jr., J. G. McAfee, and J. M. Mozley. 1960. Diagnosis of pericardial effusion by radioisotope scanning. Circulation 22:828.

Warter, J., P. Mandel, and P. Metais. 1950. Determination of cirrhotic ascitic fluid. Strasbourg Med. J. 4:244–249.

chapter 14/ Measurement of Blood Flow by the Thermodilution Technique

William Ganz/H. J. C. Swan
Department of Cardiology/Cedars-Sinai Medical Center/
and Department of Medicine/University of California/
Los Angeles, California

The measurement of blood flow by thermodilution is based on induction of a change in the intravascular heat content of flowing blood and detection of the resultant change in temperature at a point downstream. The change in heat content can be instantaneous or continuous and at constant rate, and it can be induced by adding or abstracting heat. Heat may be added directly: an electric resistance is heated by a current, the characteristics of which are known, allowing the precise calculation of the quantity of heat delivered to the blood (Afonso, 1966; Khalil, 1963; Barankay *et al.,* 1970). The abstraction of heat is performed indirectly by introduction into the blood stream of a fluid at a temperature below that of blood.

The thermodilution principle has been applied to the measurement of both cardiac output and blood flow in single blood vessels. Injection of cold fluids is the most commonly used method to alter the heat content. The discussion will, therefore, deal mainly with this application of the thermodilution technique.

MEASUREMENT OF CARDIAC OUTPUT BY THERMODILUTION

A bolus of cold fluid is injected into the central circulation before the right or the left side of the heart, and the resultant change in temperature is measured in the pulmonary artery or in the aorta. The bolus injection is similar in principle to the usual form of the dye-dilution technique and results in a similar, but not identical, temperature-time dilution curve. Cardiac output is determined from the change in the intravascular heat content and from the time course of intravascular temperature at the site of detection. It is assumed that no heat, or only a negligible amount of it, is lost between the sites of injection and detection and that even distribution of heat is attained proximal to the site of detection.

The formula used for the calculation of cardiac output by thermo-dilution is analogous to that of Stewart and Hamilton for estimation of cardiac output by dye dilution:

$$CO = \frac{V_I \times (T_B - T_I) \times S_I \times C_I \times 60}{S_B \times C_B \times \int_0^\infty \Delta T_B(t)dt}(\text{ml/min}),$$

where

CO = cardiac output in ml/minute
V_I = volume of injectate in ml
T_B = temperature of blood in °C
T_I = mean temperature of the injectate at the point of entrance into the blood stream in °C
S_I, S_B = specific gravity of injectate and blood, respectively, in g/cm³
C_I, C_B = specific heat of injectate and blood, respectively, in cal/g/°C
$\int_0^\infty \Delta T_B(t)dt$ = area of the thermodilution curve in seconds × °C

$\int_0^\infty T_B(t)dt$ can be written as A/rf, where A = area of the time-temperature curve in mm², r = recording paper speed in mm/second, f = temperature calibration factor in mm/°C.

At a hematocrit of 40% S_B = 1.045, C_B = 0.87 cal/g/°C (Mend-lowitz, 1948). The product of S_B and C_B is practically constant (about 0.91) over a wide range of hematocrit because the two values change proportionately in opposite direction (Fronek and Ganz, 1960).

Saline and 5% solution of dextrose are most commonly used for injection. The S and C values are 1.005 g/cm³ and 0.997 cal/g/°C for saline and 1.018 and 0.965 cal/g/°C for 5% dextrose, respectively.

QUANTITATION OF THERMAL INDICATOR

The effective quantity of the thermal indicator depends on the volume of injectate, the temperature difference between blood and injectate at the site of injection into the bloodstream, and the densities and specific heats of blood and injectate. Whereas the volume of injectate can be measured with minimal error, determination of the effective indicator temperature at the point of entry into the bloodstream can be a major source of error. The cold fluid is usually injected through a catheter inserted into a peripheral vein and advanced to the right atrium. The use of fluids at room temperature was advocated because it practically eliminates the problem of temperature change before injection into the catheter. However, if colder fluids are used, a proportionately smaller volume of injectate may be used to produce an appropriate thermodilution curve. In addition, measurement of fluid temperature inside a 10-cc plastic syringe showed that the temperature of ice-cold fluid (about 1°C) increased by less than 0.2°C, if injection was completed within 10 seconds after removal of the syringe from the ice bath. Since the preinjection difference between the temperature of blood and an ice-cold fluid is around 32 to 35°C, such a preinjection change in temperature is negligible (0.5%). Hence, if injection is made promptly, the use of ice-cold fluids does not introduce error of significant magnitude.

During the injection itself, indicator is lost by heat transfer between the catheter wall and the injectate. The magnitude of this heat transfer depends on the initial temperature gradient between the injectate and the blood in the vein around the catheter, the thermal conductivity of the catheter wall, the diameter of the lumen, the length of the catheter (particularly the intravascular segment), and the rate of injection. The warming of injectate is maximal initially and decreases as the catheter wall is cooled. In case of a bolus injection, the temperature of the injectate may never reach a steady state.

The effective injectate temperature and hence the quantity of thermal indicator delivered to the blood in the injected bolus is, therefore, less than indicated by the product of volume and blood-injectate temperature difference. The following methods have been used to correct such potential errors.

1. The amount of negative heat lost during injection is determined by calorimetry under conditions duplicating those of the actual measurement (Goodyer et al., 1959).

2. The time course of the injectate temperature is recorded by a fast response thermistor placed in the lumen of the catheter near the injection orifice. The mean effective injectate temperature can be obtained from this temperature-time curve on the assumption that the injection rate is practically steady (Hosie, 1962; Olsson *et al.*, 1970; Ganz *et al.*, 1971*a*).

3. The mean effective injectate temperature is determined *in vitro* under conditions duplicating those of the actual measurement. The injectate leaving the catheter is collected in a thermally insulated test tube, and its temperature is recorded by a fast response thermistor.

Most of the negative heat conducted to the catheter wall and to the blood will be absorbed through the thin venous wall into the surrounding tissues, particularly when the blood flow along the catheter is slow. Some indicator from the central portions of the catheter may, however, reach the detection site after some delay and possibly disturb the exponential course of the downslope of the thermodilution curve. Rapid injection and insulated injection lumen will minimize the effect.

After completion of injection, the cold fluid still remaining in the intravascular portion of the catheter will be rewarmed to the temperature of the blood, causing additional cooling of blood in the catheterized vein. Some of this blood may arrive at the sampling site with the injected bolus. This effect was found to be trivial in dogs (Goodyer *et al.*, 1959). However, in the experience of Olsson *et al.* (1970) and Ganz *et al.* (1971*a*) rapid withdrawal of the cold fluid from the distal part of the catheter (about 1 ml) immediately following injection is associated with a more rapid and complete return of the curve to the base line. After inscription of the curve is completed, the fluid can be injected into the patient.

Experience with a polyvinyl chloride flow-directed catheter (Ganz *et al.*, 1970; Forrester *et al.*, 1972) showed that, if the conditions of injection are partially standardized (type of catheter, 10 ml of injectate, preinjection temperature between 0 and 5°C, duration of injection between 2 to 4 seconds), the amount of heat delivered to the blood can be predicted from knowledge of the preinjection injectate temperature. The constant correcting for the heat lost during injection (c), was derived from the temperature of the injectate before the injection (T_I) and after its passage through the catheter (T_{I-E}).

$$c = \frac{T_B - T_{I-E}}{T_B - T_I}.$$

The correction constant was unaffected by the temperature of blood in the range from 35 to 39°C and negligibly changed by the length of the intravascular portion of the catheter in the range of 35 ± 10 cm from the injection orifice. The variability of the correction constant was 3%.

QUANTITATION OF THERMODILUTION CURVES

Effect of Heat Exchange

The general shape of thermodilution curves is similar to that of dye-dilution curves. There are, however, differences due to the diffusiveness of heat. When a bolus of cold fluid is introduced into the warmer vascular system, heat flows from the vessel walls to the cold blood-indicator mixture. As the cold mixture passes and warm blood follows, the heat transfer is reversed. As a consequence of this bidirectional heat exchange, the peak deflection is reduced and the downslope of the thermodilution curve is prolonged. The area under the curve will not change if the heat exchange is perfectly reversible, or it will be smaller if the heat is not fully recovered.

The magnitude of heat transfer depends on the surface area available for the heat exchange, the thickness and the thermal conductivity of the vessel wall, the thermal conductivity and heat capacity of extra-vascular tissues, and the time available for the exchange. As the vessel diameter falls, the ratio of surface area to volume increases and the velocity of blood flow decreases. The highest rate of heat transfer can be expected in small vessels, particularly in capillaries and in small, thin-walled veins.

These facts must be considered when the sites of injection and detection are chosen, since the absence of significant indicator loss is a basic condition of accurate measurement. In accordance with the above considerations, injection of cold fluid into the external jugular veins leads to an overestimation of cardiac output, as compared with right atrial injection (Mohammed et al., 1963; Wessel et al., 1971). Significant losses of indicator can also be expected when the site of temperature detection is in a peripheral artery, for instance in the carotid, iliac, or femoral, as compared with sampling in the aorta (Mohammed et al., 1963; Wessel et al., 1971).

The pulmonary capillaries differ from the other capillary beds in that they are in direct apposition to air-filled alveoli which act as a thermal insulator because of the very low heat capacity of air. Passage of the cold blood-injectate mixture through the pulmonary capillary bed leads normally to distortion of the shape of the curve due to bidirectional heat exchange (see above), but to no appreciable heat loss. Flow measurements obtained simultaneously from thermodilution curves recorded in the pulmonary artery and in the aorta after injection into the right atrium (Fegler, 1954 and 1957; Goodyer *et al.,* 1959; Klussmann, Koenig, and Lütcke, 1959; Enghoff *et al.,* 1970; Lüthy, 1961) were found to be in close agreement. A significant heat loss in pulmonary capillaries can be expected in patients with pulmonary edema.

Effect of Recirculation

The blood-injectate mixture is significantly rewarmed in the small systemic vessels. Hence, recirculation peaks, typical in dye-dilution curves, are absent in thermodilution curves (Fig. 14.1). There is,

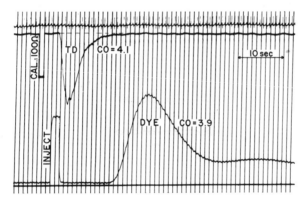

Figure 14.1 A thermodilution curve *(TD)* recorded in the pulmonary artery and a dye-dilution curve *(DYE)* obtained by sampling from the femoral artery after injection of 2.5 mg of Indocyanine green in 10 ml of cold (2°C) 5% solution of dextrose into the superior vena cava in a patient with coronary artery disease. Note the practically perfect return of the thermal curve to the base line and the distortion of the dye curve by significant recirculation. *Cal,* calibration, *CO,* cardiac output (liters/minute).

however, a difference of opinion concerning the magnitude of recirculation. Branthwaite and Bradley (1968) and Olsson *et al.* (1970) found no recirculation in curves obtained in the pulmonary artery after injection into the right atrium and no fall in pulmonary artery temperature after injection of room temperature saline into the left atrium (Olsson *et al.,* 1970). Ganz *et al.* (1971*a*) found the degree of recirculation in man negligible. According to Goodyer *et al.* (1959), the downslope of thermodilution curves in dogs has an exponential course to a very low fraction of the peak deflection, in contrast to simultaneous dye curves. Lüthy and Rutishauser (1961) and Lüthy (1961 and 1962) found, by injecting 10 patients with cold saline into the distal pulmonary artery and recording the temperature in the main trunk, that the peak of recirculation curve averaged about 3% of the peak deflection of thermodilution curves obtained by injection of the same quantity of saline at the same temperature into the right atrium. Our recent experience with a polyvinyl chloride flow-directed thermodilution catheter in 18 patients with acute myocardial infarction showed that the downslope of curves remains exponential to 5.5 ± 3.4% (mean and standard deviation; range 0 to 12) of the peak deflection, and the maximal recirculation amounts to 1.9 ± 1.0% of peak deflection (range 0 to 4). Integration of the area under the thermodilution curve with a cutoff of the downslope at 5% of the peak deflection was associated with a negligible error of −1.6 ± 1.2% (range +1.2 to −3.5), compared with the area obtained after exponential extrapolation of the downslope.

Effect of Thermistor Position

If the thermistor is wedged against the vessel wall, the thermodilution curve will be markedly distorted: the build-up time will be prolonged, the peak deflection will be reduced, and the time constant of the downslope will be increased. The area contained by the curve may be also markedly altered. Slight adjustment of the catheter position will usually eliminate the distortion (Singh *et al.*, 1970; Wessel *et al.*, 1971).

Effect of Fluctuation in Blood Temperature

There are differences between the temperature of blood in different sections of the circulation (Horvath, Rubin, and Foltz, 1950; Mellette, 1950; Eichna *et al.*, 1951; Bucher and Emmenegger, 1952; Carlsten and Grimby, 1958; Afonso *et al.*, 1962). Temperatures higher than in the pulmonary artery were found in the internal jugular and hepatic veins, in the coronary sinus, and in the renal vein (Afonso *et al.*, 1962). The temperature in the left side of the heart was found to be similar to that in the pulmonary artery (Good and Sellers, 1957; Mather, Nahas, and Hemingway, 1953; Afonso *et al.*, 1962). Experimental data and mathematical calculations (Rubenstein, Pardee, and Eldridge, 1960; Wessel *et al.*, 1971) indicate that the alveolar temperature remains virtually constant even with extreme changes in the temperature of inspired air, causing no appreciable change in the temperature of blood during its passage through the pulmonary capillary bed (Klussmann *et al.*, 1959).

The temperature in the venae cavae, the right atrium, and to a lesser extent in the right ventricle and pulmonary artery shows variations in phase with respiration. The amplitude of these variations increases with the amplitude of respiration, but is not dependent on ventilation, since it is present and even accentuated during breathing against the closed endotracheal tube (Wessel, James, and Paul, 1966*b*). It seems most likely that the respiratory variations in blood temperature at a single location are due to temperature differences between regional venous beds (the temperature in the superior vena cava is lower) and respiratory variations in the contribution of each bed to mixed venous blood.

The respiratory variation in blood temperature in the pulmonary artery in man is usually around 0.01 to 0.02°C, but can be significantly higher in patients with dyspnea or during exercise. The fluctuation of temperature represents a "physiologic noise" which makes determination of the area under the thermodilution curve less accurate. The effect of temperature fluctuation on the accuracy of measurements can be minimized by increasing the volume of injectate and the rate of injection and by using injectates at lower temperature.

Effect of Mixing

Even distribution of heat and a uniform temperature over the vascular cross-section at the sampling site is a basic requirement for the accuracy of measurements. Selection of the sites of injection and detection must be done also with regard to the mixing conditions. It is generally accepted that passage of an indicator through two cardiac chambers will ensure adequate mixing of blood and indicator. In case of a thermal indicator, mixing is enhanced by conduction of heat in addition to molecular dispersal of the injectate.

Injection into the superior or inferior vena cava at their junction with the right atrium and sampling in the pulmonary artery 4 to 5 cm from the valve seems the most suitable with respect to mixing, heat loss, and recirculation. Flow-directed catheters (Ganz *et al.*, 1970; Swan *et al.*, 1971; Forrester *et al.*, 1972) are now available which can be passed into the pulmonary artery at the bedside of the patient without fluoroscopic control, rapidly and safely, and used for the injection of cold fluid into the central venous region, recording of temperature in the pulmonary artery, and for the measurement of pressures in the pulmonary artery, pulmonary wedge position and the central venous area (Fig. 14.2).

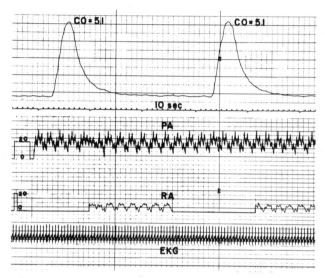

Figure 14.2 Thermodilution curves, pulmonary artery (*PA*) and right atrial (*RA*) pressures obtained by means of a single flow-directed thermistor catheter. The time interval between the two measurements of cardiac output (*CO*) was 31 seconds.

Injection into the right side of the heart with sampling in the aorta can be used only in the absence of pulmonary edema or congestion. Injection into the left atrium with sampling in the aorta is seldom practical.

MEASUREMENT OF TEMPERATURE

Thermistors, most commonly used for measurement and recording of intravascular temperature, are semiconductors with high negative temperature coefficient of resistance (3.5 to 4% change in resistance for a 1°C change in temperature). Although their temperature-resistance relationship is almost logarithmic, it can be considered linear over the small range in which the intravascular temperature changes during cardiac output determination. The relationship can be linearized over a much wider range (for instance from 0 to 40°C) by shunting the thermistor with a fixed resistance of appropriate value.

Since the temperature-resistance characteristics are individually different even in thermistors of the same type, they must be determined in each thermistor for the range of temperature in which it will be used. This can be done simply by measuring the resistance of the thermistor at different temperatures in a water bath with a sensitive thermometer. The temperature-resistance relationship is stable over long periods of time, according to James, Paul, and Wessel (1965) and also in our experience.

The resistance is measured by a Wheatstone bridge into which the thermistor is incorporated as one arm. If the thermistor is overheated, its resistance will change not only with the temperature but also with the velocity of blood; therefore, the voltage across the thermistor must be kept low (below 0.1 v). The off-balance output of the bridge (usually between 1 and 2 mv per 1°C change in temperature) is fed into a direct current amplifier with a direct current gain of at least 10^4.

Thermocouples have the advantage of an essentially linear calibration curve. However, they produce only a small electromotive force of approximately 40 $\mu v/°C$ and require, if an absolute measure of temperature is needed, a reference junction to be maintained at a very accurately controlled temperature.

Because of their small size, thermistors and thermocouples are eminently suitable for recording rapid changes in temperature. The dynamic response characteristics of a catheter-mounted thermistor depends on the size of the thermistor, the thermal conductivity and thickness of the insulating material, and the proportion of thermistor surface exposed to contact with blood (thermistors less embedded respond faster).

CALIBRATION OF THERMODILUTION CURVES

Changes in intravascular temperature, induced by injection of cold fluid, will cause corresponding changes in the resistance of the thermistor. This signal can be measured precisely by relating it to the signal obtained by introduction of a fixed resistance into the thermistor circuit. The temperature change represented by the calibrating resistance is determined from the temperature-resistance relationship of the thermistor in the actual range of temperature (usually around

37°C). A calibration factor (mm/°C) can be obtained from the ratios mm/ohm and °C/ohm. The calibration is independent of changes in bridge voltage and amplifier or recorder sensitivity (Fronek and Ganz, 1960; James *et al.*, 1965).

QUANTITATION OF THERMODILUTION CURVES BY SINGLE-PURPOSE ANALOG COMPUTERS

The thermodilution technique permits multiple determinations of cardiac output in rapid succession. However, the calculation of large numbers of curves is time consuming. For this reason and because the instant knowledge of the magnitude of cardiac output is of great value, particularly in intensive and coronary care units, single purpose electronic computers were constructed with instantaneous digital readout of the cardiac output. The area of the thermodilution curve is determined in two ways. The first way is by integration after exponential extrapolation of the downslope (Wessel *et al.*, 1966*a*). Second, because recirculation of thermal indicator is small, it is possible to eliminate the more complex extrapolation system by stopping the integration when the downslope reaches a certain value [10% of the peak concentration according to Phillips, Davila, and Sanmarco (1970)]. Small, portable battery powered computers, weighing less than 5 kg, are now available for bedside use in intensive and coronary care units and cardiac catheterization laboratories (Edwards Laboratories, Santa Ana, California).

VALIDATION OF METHOD

Studies on the accuracy of the thermodilution method include comparisons with the direct Fick method (Fegler, 1954 and 1957; Rapaport and Ketterer, 1958; Fronek and Ganz, 1960; Goodyer *et al.*, 1959; Wessel *et al.*, 1971; Evonuk *et al.*, 1961; Khalil, Richardson, and Guyton, 1966; Klussmann *et al.*, 1959; Enghoff *et al.*, 1970), with the dye-dilution method (Evonuk *et al.*, 1961; Fegler, 1954; Ganz *et al.*, 1970; Goodyer *et al.*, 1959; Heimburg *et al.*, 1964; Kochsiek *et al.*, 1964; Wessel *et al.*, 1971; Enghoff *et al.*, 1970; Olsson *et al.*, 1970; Klussmann *et al.*, 1959; Silove, Cantez, and Wells, 1971), and by extracorporeal pump (Lüthy and Galletti, 1966).

All of these studies showed a close correlation with no significant systematic deviation, except for the study of Evonuk *et al.* (1961), in which estimates of cardiac output obtained from the aortic thermodilution curves were 2% higher than those obtained from the pulmonary artery and 4% higher than those obtained by the dye-dilution method.

PATIENT SAFETY

Since the thermistor is introduced into the heart or its vicinity, the possibility of an insulation failure must be considered. Faulty insulation can be recognized from the marked, random fluctuation and drift in the base line temperature. The electric hazard to the patient in case of insulation failure can be minimized by the following measures.

1. The voltage source and the fixed resistances in the bridge are selected in such a way that the maximal current in case of leakage will not exceed 10 microamperes.

2. Isolation techniques are used in the equipment (isolation amplifier, isolation transformer, etc.).

3. The insulation of the thermistor is periodically checked (Fronek and Ganz, 1960). One terminal of the thermistor is connected through a microammeter to a source of DC voltage, whereas the second pole of the source is immersed in saline. If a deflection appears on immersing the thermistor into the saline, the insulation can be considered faulty.

4. In case of insulation failure, the thermistor is disconnected at once.

CLINICAL USEFULNESS

Although the use of heat as indicator is associated with special problems caused by its diffusiveness, comparative studies demonstrated that thermodilution is as accurate as the other generally accepted methods of cardiac output determination. From the clinical standpoint, thermodilution offers a number of advantages.

1. Intravascular sampling without withdrawal of blood from an artery or from the central circulation.

2. Simple and very accurate internal electrical calibration without removal of blood.

3. Minimal recirculation.

4. Repeatability of measurements in rapid succession (approximately twice in a minute).

5. Applicability in infants and children.

Cardiac output determinations can be performed with a single flow-directed catheter, which can be passed into the pulmonary artery without fluoroscopic control, rapidly and safely (Ganz *et al.,* 1970; Ganz and Swan, 1972; Forrester *et al.,* 1972) and permits also the measurement of the pulmonary arterial, pulmonary wedge, and central venous pressures.

MEASUREMENT OF FLOW IN SINGLE BLOOD VESSELS BY THERMODILUTION (LOCAL THERMODILUTION)

Since temperature sensors are small enough to be mounted on catheters and so inserted into vessels of relatively small diameter, thermodilution, from a technical point of view, is particularly suitable for measuring blood flow in single vessels.

The *principle* of the method is as follows. A change in intravascular heat content is induced at one point in the vessel, and the resultant change in temperature is detected downstream. Uniform temperature over the cross-section of the vessel at the point of detection is a basic requirement for the validity of the method. When cardiac output is measured, mixing is the result of the pumping action of the heart. In local thermodilution, mixing must be attained by the kinetic energy of the injectate. When a heating device is used and the kinetic energy of the injectate is therefore not available, a special stirring device must be applied for uniform distribution of the heat (Afonso, 1966). The thermal indicator can be introduced either as a single, nearly instantaneous injection or by continuous constant rate injection.

Measurement of Blood Flow in Veins by the Continuous Injection Method

When cold fluid is introduced by continuous constant rate injection, after some delay, a steady state is reached in which the amount of indicator leaving the mixing volume is equal to the amount entering the mixing volume during the same interval. Assuming that there is no heat lost from the system or that the loss is negligible, the heat gained by the injectate must be equal to the heat lost by the blood. The following heat balance equation can, therefore, be written:

$$\underset{\text{(heat lost by blood)}}{V_B \times S_B \times C_B \times (T_B - T_M)} = \underset{\text{(heat gained by injectate)}}{V_I \times S_I \times C_I \times (T_M - T_I)}$$

where T_M = temperature of blood-indicator mixture in °C, after a steady state has been reached, V_B, V_I = volumes of blood and injectate, respectively, entering the mixing volume in ml/min, $T_M - T_I$ can be written as $(T_B - T_I) - (T_B - T_M)$.

The formula for calculation of venous flow by the continuous thermodilution method is therefore:

$$V_B = V_I \times \frac{S_I \times C_I}{S_B \times C_B} \times \left(\frac{T_B - T_I}{T_B - T_M} - 1\right) \text{ml/min.}$$

Injection Rate

The kinetic energy of the injectate must be sufficiently high to induce mixing by turbulence over the whole cross-section of the vessel. The conditions of mixing have not been precisely mathematically defined; it is, however, known that the kinetic energy necessary for mixing is directly related to the velocity of blood flow and to the diameter of the vessel. The spatial orientation of the stream of injectate relative to the direction of the blood stream and the vessel wall also affect the degree of mixing.

Dynamic Response Characteristics

The continuous constant injection method can be used for measurement of transient changes in blood flow. The dynamic response of the method is characterized by the time constant of the method, that is the time interval in which the temperature of the blood-injectate mixture (T_M) reaches 63% of the maximum temperature change after a sudden change in blood flow. The time constant of the method depends not only on the time constant of the temperature sensing device, but also on hydrodynamic factors which are included in the time constant of the mixing volume:

$$\frac{\text{mixing volume (ml)}}{\text{total flow of blood and injectate (ml/sec)}}$$

(Lowe, 1968,*a* and *b*). The time constant of the sensor can be kept around 0.1 second. The time constant of the mixing volume is usually higher and is, therefore, the factor limiting the dynamic response of the method. It has been found that changes in flow 3 seconds or more in cycle length can be measured with a reasonable accuracy (Ganz *et al.*, 1971*b*).

Thermodilution Catheter

Catheters for the measurement of blood flow by the continuous thermodilution method usually carry two thermistors: one on the outside for determination of T_B and T_M, the second one in the lumen for determination of T_I. In catheters inserted downstream, the negative heat lost through the wall of the catheter to the blood may reach the mixing volume. The thermistor for measurement of T_I should be, in

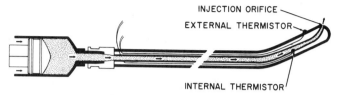

Figure 14.3 Diagrammatic representation of the catheter for measurement of coronary sinus blood flow in man.

this case, at the entrance into the vascular section from which the blood flows into the mixing volume. Thermal insulation of the injectate from blood will minimize the heat transfer and also the possible error from this heat transfer. The effective injectate temperature can also be determined by calorimetry or by external measurement of the injectate temperature under conditions duplicating the actual measurement, as discussed in connection with the measurement of cardiac output. Catheters have been constructed which can be inserted percutaneously to measure blood flow in the internal jugular, popliteal, iliac, and renal veins in human subjects (Dowsett and Lowe, 1964; Clark and Cotton, 1966; Lowe and Dowsett, 1967; Cotton and Richards, 1968; Clark, 1968; Richards, 1970; Hlavová, Linhart, and Přerovský, 1970; Hornych, Brod, and Slechta, 1971; Wilson and Halsey, 1970 and 1971). A somewhat larger catheter is employed for estimation of blood flow in the coronary sinus in man (Ganz et al., 1971b). The catheter and an actual tracing recorded in the coronary sinus are shown on Figs. 14.3 and 14.4.

Accuracy

The accuracy of the continuous constant injection thermodilution method was confirmed by comparison with direct methods in model experiments and in vivo (Pávek, Boska, and Selecký, 1964; Clark, 1968; Ganz et al., 1971b; Linzell, 1966; Reynolds, Linzell, and Rasmussen, 1968).

Measurement of Blood Flow in Veins by the Single Injection Method

If a bolus of cold fluid is injected at one point of a vein, a temperature-time curve is recorded downstream, similar in shape to that recorded during measurement of cardiac output (Fegler and Hill, 1958). The computation of blood flow is complicated by the fact that part of the temperature-time curve is inscribed during the injection, when the flow downstream to the site of injection is increased by a portion of the injectate. The second part of the curve is inscribed when the flow rate downstream has returned to its preinjection level. A complex formula for calculation of flow was derived by solving a pair of heat balance equations, which requires determination of two areas under the two parts of the curve (Fronek and Ganz, 1960; Hosie, 1962; Hornych *et al.*, 1971; White *et al.*, 1967; Warembourg *et al.*, 1969a and b).

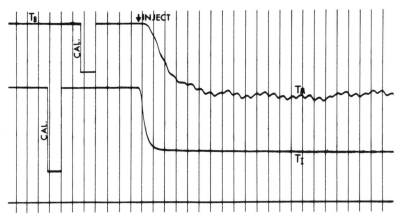

Figure 14.4 An actual tracing recorded in the coronary sinus. During the constant rate injection of the 5% dextrose solution, the thermistor in the lumen indicates the effective temperature of the injectate, T_I, whereas the external thermistor indicates fall in the intravascular temperature from the preinjection temperature of blood (T_B) to the temperature of the blood-injectate mixture (T_M). *Cal*, calibration.

Measurement of Arterial Blood Flow by the Continuous Injection Method

The continuous injection method is not suitable for measurement of arterial blood flow because the injection causes marked and unpredictable changes in blood flow. The kinetic energy of the injection tends to reduce the blood flow due to the fact that in arteries the resistance upstream is much higher than downstream: after a few seconds the reduction in blood flow is followed by a marked increase, due to arteriolar dilation [secondary probably to hemolysis (Andres *et al.*, 1954) and possibly other factors]. These changes are probably negligible, when the continuous injection method is applied in the aorta, as can be judged from the accuracy of measurements (Peterson *et al.*, 1954; Grace *et al.*, 1957).

Measurement of Arterial Flow by the Single Injection Method

Since the distance between the site of injection and the site of detection is short, in case of a bolus injection, the temperature-time curve can be recorded before the effects of the injection on the arterioles become manifest. If the kinetic energy of the injectate is not excessive (Fronek and Ganz, 1960), the injectate will displace an equal volume of blood, i.e. the flow rate downstream to the site of injection will remain unchanged during the injection. From the heat balance equation and with respect to the effect of the injection on the blood flow, a formula for estimation of arterial blood flow can be derived, which is essentially identical with that used for estimation of cardiac output (Fronek and Ganz, 1960). The single injection method has its theoretical limitations in that it requires the assumption of a constant flow (Zierler, 1962; Cropp and Burton, 1966), which is seldom present under physiological conditions. If the thermodilution curve extends over several cycles of the blood flow fluctuation, the flow estimate may be an accurate reflection of the mean blood flow (Sherman, 1960; Hosie, 1962).

As might be apparent from the preceding discussion, the local thermodilution technique is most suitable for the measurement of blood flow in veins. The continuous constant injection technique is theoretically preferable, although technically more demanding. Intravascular sampling, simple and very accurate electrical calibration, and the small size of flow probes for application in humans are the main advantages of the local thermodilution technique.

REFERENCES

Afonso, S. 1966. A thermodilution flowmeter. J. Appl. Physiol. 21: 1883–1886.

Afonso, S., G. G. Rowe, C. A. Castillo, and C. W. Crumpton. 1962. Intravascular and intracardiac blood temperatures in man. J. Appl. Physiol. 17:706–708.

Andres, R., K. L. Zierler, M. Anderson, W. N. Stainsby, G. Cader, A. S. Ghrayyib, and J. L. Lilienthal, Jr. 1954. Measurement of blood flow and volume in the forearm of man; with notes on the theory of indicator-dilution and on production of turbulence, hemolysis and vasodilatation by intravascular injection. J. Clin. Invest. 33:482–504.

Barankay, T., T. Jancsó, S. Nagy, and G. Pétri. 1970. Cardiac output estimation by a thermodilution method involving intravascular heating and thermistor recording. Acta Physiol. Acad. Sci. Hung. 38:167–173.

Branthwaite, M. A., and R. D. Bradley. 1968. Measurement of cardiac output by thermal dilution in man. J. Appl. Physiol. 24:434–438.

Bucher, K., and H. Emmenegger. 1952. Besonderheiten der Lungen-durchblutung. Arch. Kreislaufforsch. 18:94–99.

Carlsten, A., and G. Grimby. 1958. Rapid changes in human right heart blood temperature at variations in venous return. Scand. J. Clin. Lab. Invest. 10:397–401.

Clark, C. 1968. A local thermal dilution flowmeter for the measurement of venous blood flow in man. Med. Biol. Eng. 6:133–142.

Clark, C., and L. T. Cotton. 1966. Venous flow measurement in man using a local thermal dilution flowmeter. Brit. J. Surg. 53:987.

Cotton, L. T., and J. Richards. 1968. The measurement of venous blood flow by the method of local thermal dilution, p. 11–17. In W. H. Bain and A. M. Harper (eds.), Blood Flow through Organs and Tissues. Proc. Int. Conf., Glasgow, March 1967. E&S Livingstone Ltd., Edinburgh and London.

Cropp, G. A., and A. C. Burton. 1966. Theoretical considerations and model experiments on the validity of indicator dilution methods for measurements of variable flow. Circ. Res. 18:26–48.

Dowsett, D. J., and R. D. Lowe. 1964. Measurement of blood flow by local thermodilution. J. Physiol. 172:138–158.

Eichna, L. W., A. R. Berger, B. Rader, and W. H. Becker. 1951. Comparison of intracardiac and intravascular temperatures with rectal temperatures in man. J. Clin. Invest. 30:353–359.

Enghoff, E., M. Michaëlsson, K. Pávek, and S. Sjögren. 1970. A comparison between the thermal dilution method and the direct Fick and the dye dilution methods for cardiac output measurements in man. Acta Soc. Med. Upsal. 75:157–170.

Evonuk, E., C. J. Imig, W. Greenfield, and J. W. Eckstein. 1961. Cardiac output measured by thermal dilution of room temperature injectate. J. Appl. Physiol. 16:271–275.

Fegler, G. 1954. Measurement of cardiac output in anaesthetized animals by a thermodilution method. Quart. J. Exp. Physiol. 39:153–164.

Fegler, G. 1957. The reliability of the thermodilution method for determination of the cardiac output and the blood flow in central veins. Quart. J. Exp. Physiol. 42:254–266.

Fegler, G., and K. J. Hill. 1958. Measurement of blood flow and heat production in the splanchnic region of the anaesthetized sheep. Quart. J. Exp. Physiol. 43:189–196.

Forrester, J. S., W. Ganz, G. Diamond, T. McHugh, D. W. Chonette, and H. J. C. Swan. 1972. Thermodilution cardiac output determination with a single flow-directed catheter. Amer. Heart J. 83:306–311.

Fronek, A., and V. Ganz. 1960. Measurement of flow in single blood vessels including cardiac output by local thermodilution. Circ. Res. 8:175–182.

Ganz, W., R. Donoso, H. S. Marcus, J. S. Forrester, and H. J. C. Swan. 1971a. A new technique for measurement of cardiac output by thermodilution in man. Amer. J. Cardiol. 27:392–396.

Ganz, W., W. J. Mandel, D. Chonette, and H. J. C. Swan. 1970. A multipurpose flow-directed catheter for use in the acutely ill patient. Circulation 41 & 42 (Suppl. 3):65.

Ganz, W., and H. J. C. Swan. 1972. Measurement of blood flow by thermodilution in man. Amer. J. Cardiol. 29:241–246.

Ganz, W., and H. J. C. Swan. 1973. Measurement of cardiac output by thermodilution in patients with acute myocardial infarction. In E. Corday and H. J. C. Swan (eds.), New Perspectives in the Diagnosis and Management of Acute Myocardial Infarction. The Williams & Wilkins Co., Baltimore.

Ganz, W., K. Tamura, H. S. Marcus, R. Donoso, S. Yoshida, and H. J. C. Swan. 1971b. Measurement of coronary sinus blood flow by continuous thermodilution in man. Circulation 44:181–195.

Good, A. L., and A. F. Sellers. 1957. Temperature changes in the blood of the pulmonary artery and left atrium of dogs during exposure to extreme cold. Amer. J. Physiol. 188:447–450.

Goodyer, A. V. N., A. Huvos, W. F. Eckhardt, and R. H. Ostberg. 1959. Thermal dilution curves in the intact animal. Circ. Res. 7:432–441.

Grace, B. J., I. J. Fox, W. P. Crowley, and E. H. Wood. 1957. Thoracic-aorta flow in man. J. Appl. Physiol. 11:405–418.

Heimburg, V. P., P. Kadelbach, K. Kochsiek, and J. Schumacher. 1964. Vergleichende Untersuchungen zwischen dem Kälte- und dem Farbstoffverdünnungsverfahren. Z. Kreislaufforsch. 53:1230–1245.

Hlavová, A., J. Linhart, and I. Přerovský. 1970. The blood flow measurement in the human external iliac vein. Cor Vasa 12:193–203.

Hornych, A., J. Brod, and V. Šlechta. 1971. The measurement of the renal venous outflow in man by the local thermodilution method. Nephron 8:17–32.

Horvath, S. M., A. Rubin, and E. L. Foltz. 1950. Thermal gradients in the vascular system. Amer. J. Physiol. 161:316–322.

Hosie, K. F. 1962. Thermal-dilution technics. Circ. Res. 10:491–504.

James, G. W., M. H. Paul, and H. U. Wessel. 1965. Thermal dilution: instrumentation with thermistors. J. Appl. Physiol. 20:547–552.

Khalil, H. H. 1963. Determination of cardiac output in man by a new method based on thermodilution. Lancet 1:1352–1354.

Khalil, H. H., T. Q. Richardson, and A. C. Guyton. 1966. Measurement of cardiac output by thermal-dilution and direct Fick methods in dogs. J. Appl. Physiol. 21:1131–1135.

Klussmann, F. W., W. Koenig, and A. Lütcke. 1959. Über die "Thermodilution"—Methode zur Bestimmung des Herzzeitvolumens am narkotisierten und unnarkotisierten Hund. Pflügers Arch. Ges. Physiol. 269: 392–402.

Kochsiek, V. K., P. Heimburg, P. Kadelbach, and J. Schumacher. 1964. Vergleichende Untersuchungen zwischen dem Kälte- und dem Farbstoffverdünnungsverfahren. Z. Kreislaufforsch. 53:1246–1253.

Linzell, J. L. 1966. Measurement of venous flow by continuous thermodilution and its application to measurement of mammary blood flow in the goat. Circ. Res. 18:745–754.

Lowe, R. D. 1968a. Use of a local indicator dilution technique for the measurement of oscillatory flow. Circ. Res. 22:49–56.

Lowe, R. D. 1968b. Problems in the measurement of blood flow by thermal dilution, pp. 6–10. In W. H. Bain and A. M. Harper (eds.), Blood Flow through Organs and Tissues. Proc. Int. Conf., Glasgow, March 1967. E&S Livingstone Ltd., Edinburgh and London.

Lowe, R. D., and D. J. Dowsett. 1967. A catheter probe for measurement of jugular venous blood flow in man by thermal dilution. J. Appl. Physiol. 23:1001–1003.

Lüthy, E. 1961. Herzminutenvolumenbestimmungen mit Hilfe von Thermistoren. Arztl. Forsch. 15:448–454.

Lüthy, E. 1962. Die Haemodynamik des suffizienten und insuffizienten rechten Herzens. S. Karger, Basel and New York.

Lüthy, E., and P. M. Galletti. 1966. In vivo evaluation of the thermodilution technique for measuring cardiac output. Helv. Physiol. Acta 24: 15–23.

Lüthy, E., and W. Rutishauser. 1961. Die "Thermodilution" Methode. Cardiologia 38:183–189.

Mather, G. W., G. G. Nahas, and A. Hemingway. 1953. Temperature changes of pulmonary blood during exposure to cold. Amer. J. Physiol. 173:390–392.

Mellette, H. C. 1950. Skin, rectal and intravascular temperature adjustments in exercise. Amer. J. Physiol. 163:734.

Mendlowitz, M. 1948. The specific heat of human blood. Science 107: 97–98.

Mohammed, S., C. J. Imig, E. J. Greenfield, and J. W. Eckstein. 1963. Thermal indicator sampling and injection sites for cardiac output. J. Appl. Physiol. 18:742–745.

Olsson, B., J. Pool, P. Vandermoten, E. Varnauskas, and R. Wassen. 1970. Validity and reproducibility of determination of cardiac output by thermodilution in man. Cardiology 55:136–148.

Pávek, K., D. Boška, and F. V. Selecký. 1964. Measurement of cardiac output by thermodilution with constant rate injection of indicator. Circ. Res. 15:311–319.

Peterson, L. H., M. Helrich, L. Greene, C. Taylor, and G. Choquette. 1964. Measurement of left ventricular output. J. Appl. Physiol. 7:258–270.

Phillips, C. M., J. C. Davila, and M. E. Sanmarco. 1970. Measurement of cardiac output by thermal dilution. II. A new computer for rapid convenient determinations. Med. Res. Eng., pp. 25–29.

Rapaport, E., and S. G. Ketterer. 1958. The measurement of cardiac output by the thermodilution method. Clin. Res. 6:214.

Reynolds, M., J. L. Linzell, and F. Rasmussen. 1968. Comparison of four methods for measuring mammary blood flow in conscious goats. Amer. J. Physiol. 214:1415–1424.

Richards, J. B. 1970. A thermally insulated catheter for siting a local thermal dilution blood flow probe. Bio-Med. Eng. 5:65–75.

Rubenstein, E., R. C. Pardee, and F. Eldridge. 1960. Alveolar-capillary temperature. J. Appl. Physiol. 15:10–12.

Sherman, H. 1960. On the theory of indicator-dilution methods under varying blood flow conditions. Bull. Math. Biophys. 22:417–424.

Silove, E. D., T. Cantez, and B. G. Wells. 1971. Thermodilution measurement of left and right ventricular outputs. Cardiovasc. Res. 5:174–177.

Singh, R., A. J. Ranieri, Jr., H. R. Vest, Jr., D. L. Bowers, and J. F. Dammann, Jr. 1970. Simultaneous determinations of cardiac output by thermal dilution, fiberoptic and dye dilution methods. Amer. J. Cardiol. 25:579–587.

Swan, H. J. C., W. Ganz, J. Forrester, H. Marcus, G. Diamond, and D. Chonette. 1970. Catheterization of the heart in man with use of a flow-directed balloon-tipped catheter. New Engl. J. Med. 283–447–451.

Warembourg, H., Y. Houdas, M. E. Bertrand, M. Delomez, and J. Y. Ketelers. 1969a. La mesure du débit coronaire par la methode de thermodilution. Arch. Mal. Coeur 2:173–182.

Warembourg, H., Y. Houdas, M. E. Bertrand, J. Y. Ketelers, and M. Delomez. 1969b. La thermodilution en cardiologie. Ann. Cardiol. Angéiol. 18:23–31.

Warembourg, H., Y. Houdas, M. Delomez, M. Bertrand, and J. Y. Ketelers. 1968. La technique de thermo-dilution: résultats préliminaires dans la mesure de débits sanguins. Arch. Mal. Coeur 10:1401–1413.

Wessel, H. U., G. W. James, and M. H. Paul. 1966a. Quantitation of indicator dilution curves by special-purpose analog computer. Med. Res. Eng. 5:16–21.

Wessel, H. U., G. W. James, and M. H. Paul. 1966b. Effects of respiration and circulation on central blood temperature of the dog. Amer. J. Physiol. 211:1403–1412.

Wessel, H. U., M. H. Paul, G. W. James, and A. R. Grahn. 1971. Limitations of thermal dilution curves for cardiac output determinations. J. Appl. Physiol. 30:643–652.

White, S. W., J. P. Chalmers, R. Hilder, and P. I. Korner. 1967. Local thermodilution method for measuring blood flow in the portal and renal veins of the unanaesthetized rabbit. Aust. J. Exp. Biol. Med. 45:453–468.

Wilson, E. M., and J. H. Halsey, Jr. 1970. Bilateral jugular venous blood flow by thermal dilution. Stroke 1:348–355.

Wilson, E. M., and J. H. Halsey, Jr. 1971. An improved thermal dilution method for measuring jugular venous flow. Stroke 2:128–138.

Zierler, K. L. 1962. Theoretical basis of indicator-dilution methods for measuring flow and volume. Circ. Res. 10:393–407.

chapter 15/ Fiberoptics in Cardiac Catheterization I. Theoretical Considerations

Michael L. Polanyi
Research Laboratory of American Optical Company/
Framingham, Massachusetts

The *in vivo* measurements of dye concentration and oxygen saturation in free flowing blood are performed at the tip of optical fiber catheters by using light reflection techniques originated by Brinkman and Zijlstra in 1949 for *in vitro* (cuvette) measurement. The main difference between cuvette and catheter tip sensing is that when measuring in cuvette a steady blood flow pattern is created with a pump or a stirring device, whereas *in vivo* the blood flow around the sensor has an erratic pulsatile character. The flow patterns prevailing in these two systems are such that in the cuvette it is possible to correlate dye concentration (or oxygen saturation) with the variation of reflected light intensity at a single (measuring) wave length (λ_m), whereas *in vivo*, to minimize the "flow artifacts," the dye concentration (or oxygen saturation) is correlated to the ratio of two reflected light intensities, *viz.* a measuring wave length (λ_m), as in the cuvette case, and a reference wave length (λ_r). The reference wave length is chosen to be affected as little as possible by changes in dye concentration or oxygen saturation (Polanyi and Hehir, 1962).

Of late, several optical fiber *in vivo* oximeters, of relatively small physical dimension and moderate cost have been reported (Johnson *et al.*, 1971; Degonde, 1971; Krauss *et al.*, 1972); these instruments can be used for catheterization with standard type catheters and for monitoring with balloon catheters (Swan-Ganz type).

The "fiberoptics" method has the unique advantages of not requiring blood sampling for either dye or oxygen saturation determination and of giving an "on-line" fast response trace. The name fiberoptics derives from the optical fibers, or "light pipes," used to take the light to and from the tip of the catheter, the site where dye (or saturation) is sensed.

The reflection technique here presented is similar but not identical to the transmission technique used in cuvette densitometry. Since the cuvette method is well understood, the two will be compared here in some detail. In cuvette densitometry (Fig. 15.1*A*), the dye concentration, C_d, in the blood-dye mixture, is obtained from the light intensity at wave length, λ_m, transmitted by the mixture flowing through a cuvette. In the fiberoptics technique, C_d is derived by the intensity of light, at the same wave length, λ_m, scattered back or diffusely reflected by the blood-dye mixture flowing freely past the tip of the catheter (Fig. 15.1*B*).

The correlation between the transmitted light, I_t, and C_d is given by Beer's law

$$C_d = K_t \log \frac{I_0(\lambda_m)}{I_t(\lambda_m)}, \qquad (15.1)[1]$$

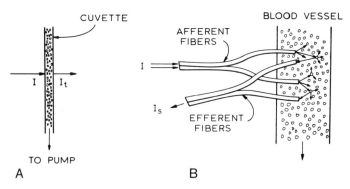

Figure 15.1 Schematic arrangements for continuous dye concentration or oxygen saturation measurements. *A*, transmission (*in vitro*); *B*, diffused reflection (*in vivo*).

[1] $K_t = (\xi_d \cdot \rho)^{-1}$, where ρ is the optical path and ξ_d is the extinction coefficient of the dye.

where K_t is constant and I_0 is the light transmitted by the blood in the absence of dye at wave length λ_m.

In the fiberoptics method, the correlation between dye concentration C_d and the intensity of the backscattered light I_s is given by

$$C_d = K_s \left(\frac{I_s^0(\lambda_m)}{I_s(\lambda_m)} - 1 \right), \qquad (15.2)^2$$

where K_s is a constant and I_s^0 is the intensity of backscattered light in the absence of dye at wave length λ_m.

Equations (15.1) and (15.2) are, as mentioned, similar. Their difference is that in transmission C_d is a linear function of the logarithm of the inverse of transmitted light intensity, whereas in reflection C_d is a linear function of the inverse of the backscattered light intensity. The instrumentation needed in the two cases is quite different, because in equation (15.1) C_d is obtained from a single measurement, whereas in (15.2) it is obtained from two measurements. In transmission, the term I_0 can be considered a constant; the stability of I_0, the base line, is due to the steady flow of the blood through the cuvette during the registration of a dye curve. In fiberoptics technique, I_s^0, the backscattered light intensity in the absence of dye, cannot be considered constant during the registration of a dye curve. This is because the flow conditions *in vivo,* at the tip of the catheter, are constantly changing, due to the pulsatile character of the blood flow, and both I_s^0 and I_s are sensitive to the concentration and orientation of the blood cells in the immediate vicinity of the detector. The erratic changes in both I_s^0 and I_s preclude the possibility of correlating C_d with the single measured quantity I_s. Using an auxiliary wave length, λ_r, which is not absorbed by the dye and in first approximation has the same "flow effects" as λ_m, a measure of C_d is obtainable which is substantially free of "flow effects." With this substitution we have

$$C_d = K_s \left[\frac{I_s(\lambda_r)}{I_s(\lambda_m)} - 1 \right], \qquad (15.3)$$

where $I_s(\lambda_r)$ has been set equal to $I_s(\lambda_m)$ for $C_d = 0$ (for Cardiogreen, $\lambda_r \approx 900$ nm and $\lambda_m \approx 800$ nm). Equation (15.3) is the basic relationship used with the fiberoptics technique to obtain C_d from the measured light intensities $I_s(\lambda_r)$ and $I_s(\lambda_m)$.

A notable additional difference between the transmission case and the reflection case is the character of the constants K_t and K_s. K_t is a

²See Appendix I.

constant independent of blood cell concentration and of the geometry of the cuvette. K_s, on the other hand, depends on both cell concentration (hematocrit) and the geometry of the optical sensor. In terms of hematocrit, H, and cross-section, α, of the optical probe, K_s can be written as follows (see Appendix I):

$$K_s = K_1 H + K_2/\alpha,$$

where K_1 and K_2 are constants. This approximate formula shows that K_s is a function of hematocrit and that the "sensitivity" of the reflection method decreases with decreasing cross-section of the optical probe.

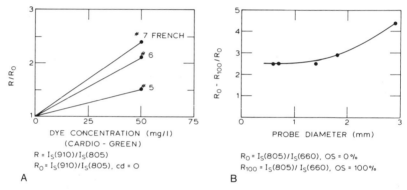

DYE CONCENTRATION (mg/l)
(CARDIO - GREEN)
$R = I_S(910)/I_S(805)$
$R_0 = I_S(910)/I_S(805)$, cd = O

A

PROBE DIAMETER (mm)
$R_0 = I_S(805)/I_S(660)$, OS = 0 %
$R_{100} = I_S(805)/I_S(660)$, OS = 100 %

B

Figure 15.2 Dependence of the change in ratio of the backscattered light intensities on the diameter of the optical probe. A, dye concentration with nos. 5, 6, and 7 French catheters; B, corresponding probe diameters.

The dependence of the sensitivity on α, as determined experimentally, is shown in Fig. 15.2. The dependence of the "sensitivity" on α imposes a lower limit on the diameter of the optical head (and on the catheter itself), below which residual flow artifacts dominate, and no useful determination of C_d is possible. This limit is about 0.5 mm for the optical probe and about 1 mm for the diameter of the catheter.

The range of validity of equation (15.3) in terms of dye concentration varies from 0 to 20 mg/liter to 0 to 100 mg/liter (Mook *et al.*, 1968), depending on the diameter of the optical probe, the diameter and the intermix of the individual afferent and efferent fibers, the ratio of afferent fibers to efferent fibers, and the optical characteristics of the fibers themselves (Fig. 15.3).

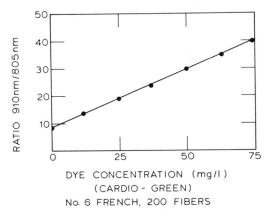

DYE CONCENTRATION (mg/l)
(CARDIO - GREEN)
No. 6 FRENCH, 200 FIBERS

Figure 15.3 Dye-dilution calibration curve. As a rule, the region of linearity between dye concentration and ratio of backscattered light intensities varies with the arrangement of the optical fibers at the tip of the catheter.

OXYGEN SATURATION

Since the same catheter and instrument that are used for dye-dilution measurement are used for oxygen saturation determination, a presentation of fiberoptics techniques would be incomplete without a discussion of fiberoptics techniques would be incomplete without a discussion of oxygen saturation (OS) measurements.

In dye dilution, as mentioned, the reference wave length, λ_r, is not absorbed by the dye, whereas the measuring wave length, λ_m, is strongly absorbed. Similarly, in OS, a reference wave length, λ_r, not effected by saturation, is used (isobestic wave length) in addition to the measuring wave length, λ_m, whose absorption is strongly influenced by saturation (Fig. 15.4). (For OS, $\lambda_r \approx 800$ nm; $\lambda_m \approx 660$ nm).

The formula used to correlate OS with backscattered light intensities, $I_s(\lambda_r)$ and $I_s(\lambda_m)$, is

$$OS = K_1 \frac{I_s(\lambda_r)}{I_s(\lambda_m)} + K_2, \qquad (15.4)[3]$$

where K_1 and K_2 are constants given in first approximation by

$$K_1 = \frac{A_1H + A_2/\alpha}{A_3H}, \qquad K_2 = \frac{A_4H + A_2/\alpha}{A_3H}.$$

$A_1, \ldots, A_4$ are constants, H is the hematocrit, and α is the diameter

[3]See Appendix II.

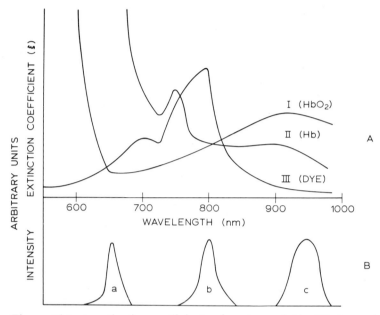

Figure 15.4 *A*, extinction coefficients of oxyhemoglobin (*I*), hemoglobin (*II*), and Indocyanine green (*III*). *B*, typical emission characteristics of light-emitting diodes used for dye dilution (*c* and *b*) and for oxygen saturation (*b* and *a*).

of the optical sensor. Both sensitivity and hematocrit effect are functions of the optical and geometrical characteristics of the probe (Fig. 15.2*B*).

The similarity of the formulas for OS in reflection (for $\alpha \to \infty$) (Polanyi and Hehir, 1960) and in transmission (e.g. Johnston, 1963) is illustrated below:

$$OS = A \frac{I_s(\lambda_r)}{I_s(\lambda_m)} + B \quad (\text{reflection } \alpha \to \infty)$$

$$OS = A \frac{\log I_t(\lambda_r)}{\log I_t(\lambda_m)} + B \quad (\text{transmission})$$

where

$$A = \frac{\bar{\xi}(\lambda_r)}{\xi_0(\lambda_m) - \xi_r(\lambda_m)}, \quad B = \frac{\bar{\xi}_r(\lambda_m)}{\xi_0(\lambda_m) - \xi_r(\lambda_m)}.$$

$\bar{\xi}(\lambda_r)$ is the common extinction coefficient at λ_r of oxyhemoglobin and reduced hemoglobin; $\xi_0(\lambda_m)$ and $\xi_r(\lambda_m)$ are the extinction co-efficients of oxyhemoglobin and reduced hemoglobin at λ_m.

In practice the constants K_1 and K_2 of equation (15.4) are determined experimentally; a typical calibration curve is shown in Fig. 15.5 (Enson, 1962.)

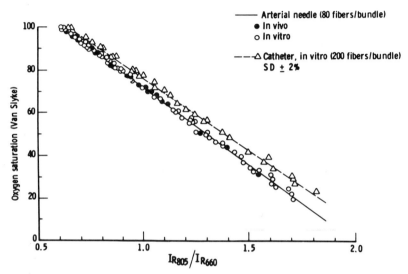

Figure 15.5 Typical oxygen saturation calibration curves. (Reprinted, with permission, from Enson *et al.* 1962. In vivo studies with an intravascular and intracardiac reflection oximeter. J. Appl. Physiol. 17:553).

FIBEROPTICS CATHETER

A typical fiberoptics catheter (Fig. 15.6) consists of a standard woven cardiac catheter (no. 6 or 7 French), through which are threaded a number of optical fibers and a plastic tube (0.022-inch inside diameter) for pressure measurements. Both the optical fibers and the pressure tube terminate at the distal end, where they are cemented, ground, and polished in a plane normal to the axis of the catheter (Fig. 15.7*A*). A cage is attached to the tip of the catheter to keep the tip about 3 mm axially from the vessel walls (insuring a view of the blood rather than the vessel walls) and to minimize any interference of the blood flow at the catheter tip. At the proximal end the optical fibers are cemented in a plug and divided into two parts: the afferent fibers, through which the light travels to the tip, and the efferent fibers, through which the light backscattered by the blood cells is returned. The pressure tube is terminated with a standard Luer-lok fitting. The overall length of the catheter is about 150 cm. Smaller diameter catheters, no. 5 French without pressure lumen and no. 3 French without pressure lumen or cage, are also made.

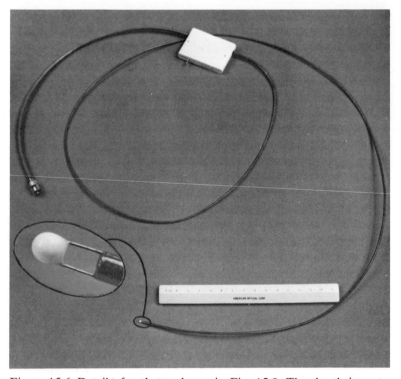

Figure 15.6 Detail of catheter shown in Fig. 15.8. The sheath is a standard no. 6 French cardiac catheter. The optical connections are made at the white plug-in box, with the pressure lumen continuing on to the Luer-lok connection. The caged tip is shown in detail (see text for explanation).

The optical fibers used in these catheters can be either glass or plastic. As a rule, to minimize breakage due to inadvertent sharp bending, the glass fibers used are of relatively small diameter, about 0.001 inch. The plastic fibers on the other hand can be as large as 0.01 inch without danger of damage by sharp bending. Two distinct

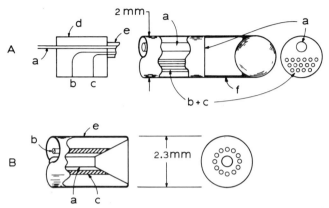

Figure 15.7 Schematic catheter tip arrangements. In *A*, the fibers end in a plane normal to the catheter. Contact between the fibers and the vessel walls is prevented by a cage which terminates in a plastic sphere. This sphere acts as a reflector to standardize the catheter before insertion in the vessel. In *B*, contact with the walls is prevented by receding the end of the fibers within the catheter. Symbols: *a*, pressure lumen; *b* and *c*, afferent and efferent optical fibers; *d*, proximal terminal assembly; *e*, standard cardiac catheter; *f*, cage.

considerations limit the minimum diameter of the catheters: (*a*) the decrease in sensitivity (which has already been mentioned), and (*b*) the decrease of light intensity available for the measurements (a technical limitation). In our experience the minimum number of fibers (for a 150-cm long catheter) are three 0.010-inch plastic fibers or about 150 of the 0.001-inch glass fibers.

As mentioned, the catheter tip must be prevented from contacting the vessel walls. This can be accomplished by so positioning the tip within the vascular system to avoid contact or by using a cage (Fig. 15.7*A*). Another method is to recess the fibers within the catheter (Johnston *et al.*, 1972) (Fig. 15.7*B*). When optical fibers are used with a balloon catheter (Swan-Ganz type), contact with the wall can be avoided by partial inflation of the balloon.

Various arrangements of the efferent and afferent fibers at the distal end can be used. These various arrangements may cause (*a*) different sensitivity to dye and oxygen saturation, (*b*) different linearity for dye and oxygen saturation, and (*c*) different sensitivity as a function of hematocrit. Sensitivity and linearity are also affected by minor changes in λ_r and λ_m.

THE INSTRUMENT (Fig. 15.8)

The dye concentration and the oxygen saturation are derived as a linear function from the ratio of two backscattered light intensities, $I_s(\lambda_r)$ and $I_s(\lambda_m)$. In practice, these two light intensities are not mea-

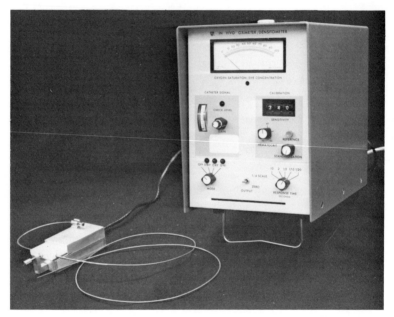

Figure 15.8 Solid state *in vivo* oximeter densitometer.

sured concurrently but are sampled sequentially in rapid succession. The advantage of sequential sampling is that a single set of fibers and a single light detector can be used for both wave lengths. In this way, possible drift of the detector will affect both wave lengths equally and have no influence on the ratio; the construction of the catheters is also simplified.

The light of the two wave lengths, λ_r and λ_m, can be obtained by using a tungsten lamp and two light filters (Gamble *et al.,* 1965), or by sequentially flashing two light-emitting diodes of appropriate wave lengths. The instrument used in most of the work described in the second part of this chapter uses a tungsten filament lamp and wave length isolating interference filters. The use of light-emitting diodes, which of late have become commercially available, greatly simplifies the construction and substantially decreases both the bulk and the cost of the instrumentation. Since the use of light-emitting diodes does not change the principle or the logic of the system, the instrument here described is of the light-emitting diode type, of which several have been built in our laboratory and elsewhere (Johnson *et al.,* 1971; Krauss *et al.,* 1972) and have proved viable in actual clinical work.

The instrument consists of an *optical head* which is relatively small and contains the light-emitting diodes, the light detector, and the preamplifiers (Fig. 15.9). In addition, the optical head has provisions for optical and mechanical coupling of the proximal end of the catheter to the sources of light and to the detector. The light from the

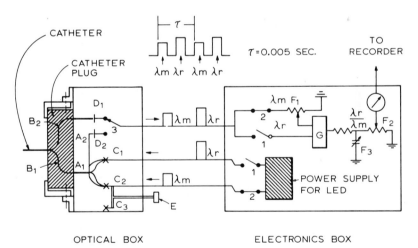

Figure 15.9 Block diagram of the electronics and optical head C_1, C_2, and C_3 are light-emitting diodes (*LED*); A_1 is a bifurcated optical fiber bundle which illuminates the afferent fibers (B_1) of the catheter. As shown the light-emitting diodes, C_1 and C_2, are used (say dye mode). By turning knob *E,* diodes C_1 and C_3 would be used (oxygen saturation mode). D_1 is the measuring photocell. D_2 is an auxiliary photocell which "looks" at the light-emitting diodes by means of the auxiliary fibers, A_2. By using switch *3,* the instrumental drift can be checked. A_2, D_2, and switch *3,* constitute the "by-pass" arrangement (see text).

light-emitting diodes is coupled to the catheter by means of a bifurcated optical fiber bundle. A mechanical linkage permits the interchange of light-emitting diodes from the dye mode wave lengths ($\lambda_r \approx 900$ nm, $\lambda_m \approx 800$ nm) to the OS mode wave lengths ($\lambda_r \approx 800$ nm, $\lambda_m \approx 660$ nm). The same interchange can be accomplished, at the cost of some loss of light intensity and without mechanized linkage, with a trifurcated fiber bundle.

A "by-pass" arrangement permits the operator to check and correct instrumental drifts at anytime during a run.

The *electronic package* contains, besides power supplies and the controls for calibration, a meter to read dye concentration or oxygen saturation, a second meter which indicates $I_s(\lambda_r)$ (used to monitor the performance of the catheter), and appropriate outlets for recorders and cardiac output computers.

The calibration controls, F_1 and F_2, are used to select the proper intercept and slope. In Fig. 15.9, F_1 is labeled "sensitivity" and F_2 "calibration." In addition, a mode selector (C_d or OS), a frequency response control F_3, and a "hematocrit" control are provided.

The meter which indicates $I_s(\lambda_r)$ provides an indication of the possible presence of clots or of "wall effect" which give rise to spurious readings. If the catheter tip is free of clots, the "flow effect" appears as meter oscillations of some amplitude in synchronism with the heart cycle. In case of clot or fiber formation on the catheter tip, these oscillations are damped or disappear all together. On the other hand, wall effect shows up either as lack of oscillation (if the tip is continuously against the vessel wall) or as a very large oscillation if the tip periodically hits the vessel wall.

CALIBRATION PROCEDURE

The calibration procedure of a catheter is done, as a rule, in two stages: a primary, nonsterile calibration and a secondary sterile calibration.

Since the output of the electronic ratio determining circuit is a linear function of the dye concentration C_d (or OS), the primary calibration procedure consists of adjusting F_2 and F_1 so that the meter reads correctly for two blood samples of known C_d (or OS). The secondary calibration consists of reproducing the setting as determined in the primary one. This is accomplished by setting F_2, the overall gain, at the value determined in the primary calibration, and setting F_1, with the catheter in a sample of blood of known C_d or OS, to read at the proper value on the meter. In the case of dye calibration, the blood within the circulation system, for which $C_d = 0$ before the dye injection, can be used. The second step of dye calibration consists of setting the meter to read zero, using F_1, with the catheter in the circulatory system, before the first dye injection. In the case of O.S. calibration, instead of using a blood of known saturation, an involved procedure, the second step can be accomplished with an auxiliary sterile standard, such as the white ball of the cage. The second step then, for O.S. calibration, consists of setting the meter to read the "equivalent" saturation of the auxiliary standard, using F_1, prior to the insertion of the catheter itself in the circulatory system (Fig. 15.4). In this case, the equivalent "saturation" reading of the auxiliary reflection standard must be established concurrently with the primary calibration. It should be noted that all *in vitro* measurements with blood must be performed with the blood in motion (magnetic stirrer or similar arrangement). Details of the procedure are given in Part II of this chapter.

APPENDIX I (DYE)

Cuvette or Transmission Technique

To calculate the light intensity, I_t, transmitted by a blood-dye mixture, the validity of Beer's law is assumed.

This assumption, due to the optically inhomogeneous character of unhemolyzed blood, is valid only for a limited dye concentration range (for Cardio-green, less than 45 mg/liter). From Beer's law,

$$I_t = I_0' e^{-\rho(\bar{\xi}HK' + \xi_d C_d)},$$

Where I_0' is the light intensity transmitted by the cuvette in the absence of blood; $\bar{\xi}$ and ξ_d are the extinction coefficients of hemoglobin and of dye, respectively, at the measuring wave length, λ_m; H is the hematocrit; K' is the mean cell hemoglobin concentration; ρ is the optical path; and C_d is the dye concentration ($\lambda_m \sim 800$ nm). Solving for C_d, we get

$$C_d = \frac{1}{\xi_d \cdot \rho} \left(\log \frac{I_0'}{I_t} - \rho \bar{\xi} K' H \right).$$

The sensitivity of this measurement is proportional to optical path ρ.

Substituting for I_0', the light intensity transmitted by the blood in the absence of dye, I_t^0, ($I_t^0 = I_t$ for $C_d = 0$), we have

$$C_d = K_t \log \frac{I_t^0}{I_t},$$

where K_t is a calibration constant ($K_t = 1/\xi_d \cdot \rho$).

Fiberoptics or Reflection Technique

The calculation of the backscattered light intensity, I_s, is based on the following approximations: (a) Light intensity, I_0, entering into the blood stream from the afferent fiber (A_1, Fig. 15.9), propagates in a straight line and is attenuated according to Beer's law. (b) The light intensity, $dI_s'(x)$, scattered by a layer, dx, is proportional to the thickness, dx, of the layer; to the concentration, H, of blood cells; to the intensity, $I_s(x)$, of the light reaching the scattering layer; and to a scattering coefficient, σ. (c) Of this scattered light (since the blood cells scatter light in all directions), the efferent fiber (A_2, Fig. 15.9) "sees" only a fraction proportional to $e^{-2Kx/\alpha}$, where α is the cross-section of A_2[4] and K is a constant. On the basis of these assumptions, the light intensity dI_s backscattered by layer dx and picked up by A_2 is

$$dI_s(x) \approx \sigma H I_0 e^{-2x(\bar{\xi}K'H + \xi_d C_d + K/\alpha)} dx.$$

The total backscattered light intensity picked up by A_2 is

$$I_s(\lambda_m) \approx \int_{x=0}^{x=\infty} dI_s(x) = \frac{H\sigma I_0}{2(\bar{\xi}K'H + \xi_d C_d + K/\alpha)}. \tag{15.5}$$

In the absence of dye ($C_d = 0$), the backscattered light intensity I_s^0 is

$$I_s^0 \approx \frac{\sigma I_0 H}{2(\bar{\xi}K'H + K/\alpha)}. \tag{15.6}$$

Solving (15.5) and (15.6) for C_d,

$$C_d = K_s \left(\frac{I_s^0}{I_s} - 1 \right)$$

where

---

[4] α can also be considered as the sum of the cross-sections of all efferent fibers, i.e. a linear function of the cross-section of the optical probe.

$$K_s = \frac{1}{\xi_d}\left(\frac{K}{\alpha} + \bar{\xi}K'H\right).$$

Here $\bar{\xi}$, ξ_d, C_d, K', and H, have the same meaning as in the transmission.

APPENDIX II (OXYGEN SATURATION)

It will be assumed that the only two pigments present in the blood are oxyhemoglobin and reduced hemoglobin. Referring to equation (15.6), the backscattered light intensity in the absence of dye is given by

$$I_s(\lambda_m) \approx \frac{\sigma H I_0(\lambda_m)}{2(\bar{\xi}K'H + K/\alpha)}, \quad \lambda_m \approx 800 \text{ nm},$$

where $\bar{\xi}$ is the extinction coefficient of hemoglobin at the isobestic wave length (~ 800 nm).

Choosing a measuring wave length for which the extinction coefficients of oxyhemoglobin and reduced hemoglobin are different, e.g. 660 nm (Fig. 15.4), we have in absence of dye

$$I_s(\lambda_m) \approx \frac{\sigma H I_0(\lambda_m)}{2[(C_0\xi_0 + C_r\xi_r)HK' + K/\alpha]} \tag{15.7}$$

$$\lambda_m = 660 \text{ nm}$$

where $I_0(\lambda_m)$ is the light intensity impinging upon the blood of the measuring wave length, C_0 and C_r are the concentrations of oxyhemoglobin and reduced hemoglobin ($C_0 + C_r = 1$), ξ_0 and ξ_r are the extinction coefficients of the two forms of hemoglobin, at the measuring wave length.

Using as reference wave length the isobestic wave length ($\lambda_r \simeq 800$ nm) we have

$$I_s(\lambda_r) \approx \frac{\sigma H I_0(\lambda_r)}{2(\bar{\xi}HK' + K/\alpha)}. \tag{15.8}$$

From equations (15.7) and (15.8), the oxygen saturation, C_0, is given by

$$C_0 = OS = K_1\frac{I_s(\lambda_r)}{I_s(\lambda_m)} + K_2,$$

where

$$K_1 = \frac{K'H\bar{\xi} + K/\alpha}{K'H\Delta\xi},$$

$$K_2 = \frac{K'H\xi_r + K/\alpha}{K'H\Delta\xi},$$

and $\Delta\xi = \xi_0 - \xi_r$ and $I_0(\lambda_m) = I_0(\lambda_r)$ (Rodrigo, 1953; Polanyi and Hehir, 1960).

ACKNOWLEDGMENTS

The author gratefully acknowledges the valuable collaboration and technical assistance of the following members of the research laboratory of the American Optical Corporation: R. E. Bober, A. DeSourdy, R. E. Innis, D. Ostrowski, and P. F. Putis.

REFERENCES

Brinkman, R., and W. G. Zijlstra. 1949. Determination and continuous registration of the % oxygen saturation in clinical conditions. Arch. Chir. Nedeerlandicum. 1:177–183.

Degonde, J. 1971. Oxymètre opttoelectronique a fibres de verres. Ann. Anesth. 13:525–529.

Enson, Y. 1962. In vivo studies with an intravascular and intracardiac reflection oximeter. J. Appl. Physiol. 17:552–558.

Gamble, W. J., P. G. Hugenholtz, R. G. Monroe, M. L. Polanyi, and A. S. Madas. 1965. The use of fiberoptics in clinical cardiac catheterization. I. Intracardiac oximetry. Circulation 31:328–343.

Johnson, C. C., R. D. Balm, D. D. Stewart, and W. E. Martin. 1971. A solid state fiber optics oximeter. J. Ass. Adv. Med. Inst. 5:77–83.

Johnson, C. C., P. W. Chung, and W. E. Martin. 1972. An improved fiberoptic catheter oximeter system. Proc. 25 ACEMB. Bal Harbour, Florida. 86.

Johnston, G. H. 1963. Standard Methods of Clinical Chemistry, D. Seligson (ed.). Academic Press, New York, p. 188.

Krauss, X. H., P. D. Verdouw, P. G. Hugenholtz, F. Hagemeijer, and M. L. Polanyi. 1972. Continuous monitoring of O_2 saturation in the intensive care unit. Circulation 45 (Suppl. II): 178.

Mook, G. A., P. Osypka, R. E. Sturm, and E. H. Wood. 1968. Fiberoptics reflection photometry in blood. Cardiovasc. Res. 2:199–209.

Polanyi, M. L., and R. M. Hehir. 1960. A new reflection oximeter. Rev. Scient. Instr. 31:401–403.

Polanyi, M. L., and R. M. Hehir. 1962. In vivo oximeter with fast dynamic response. Rev. Scient. Instr. 33:1050–1054.

Rodrigo, C. A. 1953. Determination of the oxygenation of blood in vitro by using reflected light. Amer. Heart J. 45:809–822.

chapter 15/ Fiberoptics in Cardiac Catheterization
II. Practical Applications

P. G. Hugenholtz/P. D. Verdouw/G. T. Meester
Thorax Center, Medical Faculty and University Hospital/
Erasmus University, Rotterdam, The Netherlands

Since the first report of fiberoptic catheterization in a clinical setting (Gamble and Hugenholtz, 1964), an extensive literature has accumulated on the subject (Hugenholtz *et al.*, 1965; Frommer *et al.*, 1965; Wagner *et al.*, 1968; Hugenholtz *et al.*, 1969*a,b;* Singh *et al.*, 1970). Enson *et al.* (1964) demonstrated the applicability of the fiberoptic principle to *in vivo* oximetry, but the complexity and cost of the method delayed its general acceptance in the laboratories of Europe and America. However, further fundamental work (Mook *et al.*, 1968; Mook *et al.*, 1971) and recent advances in technology, such as microcircuitry, solid state electronics, light-emitting diodes, and improved glass fiber design with float-in catheters have now eliminated the practical objections. As described at the end of the previous section, the most recent instrument incorporates all of the prerequisites of a simplified dye-dilution system for cardiac output providing a practical and useful tool even at the bedside of the critically ill patient.

THE CALIBRATION PROCEDURE

The following procedures apply to the equipment most frequently used. Only slight modifications are required for the most recent instrument. The components necessary for the calibration procedure consist of a sterilizable 5-inch nylon disc in the center of which a 5-ml glass cuvette containing a ½-inch long Teflon-coated magnetic stirrer is placed. The entire assembly may be autoclaved and attaches by adjustable screws to a Fisher Scientific model 14–511–2 magnetic stirring device. A metal, adjustable arm with a clamp permits the placement of the fiberoptic catheter tip inside the cuvette. Three milliliters of arterial blood are required to fill the cuvette to approximately 3 mm above the fiberoptic sensing tip. The stirrer is set to rotate at about 350 revolutions per minute. A 50-μl micropipette (no. 701; Hamilton Precision Company, Whittier, California) completes the equipment. All components, with the exception of the stirring device, can be assembled under sterile conditions by the operator before the procedure. All parts may be sterilized, preferably by the slow-cycle ethylene oxide gas method.

Calibration of the system is made in two steps. First, without dye in the 3-ml blood sample, the apparatus is set to zero with the "standardization" control. Next, 20 μl of a mixture containing approximately 4.8 mg of the Indocyanine green per ml of diluent is added to the 3 ml of the blood sample. The Cardio-green mixture utilized throughout the study is made by mixing 20 ml of distilled water and 1 cc of human albumin with 100 mg of Cardio-green powder. Previous experience has shown that the addition of albumin stabilizes the mixture for over a 4-hour period, the time usually necessary for a complete cardiac catheterization, and the amount, including spillage, suffices for up to 15 cardiac output determinations. After thorough mixing in the agitating chamber (which acts as a flow simulator), the deflection from the base line is recorded, preferably on a direct writing instrument, and corresponds to approximately 32 mg of Cardio-green per liter at 80% of the scale. The indicator needle on the oximeter is now set to one-half scale with the span control, to provide adequate range for high curves. This reading and the control value must again be recorded. Since the dye mixture added to the sample is the same as that which will be used for injection into the patient, a full scale deflection on the meter will correspond to a concentration of approximately 40 mg of Indocyanine green per liter of blood. This two-point calibration is considered adequate on the basis of previous *in vitro* calibration results, showing a virtually linear calibration curve for doses of Indocyanine green as high as 50 mg per liter of blood and an unchanged calibration with desaturation or when the hematocrits varied from 40 to 55%. An example of such a calibration curve is shown in Fig. 15.3 of Part I.

In all other respects, the calibration procedure corresponds to that described in previous studies utilizing the formula

$$\text{cardiac output (liters/minute)} = \frac{60 \times \text{calibration deflection} \times I}{\text{curve area}},$$

where calibration deflection = the average deflection from the base line in millimeters per microliter of dye added to 1 ml of blood (mm/μl of dye/liter of blood), curve area = the planimetrically obtained area of the inscribed curve (mm of deflection × seconds), 60 = seconds per minute, I = the volume of Indocyanine green mixture injected into the patient at the time of the dye curve (ml of dye).

THE INJECTION AND SAMPLING PROCEDURE

After the sensing catheter has been removed from the calibration device and placed in the desired location inside the cardiovascular system and after the instrument has been reset for zero dye reading (0 on the scale), injection of the same Cardio-green mixture in quantities of 1 ml for each cardiac output determination may be carried out at 3-minute intervals provided normal liver function is present. In most cases curves may be obtained much sooner since the dye is promptly taken up in the liver and excreted.

Injection is carried out either by hand, employing a volumetric pipette and 5 ml of saline as flush, or by a power injection, using a mixture of 1 ml of the dye in 4 ml of saline. In the latter instance, the catheter remains connected to the injecting syringe throughout the study period, assuring delivery of virtually the same amounts of indicator each time injection is carried out. Since the injecting pressure varies and different lengths of catheter may be used, the exact volume delivered during the injections has to be verified after completion of the study by means of volumetric measurements of the injection system employed. An improved system utilizes precalibrated Becton-Dickinson glass tubes connected by stopcocks. The power injector is employed for injection and flushing.

THE CURVE AREA CALCULATION

The traditional difficulty in conventional sampling systems which employ withdrawal pumps is caused by particles of dye which remain behind in the eddies of the sampling system, such as the catheter, the tubing, and the sensing chamber of the densitometer. These "trailing" particles have always been thought to indicate early recirculation which in fact may be obscured by them. This distortion of the downslope in turn has led to the development of elaborate mathematical methods and even of cardiac output computers to "correct" curves obtained by these systems (Karlsson et al., 1971).

In view of the fast response time of the fiberoptic system and the absence of any delay in the sampling circuit, the primary curve returns to the base line immediately after its first circulation through the system (Fig. 15.10). This makes mathematical correction of the obtained curve through extrapolation of the downslope unnecessary, provided intracardiac shunts are absent.

A failure of dye curves derived from a fiberoptic system to return to the base line was observed only when distal injection (such as upstream from the superior vena cava) was used in association with far downstream sampling such as in the brachial artery. This finding corresponds to that described by Ryan et al. (1961), who found that the more distal was the injection of indicator, the wider the spread of the indicator concentration became and, hence, the greater the

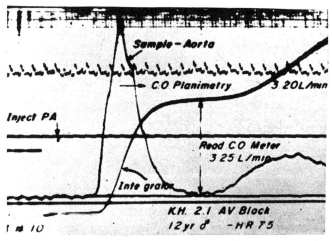

Figure 15.10a Fiber optic (FO) dye dilution curves recorded in the aorta. The site of injection is in the pulmonary artery (PA). Notice that the FO curve returns to zero before recirculation appears. This makes it possible to obtain a direct read-out of the cardiac output (CO) by using a calibrated integrator (By permission from The American Heart Association, American Heart Journal, Vol. 77, 1969, page 182, Fig. 6).

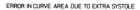

ERROR IN CURVE AREA DUE TO EXTRA SYSTOLE

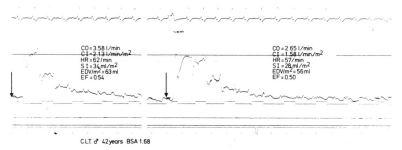

CLT ♂ 42years BSA 1.68

Figure 15.10b Errors introduced in the cardiac output determination at-tributable to an extra systole. On the left-hand side is a normal curve, while on the right-hand side is a curve disturbed by an extra systole. Both curves were obtained in the same patient, within two minutes of each other.

chance of an overlap of the primary circulation with the first recircula-tion (see Chapter 4).

When early recirculation is a real event such as in the case of intra-cardiac shunts, the downslope of the curve clearly shows a different contour. By selection of alternate sampling sites (for example, in ven-tricular septal defects by injecting in the superior vena cava and sampling in the right ventricle just beyond the tricuspid valve or in atrial septal defects by injection in the left ventricle and sampling in the aorta), the exact magnitudes of the systemic cardiac output and pulmonary blood flow may readily be obtained.

Since there is no need for logarithmic extrapolation, simple plani-metry suffices for the derivation of the area under the curve. Only premature beats or other occasional irregularities in cardiac rhythm will spuriously enlarge the area and disturb the calculations. Such curves should be discarded since the procedure is very easily repeated, without the need for recalibration. If one does not wish to allow a few minutes for complete excretion of the dye after the last injection, adjustment of the base line reading is adequate. An integrating circuit has been developed to remove the need for manual planimetry and to take advantage of the return of the curve to the base line before true recirculation begins. This integrating circuit requires electrical calibration only. The circuit is first balanced in the "hold" position with the zero adjust control; after immersion of the fiberoptic tip in the calibrating device (blood without dye), adjustment for drift is made in the "run" position with the offset control until the needle remains stationary at any position on the scale. By means of a calibrating signal which gives a series of readings on the meter between zero and 30 liters per minute and a series of deflections from the base line on the recorder, subsequent readings on the meter during the actual catheterization can be made directly. Later they can be compared with recorded deflections. Calibration was usually carried out before the fiberoptic catheter was inserted in the cuvette and required at most a 5-minute effort. The saving in time by not having to carry out planimetry of the curve, and the attractiveness of immediate readout is evident. In a study to evaluate the accuracy of this system, two series of data were obtained (Hugenholtz et al., 1969b). In the first, the direct reading from the meter in 124 curves was obtained by stopping further integration of the curve points at that instance where, by visual inspection of the curve on the oscilloscope, its first return to the base line was noted. This result was compared with the reading obtained from the simultaneously recorded and subsequently measured deflection from the base line of the integrator output at that instant. In the second, an analysis was carried out in 134 observations of the relationship of this direct reading to the results after conventional calculation of the dye-dilution curve employing planimetry and the formula given earlier. For cardiac outputs over a range of 1 to 16 liters per minute, a highly significant linear agreement ($r = 0.99$) existed between the direct reading of the integrator meter and that measured subsequently from the recorded integrator output. However, calculation of the regression equation indicated an overestimate by direct reading of 3.3%, presumably as a result of too early arrest of the integrating procedure after direct visual interpretation. In the second analysis, the comparison of the direct integrator output with the conventional calculation of cardiac output by planimetry of the curve, a similar high degree of correlation was found ($r = 0.99$). However, regression analysis again indicated that the results by direct integration overestimated the subsequently measured data. In this case, the average overestimate was 4.2%, presumably from instrumentation error. Both errors indicate that the "automated" cardiac output may exceed the normally computed result by an average of 7.3%. However, both may be readily introduced as correction factors in any scheme for automation of the procedure.

INJECTION AND SAMPLING SITES

Injection for cardiac output determination is conventionally carried out in the pulmonary artery through a no. 7 French catheter with side holes to avoid recoil during injection. Sampling preferably takes place in the central aorta through a standard no. 6 French fiberoptic catheter. In this combination, mixing of the indicator in the pulmonary capillary bed is complete, and sampling takes place before recirculation has occurred. On the other hand, when the dye-dilution system is utilized in an effort to localize shunts, injection has to take place upstream and sampling downstream of the site of shunting. For example, in a suspected coronary fistula where the shunt had already been identified by a sudden increase in oxygen saturation, the site of drainage of the fistula could be exactly located in the right ventricle by moving the tip of the fiberoptic catheter from the atrium to the ventricle along the ventricular septum. It is clear that the catheter tip sampling system completely obviates the complex sampling and calculation procedures needed to separate "double hump" curves recorded downstream in cases of intracardiac shunts. Furthermore, the high sensitivity and the rapid response time of the system (0.04 second) in oxygen saturation mode has all but obviated the use of indicator dyes in the localization of shunts since small (up to 1%), phasic and cyclical increases in oxygen saturation, often with characteristic patterns (Fig. 15.11) have returned to oxyhemoglobin its traditional role of the "cheapest indicator" available.

Although pulmonary artery injection and aortic root sampling is conventional during a routine cardiac catheterization, the present interest in left ventricular function studies requires the following modifications. The injecting catheter for left ventricular angiocardiography, usually a no. 8 French catheter in adults or a no. 7 French in children, is advanced in a retrograde manner to the left ventricular cavity and placed in the midcavity position at a point between the apex and the entrance of the left ventricular outflow tract. Power injection is preferred with between 200 and 400 psi injection pressure, although it may be reduced toward 200 psi whenever the injection results in premature ventricular contractions. With power injection a variable delay between the peak of the R wave in the electrocardiogram and the time of the injection should be selected related to the heart rate. This delay usually varies from 150 to 300 milliseconds and permits delivery of the indicator mixture within the last half of diastole, with a maximal chance for adequate mixing and a minimal chance of ventricular irritability.

The fiberoptic catheter is advanced into the ascending aorta under fluoroscopic observation so that the tip is placed about 2 cm above the aortic valve. It should be pointed toward the anterior wall of the aorta so that it is within the direct path of the blood ejected from the left ventricle. Exact placement of the catheter is essential, since a position

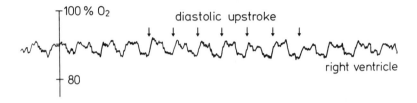

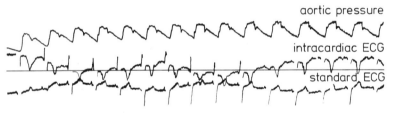

Figure 15.11 Continuous oxygen saturation, recorded in the right ventricle of a patient with a coronary artery fistula.

in the sinus of Valsalva or higher in the aortic root may lead to significant admixture of the ejected volume with blood in the ascending aorta. Placement of catheters in this wall still permits the routine determination of cardiac output while left ventricular injection and aortic sampling also makes available the calculation of ejection fraction, end-diastolic volume, and end-systolic volume from the same curve.

RESULTS FOR CARDIAC OUTPUT AND STROKE VOLUME

The mean stroke volume derived by fiberoptics in a subgroup of 40 patients without volume overloads was 46.8 ± 9.1 ml/m^2 (Wagner *et al.,* 1968; Hugenholtz *et al.,* 1968). Comparison of the fiberoptic with the standard dye-dilution technique (densitometer and withdrawal pump) shows that the results for cardiac output and stroke volume agree well with one another (Fig. 15.12). The fiberoptic values correspond closely with the 45.0 ± 6.0 ml/m² value reported by Levinson *et al.* (1964, 1966) in normal volunteers, obtained by conventional dye-dilution technique. After the first studies from our laboratory (Gamble and Hugenholtz, 1964; Hugenholtz *et al.,* 1965) other workers such as Singh *et al.* (1970) did extensive work confirming these results in 176 sets of curves (95% confidence limits $0.95 \geqslant r \geqslant 0.93$, fiberoptics $= 0.97$ dye $- 0.06$).

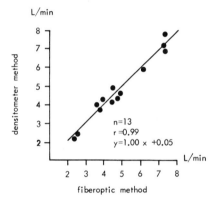

Figure 15.12 Comparison of indicator-dilution cardiac output by densitometric (integrated sample) and fiberoptic techniques. Injection was in superior vena cava in both circumstances. Sampling was in the pulmonary artery for densitometric techniques and in the aorta for fiberoptic techniques.

In a separate study, our group reported that results for stroke volume by the fiberoptic method agreed well with both the Fick and the angiocardiographic methods, although small systematic errors affecting each of the three techniques were found (Wagner *et al.,* 1968). Bussmann and co-workers (1971) have carried out extensive studies with simultaneous determination of cardiac output and stroke volume by the fiberoptic and thermodilution methods. Very acceptable agreement between both methods was achieved, despite the fact that they have inherent differences in terms of the mixing characteristics and the sampling system (Fig. 15.13).

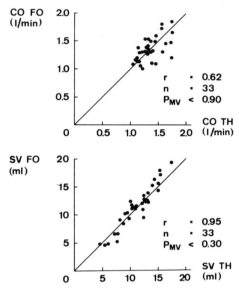

Figure 15.13 Simultaneous determination of cardiac output and stroke volume by fiberoptic and thermodilution methods. There is a high degree of correlation. (From Bussmann, W. D., H. P. Krayenbühl, and W. Rutishauser, 1971. Simultaneous determination of the stroke volume and left ventricular residual fraction with the fiberoptic and thermodilution method. Cardiovasc. Res. 5:136–140.)

RESULTS FOR EJECTION FRACTION AND END-DIASTOLIC VOLUME

The end-diastolic volume (EDV) can be calculated from the ascending aortic dilution curve after left ventricular injection of dye, utilizing the rapid response time of the fiberoptic reflectometer to recognize the stepwise increases and decreases of the indicator concentration (Fig. 15.14). In our laboratory the calculation of end-diastolic volume was evaluated by two methods, originally proposed by Holt (1956). Technique A is based on the measurement of the concentration of indicator during the very first diastolic period after injection of the indicator. The method solves for end-diastolic volume by the relationship

$$EDV = \frac{Q}{C_1},$$

(15.6)

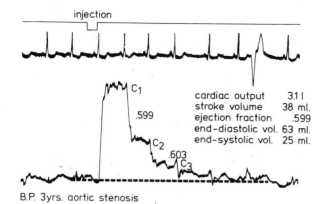

injection

cardiac output	3.1 l
stroke volume	38 ml.
ejection fraction	.599
end-diastolic vol.	63 ml.
end-systolic vol.	25 ml.

C_1

.599

C_2

.603

C_3

B.P. 3yrs. aortic stenosis

Figure 15.14 Fiberoptic indicator-dilution curve—left ventricular injection in late diastole, aortic sampling. The curve permits calculation of cardiac output, stroke volume, ejection fraction, and ventricular volumes. Note the phasic changes in the base line (flow effect) and the return of the curve to the base line before recirculation occurs.

where Q = the injectate (in ml) $\times$ the calibration deflection (in mm) and C_1 = the deflection of the first step after injection, measured at the time the QRS complex was inscribed (in mm). When technique A is employed, the dye has to be injected precisely in the last two-thirds of diastole to guarantee maximal mixing. End-diastolic volume in patients with mitral regurgitation or a shunt lesion can be calculated by this method only.

Technique B is based on the definition

$$EDV = \frac{FSV}{1 - \left(\frac{C_{n+1}}{C_n}\right)},$$
(15.7)

where FSV = the forward stroke volume and C_n and C_{n+1} = reflect the dye concentration in the aortic blood at steps n and $n + 1$, respectively. The concentration $(C_n \ldots, C_{n+1})$ again is measured at the time of inscription of the QRS complex. This technique was utilized in most of the patients studied except those with mitral regurgitation. Beginning with the first clearly distinguishable step on the downslope of the curve, usually the second beat after the peak of the curve, the ratio C_{n+1}/C_n was determined for three to five successive beats, depending on the heart rate. These ratios were then averaged: $1 - (C_{n+1}/C_n)$ expresses the ejection fraction (EjF). End-systolic volume

(ESV) is derived by subtraction of forward stroke volume from end-diastolic volume (EDV − FSV = ESV) in all instances where no valvular regurgitation is present. Further discussion of the methodology can be found in Chapter 14.

Comparison of methods A and B showed excellent agreement in nonshunt and nonregurgitant lesions (Hugenholtz et al., 1968). Since method A requires very exacting conditions for injection of the indicator, method B is generally preferred and mostly employed in this and other studies referred to in this chapter.

Some disagreement exists in the literature among different authors when the ejection fraction, end-diastolic, and end-systolic volumes are considered. Considerable differences in results have been described depending on the technique utilized and the type of patient studied. In the previous chapter this topic was covered in detail, and the confusing, alleged differences were elucidated. Accordingly, only a few points pertinent to the fiberoptic system will be detailed here. When the same patient is studied at close intervals by two different techniques, the reported differences seem to be more apparent than real (Singh et al., 1970; Hugenholtz et al., 1968). Similar data were shown by Bussmann and his group (1971) when ice-cold Cardiogreen was injected into the LV and sampled by thermistor and fiberoptic systems simultaneously at the same site in the aorta (Fig. 15.15). The conclusion can be reached from these studies that the fiberoptic indicator-dilution technique is the most accurate system if one wishes to measure the ejection fraction and the stroke volume from the same curve. The results of these studies and of our own (Hugenholtz et al., 1968; Hugenholtz et al., 1969a) indicate excellent agreement with those reported by Levinson et al. (1966), who also employed Indocyanine green as the indicator but utilized a fast withdrawal system and a standard cuvette densitometer.

The reason for differences in results of different washout techniques reported in the literature may reside in the species, the age or diseases of the individuals studied, or the characteristics of the instrumentation and techniques utilized. In this respect, the application of fiberoptics to the study of instantaneous changes in indicator concentration holds particular promise. This may be ascribed to the following factors: (1) the instrumentation is very stable, is calibrated with ease, and provides a more adequate response time for this application (95% of the response is reached in less than 0.1 second), (2) the same curve from which cardiac output is determined permits the calculation of washout

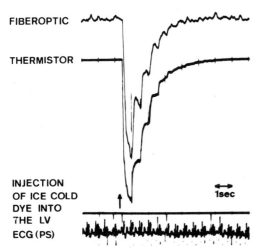

FIBEROPTIC

THERMISTOR

INJECTION
OF ICE COLD
DYE INTO
THE LV
ECG (PS)

1sec

Figure 15.15 Simultaneous left ventricular "washout" curves by fiberoptic and thermodilution techniques. Both curves were recorded from the ascending aorta. The different contour of the steps results from the faster response time of the fiberoptic system. (Reprinted, as in Fig. 15.13, with permission.)

fraction so that the variations in forward stroke volume are minimized, and (3) the indicator utilized is Indocyanine green, a colorimetric substance not suffering from loss to surrounding tissues, such as may occur with heat in the thermodilution technique. Difficulties in calibrating thermodilution systems have led several authors to employ the Fick principle for the determination of cardiac output. Their utilization of ejection fraction with nonsimultaneous determination of forward stroke volume and our observations of frequent variations within short intervals in stroke volume and heart rate during cardiac catheterization lead us to believe that errors incurred by this practice may be a part of the explanation for the observed discrepancies. Loss of heat during injection via the injecting catheter wall and, after the mixing, from the blood to the ventricular walls surrounding the end-diastolic volume may be an even more important factor, as discussed in the preceding chapter. These possible errors have also been discussed by Singh *et al.* (1970) and Salgado and Galletti (1966) in their excellent studies on the thermodilution principle in dogs and in the model. A further factor may be the difficulty in determining the fraction C_{n+1}/C_n from thermodilution or polarographic curves. Inspection of some published data using these techniques demonstrates a downward sloping of the steps on the descending limb of the cardiac output curve. Consequently, it appears difficult to select from these curves the proper point in diastole for the calculation of the fraction. In contrast, such apparent changes in concentration of the indicator during diastole were not observed in our series except in the presence of aortic regurgitation (LaFarge et al., 1965). Of course, flow artifacts, so well described by Mook et al. (1968) in their fundamental study on fiberoptic reflection photometry in blood, must at all times be looked for and inadequate curves rejected.

Measurement of the deflection at the time when no aortic flow occurs (during inscription of the QRS complex) gives the most accurate values. Indeed, variance from the characteristic horizontal plateau of indicator from the beginning of systole to the end of diastole has been utilized in the assessment of aortic regurgitation since the slope of the deviation from the horizontal plateau was found· to be directly related to the amount of regurgitation (LaFarge et al., 1965).

Holt (1956) has summarized the critical problems affecting the measurement of the ejection fraction as those related to the method of injection of the indicator, the dynamic response time of the system employed, and the degree of mixing achieved. Since evidence exists that, at comparable heart rates, fiberoptic stroke volumes after PA or LV injection are identical (Wagner et al., 1968) and that stroke volumes by angiographic and fiberoptic methods are similar (Hugenholtz et al., 1968), the calculation of stroke volume per se does not appear to be the major cause for difference between the systems utilized. Also, the dynamic response time of the systems appears very adequate. Thus, of the three factors mentioned, the degree of mixing is the remaining source of contention affecting the measurement of the ejection fraction.

Ideally for the determination of end-diastolic volume, the entire amount of indicator should be mixed in one diastolic period in the left ventricle or, if incompletely mixed in the first diastolic period, no further indicator should be added during subsequent cycles. Of the two methods evolved from these concepts, the one advocated by Lüthy (1962) and Weigand and Jacob, (1965) (technique A) assumes complete mixing in one diastolic period, and the other (technique B) advocated by Holt (1956) assumed constant mixing in subsequent steps. Adherence to a protocol employing a side-hole catheter in the midcavity of the left ventricle and, with technique A, the injection of the bolus under pressure precisely within the second two-thirds of the diastolic filling period—avoiding excessive amounts of "flushing" volume—appear to fulfill most of the requirements for optimal mixing. In fact, angiographic analysis of sudden injection of small amounts of Renovist in some of our cases has shown complete mixing throughout the cavity, provided the timing of the injection was within diastole. Yet, the finding that over a wide range of volumes (47 to 375 ml) methods A and B agreed well ($r = 0.95$; $r = 0.98$ after power injection, average by method A, 116.9 ml, average by method B, 118.2 ml) with an insignificant difference between the results and the fact that both methods gave similar agreement with the angiographic volumes indicate that, even with method B, mixing is about as adequate as it is under the "optimal" conditions of method A.

Although the numbers are small ($n = 29$), the marked decrease in variance in the results by method A would still further suggest that it is the technique to be preferred despite the relatively more complex manner of injection. These conclusions are in keeping with the results by Salgado and Galletti (1966), who showed in their model that when a "spray nozzle" catheter was used with forced injection over a short period of time, measurements of stroke volume and end-diastolic volume over a range of 60 to 300 ml gave no systematic error when compared to known volumes. They also found a standard deviation for end-diastolic volume of 11%. Although they emphasized the advantages of method A, since it proves to be applicable to patients with mitral valve regurgitation, they also found no major differences with method B. Recently, Swan et al. (1968) have once more critically reviewed the accuracy of the determination of end-diastolic volume by the indicator-dilution and angiocardiographic methods. Starting from a position of considerable doubt, given the conflicting statements in the literature, they showed conclusively in animals that both approaches gave results in close agreement. Their data have apparently ended a scientific argument which has raged over a decade.

The ejection fraction in patients with aortic or pulmonic stenosis or subaortic stenosis in our study was 0.58 ± 0.08, whereas in Levinson's volunteers (1966) the ejection fraction was 0.55 ± 0.08. The end-diastolic volume in our patients was 82 ± 18 ml/m^2 versus 82 ± 12 ml/m^2 in the other series. Holt (1956) indicated that in animals the end-diastolic volume had an average of 2.2 ml/kg, whereas our data in the "non-volume overloaded" group showed 2.7 ± 0.8 ml/kg. The average of the two animal species closest in weight to man was also 2.7 ml/kg of body weight. Even in children with different body weights, without shunts or regurgitation, the same basic relation between end-diastolic volume and body weights exists as Holt found for different animal species with different weights.

STROKE VOLUME, EJECTION FRACTION, AND END-DIASTOLIC VOLUME BY THE FIBEROPTIC TECHNIQUE COMPARED TO THE ANGIOCARDIOGRAPHIC TECHNIQUE

Although it has been indicated that left ventricular volumes may be measured by a variety of methods, the "direct" method through the use of rapid biplane angiocardiography or cineangiocardiography for visualization of the left ventricular cavity has now found acceptance in most centers as the best reference (Hugenholtz et al., 1968; Dodge et al., 1960b; Dodge et al., 1966). It permits the measurement of serial changes in intracardiac volume during systole and diastole usually over several cycles. Although some question persists in regard to the reliability of the determination of end-systolic volume, values for either end-diastolic volume or for stroke volume have agreed well with those obtained by washout methods. However, the angiocardiographic method is time consuming, involves the use of expensive equipment and material, and does not lend itself to serial observations of end-diastolic volume during rapidly altering hemodynamic states. Thus indirect methods based on the principle of rapid sensing of the washout of indicator substances have been preferred ever since the early work of Bing et al. (1951) and Holt (1956).

Data from angiocardiographic films calculated by the area-length method can be used as an *in vivo* standard against which to compare the fiberoptic indicator-dilution method (Hugenholtz et al., 1968). The estimates of end-diastolic volume by fiberoptics correlate well with those of end-diastolic volume by angiography ($r = 0.97$, 95% confidence limits $0.93 < r < 0.98$). The regression equation describing this relationship is $EDV_A = 0.97 \, EDV_{FO} - 10.7$ (in ml/m^3). Analysis of the 49 patients in whom heart rates differed 10 beats or less between the two determinations and in whom no arrhythmia occurred gave similar results ($r = 0.97$, $EDV_A = 0.97 \, EDV_{FO} - 12$). Heart rates differed by more than 10 beats in 14 instances, and volumes calculated in these patients were chiefly responsible for the degree of scatter observed in the total series. For all 63, the average

EDV_{FO} was 106.0, and the average EDV_A was 93.0 ml/m², a difference of 13.0 ml/m². Standard error (SE) was 2.3 ml/m². In view of these findings, it would seem advisable to correct end-diastolic volume derived by fiberoptics by the regression equation given above. Considering the area-length angio method as standard, the fiberoptic technique appears to be the first indicator-dilution method to give results in such close agreement.

APPLICATION IN STRESS TESTING AND MONITORING

The dynamic and flexible nature of the fiberoptic system makes its use particularly attractive in studies where hemodynamic conditions change rapidly. Its application to the study of ventricular performance under conditions of stress, such as exercise and isoproterenol infusion, has been described (Hugenholtz and Wagner, 1967; Hugenholtz et al., 1970b). It was concluded that when an increase in end-diastolic and end-systolic dimensions occurred in the face of unchanged or decreased "afterload," impaired myocardial reserve was present. A change in the ejection fraction or the end-diastolic volume may result from a variation in each of three factors, the "preload," the "afterload," or the myocardial force-velocity relations (Hugenholtz et al., 1970a; Taylor et al., 1966). In recent studies, Urschel and co-workers (1968) found by suddenly altering aortic compliance that the performance of the left ventricle was dependent, at any instant during ejection, on the interrelation between fiber length, tension, velocity, and the impedance to ejection. Since force and velocity under clinical circumstances cannot be derived directly, the interrelationship may be rephrased as follows:

> If in response to a decrease in afterload, the end-diastolic volume does not decrease or the ejection fraction does not increase, impaired myocardial function is present. A normal response to such a stimulus is illustrated in Fig. 15.16 and an abnormal response in Fig. 15.17.

These illustrations and other detailed studies into the performance of the left ventricle demonstrate clearly that the washout techniques have provided the means by which such studies can be carried out during clinical cardiac catheterization (Hugenholtz and Nadas, 1963; Astrand and Rhyming, 1954; Braunwald et al., 1967; Hugenholtz et al., 1970a; Taylor et al., 1966; Urschell et al., 1968; Dodge et al., 1960a; Bristow et al., 1963 and 1966).

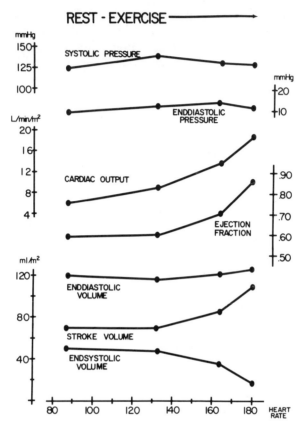

Figure 15.16 Normal exercise response in an anemic youth. Data at rest and at increasing levels of exercise are shown. The increase in stroke volume resulted from an increase in ejection fraction (58 to 87%) with relatively constant end-diastolic volume and pressure.

Fiberoptics have also been used in the intensive care unit setting since the techniques particularly lend themselves to both long-term monitoring of oxygen saturation and intermittent determination of cardiac output from the same catheter. Twenty-one patients have been studied either immediately after cardiac surgery or in the coronary care unit, by means of the instrument described in Part I and a modified no. 7 French Swan-Ganz balloon catheter (Swan *et al.*, 1970) which contains a fiberoptic bundle in addition to a pressure channel.

The catheter was inserted into the brachial vein and the balloon inflated in the right atrium. The tip of the catheter then floated to the pulmonary artery (PA) in 12 patients. It remained in the right atrium in nine other patients despite attempts to manipulate it to the PA under fluoroscopy and ECG control. The failure to advance the catheter may have been caused by the increased weight of the catheter. In seven of the patients in which the fiberoptic catheter was positioned in the pulmonary artery, a second catheter was inserted in the same

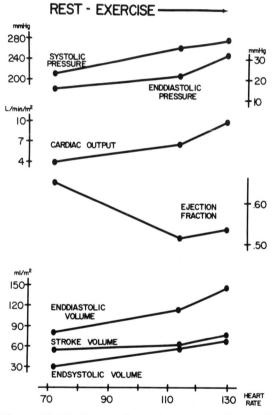

Figure 15.17 Abnormal exercise response in a patient with severe aortic valvular stenosis. Data as in Fig. 15.16. The increase in cardiac output was largely achieved by increase in stroke volume (51 to 77 ml/m²). However, ejection fraction decreased from 67% to an abnormal 52%, and end-systolic and end-diastolic volumes increased more than twofold. The concomitant rise in end-diastolic pressure from 16 mm Hg to 32 mm Hg completed this abnormal profile.

GRAFIEK VAN O2 VERZ. VOOR PATIENT S   (4.1) OVER 1 UUR TOT 17:36   21/12/72

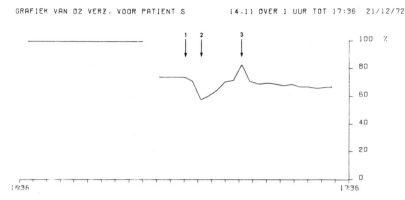

Figure 15.18 Graphic display of 1 hour of monitoring of pulmonary artery oxygen saturation. At time *1*, the patient was turned on the left side; at time *2*, tracheal suction was performed; and at time *3*, intranasal oxygen flow was increased. The initial straight line represents a calibration procedure.

vein and positioned in the superior vena cava or in the right atrium for the injection of Indocyanine green. All patients received some anticoagulant therapy during the monitoring period, either because of their myocardial infarction or as an after effect of the cardiopulmonary by-pass procedure. Signals were recorded either on a direct writing recorder or processed by a digital computer for immediate recall of events of the preceding 24 hours (Mook *et al.*, 1971; Meester *et al.*, 1971). An example of this kind of display of pulmonary artery O_2 saturation is given in Fig. 15.18. The monitoring periods varied from 15 to 46 hours with an average of 22 hours, during which time no serious complications were observed as a result of the in-dwelling catheter. At times inadequate inflation of the balloon brought about artifacts in the O_2 saturation readings. Slight adjustment in the position of the tip or partial reinflation of the balloon up to 0.5 ml of air readily corrected for this. In two instances dye-dilution curves became unsatisfactory due to low gains after a period of 12 hours, whereas O_2 saturation readings remained accurate. A likely cause for this observation was the development of a light fibrin deposit on the catheter tip. It appears that this thin layer will selectively decrease transmission of light at 805 nm. Actual clot formation at the tip of the catheter was not encountered in these 21 patients.

In this study, the previously described calibration procedure was further simplified as follows. Two cuvettes are each filled with 3 ml of blood with a hematocrit between 40 to 50%. Twelve microliters of the previously described solution of Indocyanine green dye is added to one of the cuvettes. By means of the standardization control a reading of zero dye is obtained for the first curvette, and by means of the calibration control a reading of 50% meter deflection is obtained for the second cuvette. This procedure is repeated twice or until no further adjustment is necessary. When the procedure is carried out under sterile conditions the catheter is directly available for use in the patient. Alternatively, this calibration procedure may be carried out well in advance under nonsterile conditions, and the calibration factor for this particular catheter and dye mixture may be retained for use after the catheter has been inserted into the vein. Calibration for oxygen saturation may also be carried out in advance by a two-point procedure. One-hundred per cent and 50 to 60% saturated blood samples are obtained, and a calibration point is recorded by adjustment of the calibration control for the saturation values. Usually the independent method employs a spectrophotometer (Nahas, 1951; Wood, 1950) or the hemoreflectometer. Again the calibration factor for that particular catheter must be retained for subsequent use. Finally, the calibration may be carried out after the catheter has been primarily inserted in the patient by withdrawal of venous blood through the pressure channel and appropriate adjustment of the standardization control to equal the reading has been obtained on that sample by an independent method. The apparatus, designed for monitoring purposes and described in Part I, also contains an internal reference circuit. Thus a previously known reference value can at all times be obtained. This guards against undetected drift in the electrical circuitry.

Reliability of instrument and catheter were checked by comparison of the fiberoptic readings with O_2 saturations from samples drawn simultaneously through the pressure lumen of the fiberoptic catheter and measured with a hemoreflectometer. Fig. 15.19 shows the relation between the fiberoptic and the sample readings. Eighty-eight comparisons were made in which the oxygen saturations ranged from 35 to 82%. The correlation coefficient was $r = 0.98$ with confidence limits $0.92 < r < 0.98$; SE equaled 2.1%. Comparison of the sample values with the fiberoptic readings also showed that there was no systematic pattern in the differences of the two measurements in terms of the duration of monitoring.

A number of interesting observations were made, many of which were in agreement with the findings of a similar study by Taylor *et al.*

(1972). At rest, small fluctuations of mixed venous oxygen saturation occurred and these were exaggerated by simple positional changes. Tracheal suction caused abrupt and severe (up to 20%) decreases in pulmonary artery oxygen saturation. In some patients such desaturation repeatedly preceded bouts of arrhythmia and consequently were of prognostic value. In agreement with the observations of Scheinman

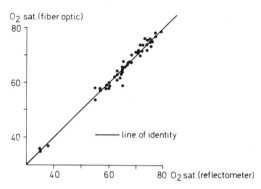

Figure 15.19 Simultaneous determination of oxygen saturation by intravascular fiberoptic and extravascular reflectometric methods. The later analysis was made on blood samples drawn from the pulmonary artery through the lumen of the fiberoptic catheter.

et al. (1969), patients with cardiogenic shock and inadequately treated failure after myocardial infarction were found to have depressed mixed venous oxygen saturation. Whalen et al. (1971) have proposed an index which includes mixed venous oxygen saturation, to predict the occurrence of congestive heart failure and of mortality in acute and chronic heart disease. Although data are limited, they appear to indicate that O_2 saturation in excess of 60% monitored in the pulmonary artery indicate a normal or above normal cardiac output. Below 60% saturation, it would appear that cardiac output cannot be estimated in this manner (Oeseburg and associates, 1972).

It is in the area of intensive care monitoring that the combination of fiberoptic dye dilution and oximetry with a small digital computer for interactive control interpretation and display appears to have its greatest potential (Meester et al., 1971). Coronary blood flow measurements during stress-testing (Sigwart et al., 1972) (Fig. 15.20) and assessment of liver function (McCarthy et al., 1967) provide further avenues for the utilization of fiberoptic techniques.

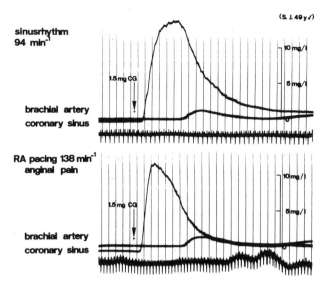

Figure 15.20 Measurement of coronary sinus blood flow by fiberoptics during rest and during pacing-induced angina pectoris. By courtesy of Drs. Sigwart and Rutishauser.

ACKNOWLEDGMENTS

The earlier data were obtained in collaboration with Drs. Gamble, Wagner, and Ellison at the Children's Hospital Medical Center in Boston, Massachusetts, and the more recent data with Drs. J. Roelandt, K. Meeter, and M. van den Brand at the Thoraxcenter of the University Hospital at Rotterdam, The Netherlands.

Our thanks are due to Drs. J. Nauta, X. H. Krauss, and F. Hagemeijer for permission to include case reports on their patients. Work carried out in other laboratories has also been included wherever possible. Particular mention goes to Drs. Rutishauser and Krayenbühl of the Kantonsspital in Zürich, Switzerland, for making available their work in comparing the thermodilution and fiberoptic methods.

REFERENCES

Astrand, P. O., and J. Rhyming. 1954. A nomogram for calculation of aerobic capacity (physical fitness) from pulse rate during sub-maximal work. J. Appl. Physiol. 7:218–219.

Bing, R. J., R. Heimbecker, and W. Falholt. 1951. Estimation of the residual volume of blood in the right ventricle of normal and diseased human hearts in vivo. Amer. Heart J. 42:483–502.

Braunwald, E., E. H. Sonnenblick, J. Ross, Jr., G. Glick, and S. E. Epstein. 1967. An analysis of the cardiac response to exercise. Circ. Res. 20 (Suppl. 1): 44–58.

Bristow, J. D., R. E. Ferguson, F. Mintz, and E. Rapaport. 1963. Thermodilution studies of ventricular volume changes due to isoproterenol and bleeding. J. Appl. Physiol. 18:129–133.

Bristow, J. D., F. E. Kloster, C. Farrehi, M. T. H. Brodeur, R. P. Lewis, and H. E. Griswold. 1966. The effects of supine exercise on left ventricular volume in heart disease. Amer. Heart J. 71:319–329.

Bussmann, W. D., H. P. Krayenbühl, and W. Rutishauser. 1971. Simultaneous determination of the stroke volume and the left ventricular residual fraction with the fiberoptic and thermodilution method. Cardiovasc. Res. 5:136–140.

Dodge, H. T., J. D. Lord, and H. Sandler. 1960a. Cardiovascular effects of isoproterenol in normal subjects and subjects with congestive heart failure. Amer. Heart J. 60:94–105.

Dodge, H. T., H. Sandler, D. W. Ballew, and J. D. Lord. 1960b. The use of biplane angiocardiography for measurement of left ventricular volume in man. Amer. Heart J. 60:762–776.

Dodge, H. T., H. Sandler, W. A. Baxley, and R. R. Hawley. 1966. Usefulness and limitations of radiographic methods for determining left ventricular volume. Amer. J. Cardiol. 18:10–24.

Enson, Y., A. G. Jameson, and A. Cournand. 1964. Intracardiac oximetry in congenital heart disease. Circulation 29:499–507.

Frommer, P. L., J. Ross, Jr., D. T. Mason, J. H. Gault, and E. Braunwald. 1965. Clinical application of an improved rapidly responding fiberoptic catheter. Amer. J. Cardiol. 15:672–679.

Gamble, W. J., and P. G. Hugenholtz. 1964. In vivo cardiac output determination by means of a fiberoptic catheter. Clin. Res. 12:441.

Holt, J. P. 1956. Estimation of the residual volume of the ventricle of the dog's heart by two indicator dilution techniques. Circ. Res. 4:187–195.

Holt, J. P., E. A. Rhode, S. A. Peoples, and H. Kines. 1962. Left ventricular function in mammals of greatly different size. Circ. Res. 10:798–806.

Hugenholtz, P. G., R. C. Ellison, C. Urschel, I. Mirsky, and E. H. Sonnenblick. 1970a. Myocardial force-velocity relations in clinical heart disease. Circulation 41:191–202.

Hugenholtz, P. G., W. J. Gamble, R. G. Monroe, and M. L. Polanyi. 1965. The use of fiberoptics in clinical cardiac catheterization. II. In vivo dye-dilution curves. Circulation 31:344–355.

Hugenholtz, P. G., and A. S. Nadas. 1963. Exercise studies in patients with congenital heart diseases. Pediatrics 32: 769–775.

Hugenholtz, P. G., and H. R. Wagner. 1967. Assessment of myocardial function in congenital heart disease. Proceedings of the Symposium on Pathophysiology of Congenital Heart Disease, Los Angeles, July.

Hugenholtz, P. G., H. R. Wagner, R. C. Ellison, and E. Hull. 1969a. Application of fiberoptic dye dilution technique to the assessment of myocardial function. I. Description of technique and result in 100 patients with congenital or acquired heart disease. Amer. J. Cardiol. 24:79–94.

Hugenholtz, P. G., H. R. Wagner, W. J. Gamble, and M. L. Polanyi. 1969b. Direct read-out of cardiac output by means of the fiberoptic indicator dilution method. Amer. Heart J. 77:178–186.

Hugenholtz, P. G., H. R. Wagner, W. H. Plauth, Jr., and E. Hull. 1970b. The application of fiberoptic indicator-dilution technique to the assessment of myocardial function. II. The interaction of the ejection fraction and end-diastolic volume during exercise and isoproterenol infusion. Amer. J. Cardiol. 26:490–504.

Hugenholtz, P. G., H. R. Wagner, and H. Sandler. 1968. The in vivo determination of left ventricular volume. Circulation 37:489–508.

Karlsson, H., G. J. Rosenhamer, and O. Wigertz. 1971. One stage digital correction of distorted dye-dilution curves. Scand. J. Clin. Lab. Invest. 27:287–291.

LaFarge, C. G., W. J. Gamble, P. G. Hugenholtz, R. S. Replogle, and R. G. Monroe. 1965. A new technique for the quantitation of aortic valvular regurgitation. Trans. Amer. Assoc. Physicians 78:278–281.

Levinson, G. E., M. J. Frank, and H. K. Hellems. 1964. The pulmonary vascular volume in man. Measurement from atrial dilution curves. Amer. Heart J. 67:734–741.

Levinson, G. E., A. D. Pacifico, and M. J. Frank. 1966. Studies of the cardiopulmonary blood volume. Measurement of the total cardiopulmonary volume in normal human subjects at rest and during exercise. Circulation 33:347–356.

Lüthy, E. 1962. Die Haemodynamik des suffizienten und insuffizienten rechten Herzens. Bibl. Cardiol. II:1–168.

McCarthy, B., W. B. Hood, Jr., and B. Lown. 1967. Fiberoptic monitoring of cardiac output and hepatic dye clearance in dogs. J. Appl. Physiol. 23:641–645.

Meester, G. T., C. Zeelenberg, and N. Bernard. 1971. Directe verwerking van gegevens door middel van de computer tijdens hartcatheterisatie. Heart Bull. 2:76–82.

Mook, G. A., N. Knop, and M. L. J. Landsman. 1971. Further developments in fiberoptic reflection oximetry and densitometry. Pflügers Arch. 328:253.

Mook, G. A., P. Osypka, R. E. Sturm, and E. H. Wood. 1968. Fiberoptic reflection photometry on blood. Cardiovasc. Res. 2:199–209.

Nahas, G. G. 1951. Spectrophotometric determination of hemoglobin and oxyhemoglobin in whole hemolyzed blood. Science 113:723–725.

Oeseburg, B., M. L. J. Landsman, G. A. Mook, and W. G. Zijlstra. 1972. Direct recording of oxyhaemoglobin dissociation curve in vivo. Nature 237:149–150.

Ryan, T. C., I. H. Williams, and W. H. Abelmann. 1961. Influence of injection site upon observed flow relationship between right and left ventricle. Circulation 24:1028.

Salgado, C. R., and P. M. Galletti. 1966. In vitro evaluation of the thermodilution technique for the measurement of ventricular stroke volume and end-diastolic volume. Cardiologia 49:65–78.

Scheinman, M. M., M. A. Brown, and E. Rapaport. 1969. Critical assessment of use of central venous oxygen saturation as a mirror of mixed venous oxygen in severely ill cardiac patients. Circulation 40:165–172.

Sigwart, U., H. Hirzel, W. Burian, and W. Rutishauser. 1972. Transfer Funktion des Coronairgefassbettes beim Menschen. Personal communication.

Singh, R., A. J. Ranieri, Jr., H. R. Vest, Jr., D. L. Browers, and J. F. Dammann, Jr. 1970. Simultaneous determinations of cardiac output by thermal dilution, fiberoptic and dye-dilution methods. Amer. J. Cardiol. 25:579–587.

Swan, H. J. C., W. Ganz, J. Forrester, H. Marcus, G. Diamond, and D. Chonette. 1970. Catheterization of the heart in man with use of a flow-directed balloon-type catheter. New Eng. J. Med. 283:447–451.

Swan, H. J. C., V. Ganz, J. C. Wallace, and K. Tamura. 1968. Left ventricular end-diastolic volume by angiographic and thermal methods in a single diastole. Circulation 37 (Suppl. 6): 193.

Taylor, J. B., B. Lown, and M. Polanyi. 1972. In vivo monitoring with a fiberoptic catheter. J. A.M.A. 221:667–673.

Taylor, R. R., H. E. Cingolani, and R. H. McDonald, Jr. 1966. Relationships between left ventricular volume, ejected fraction and wall stress. Amer. J. Physiol. 211:674–680.

Urschel, C. W., J. W. Covell, E. H. Sonnenblick, J. Ross, Jr., and E. Braunwald. 1968. Effects of decreased aortic compliance on performance of the left ventricle. Amer. J. Physiol. 214:298–365.

Wagner, H. R., W. J. Gamble, W. H. Albers, and P. G. Hugenholtz. 1968. Fiberoptic dye dilution method for measurement of cardiac output. Circulation 37:694–708.

Weigand, K. H., and R. Jacob. 1965. Zur Frage des Restvolumenbestimmung des linken Ventrikels im natürlichen Kreislauf. Arch. Kreislaufforschung. 46:97–114.

Whalen, R. E., B. W. Ramo, and A. G. Wallace. 1971. The value and limitations of coronary care monitoring. Prog. Cardiovasc. Dis. 13: 422–436.

Wood, E. H. 1950. The oximeter. In O. Glasser (ed.), Medical Physics, Ed. 2. Year Book Publisher, Inc., Chicago.

chapter 16/ Roentgen Videodensitometry

Joachim H. Bürsch/Erik L. Ritman/Earl H. Wood/
Ralph E. Sturm
Mayo Clinic and Mayo Foundation/Rochester, Minnesota

Videodensitometry is a technique for obtaining objective measurements of the dilution of roentgen contrast media in the circulatory system. The basis of this technique is the measurement of brightness at selected sites in video X-ray images. Since initial development of the technique in 1964 (Sturm, Sanders, and Wood, 1964; Wood, Sturm, and Sanders, 1964), great effort has been made to assess the general usefulness and limitations of roentgenologic indicator techniques in the field of cardiovascular diagnostics. Detailed descriptions of fundamental investigations performed with videodensitometric equipment up to 1971 were published recently in an international symposium volume entitled *Roentgen-, Cine- and Videodensitometry,* edited by P. H. Heintzen and published by Georg Thieme Verlag. Similar densitometric results have been obtained by other methods, especially cinedensitometry (Sinclair *et al.,* 1960; Heintzen and Halsband, 1965; Heintzen and Bürsch, 1966; Heintzen *et al.,* 1967; Rutishauser *et al.,* 1967 and 1970). Roentgen kymography is a related technique (Moros, Neri, and Villagordoa, 1953; Stauffer *et al.,* 1957; Marchal, Marchal, and Kourilsky, 1963).

These methods make use of brightness measurements with photocells of images being displayed on fluoroscopic screens or recorded on cinefilms such as were used rather early and extensively in angio-cardiographic studies of patients (Rutishauser 1969). The investiga-tions were helpful both in testing special applications and in extending understanding of the problems encountered in roentgen photometric and quantitative roentgen contrast-medium dilution methods.

The videodensitometric technique has the greatest versatility for dynamic density measurements in roentgenograms with or without injection of a contrast medium. This is apparent because of (1) ready manipulation of the image format (e.g. logarithmic conversion of video signals), (2) freedom to vary the shape and size of the sampling window (that area of the image used to produce a single densitometric value), (3) sampling simultaneously with high temporal and spatial resolution from any region of roentgen image, and (4) convenience, speed, and reproducibility with which the resulting electromagnetic transformation of the data can be analyzed, using modern electronic data processing and computing techniques when desired. For these reasons this chapter will be restricted to videodensitometry, although some of the applications are equally relevant to cinedensitometry.

PRINCIPLE AND GENERAL CONSIDERATIONS

As is known from conventional indicator-dilution techniques, a densi-tometric system consists of three functional entities: (1) a generator of radiation of known physical properties, (2) an indicator that specifically absorbs the radiation, and (3) a photometric system with which the intensity of the radiation is measured.

Radiation

In the case of roentgenologic investigations, the radiation is generated by a conventional roentgen tube used in diagnostic X-ray equipment. Constant radiation intensity and spectral distribution must be main-tained. Stability of radiation intensity is normally achieved with con-tinuous radiation. If measurements with pulsed radiation are used, the frequency of the X-ray pulses must be synchronized with the video field repetition rate, i.e. 60 cps (Heintzen, 1971*b*). For quantitative density measurements, monochromatic roentgen radiation is desirable. Although it is impossible to generate monochromatic X-ray radiation with conventional X-ray equipment, the use of a suitable copper filter greatly narrows the spectral distribution of the radiation (Bassingthwaighte and Sturm, 1971; Bürsch, Johs, and Heintzen, 1971a; Heintzen and Moldenhauer, 1971b).

Indicator

Conventional angiocardiographic contrast media are used to tag the flow and confines of the blood in which the indicator is diluted. The fractional X-ray transmission depends on two physical factors: (1) amount (concentration times depth) of transradiated contrast medium and (2) frequency of radiation (dependent on X-ray tube kilovoltage and the degree of filtering) as well as atomic weight of the absorbing medium (normally iodine), all of which determine the specific absorption coefficient.

The physical relationship between these factors and the fractional X-ray transmission is given by the law of Lambert and Beer for monochromatic radiation:

$$T = e^{-k \cdot C \cdot D}, \tag{16.1}$$

where T = transmission (fractional), k = coefficient of absorption, C = concentration of contrast medium, D = depth of solution of contrast medium along the path of the X-ray beam. Because the specific absorption coefficient can be considered constant during one angiographic procedure, measurements of changes in X-ray intensity give information about the concentration and depth of the contrast medium-blood mixture being transradiated.

Calibration of indicator concentrations in the circulation may be performed by comparing the fractional X-ray absorption of the injected dye with that of solutions of known concentrations and depth (Heintzen et al., 1967). The calibration should be performed at the time of angiocardiography to assure similar conditions.

Densitometer

The photometric system is the most complex part of the whole X-ray videodensitometric system. It includes (1) conversion of X-ray radiation into visible light with concomitant intensification (normally performed by an image intensifier), (2) conversion of the image produced by the image intensifier into video signals (voltages) by a video camera, and (3) measurements (photometric) of voltage by the videodensitometer within selected sampling regions of any size and shape within the video picture. Technical problems associated with videodensitometry are caused mainly by the image intensifier video chain (Bürsch *et al.,* 1971*b*). The input-output relationship of the video chain (video camera and densitometer) is virtually proportional over the operational range provided the circuitry is adequately clamped in relation to ground (zero) potential.

The image intensifier, however, has a *linear* input-output relationship with a significant residual output at zero input (localized to a small part of the image) due to internal photon scattering. The contribution of this residual output to the videodensitometric value depends on the size of the individual sampling window selected and the size of the roentgen image projected on the image intensifier. In quantitative density measurements, and especially if calibration procedures are performed, the contribution of this radiation scattering has to be considered. This can be done either by a separate nonangiographic brightness measurement or by comparative photometric measurements at the time of angiocardiography when the same amount of light scattering may be assumed for two sampling fields.

Second, the law of Lambert and Beer indicates that the image intensity converted to logarithmic form has a proportional relationship with the product of concentration and the depth of transradiated contrast medium:

$$\ln T = - k \cdot C \cdot D. \tag{16.2}$$

When large sampling windows are used in videodensitometry, it is imperative that the logarithmic conversion be completed before the videodensitometer sampling (Heintzen and Moldenhauer, 1971*a*). This can be achieved by a videologarithmic amplifier with a frequency response of $f > 4$ MHz. If minute sampling windows are used, logarithmic conversion after densitometric sampling is generally satisfactory.

It is possible to sample concentration patterns within distinct regions of the bloodstream (small sampling windows) as well as to measure amounts of the indicator in entire sections of the circulatory system (large sampling windows) (e.g. over the entire silhouette of a cardiac chamber such as the end-diastolic silhouette of the left ventricle). In fact, it is possible to adjust electronically the sampling field to the actual area of the ventricular silhouette described as a contoured window (Heintzen and Pilarczyk, 1971) or to use the automatic border recognition circuitry devised for roentgen videometry

(Ritman, Sturm, and Wood, 1971). The use of large area video-density sampling windows is based on the relationship:

$$\ln T \sim M, \tag{16.3}$$

where M = amount of contrast medium.

EQUIPMENT AND METHOD

The equipment chain and the procedural sequence of videodensi-tometry may be divided into two parts: the first deals with adequate generation and recording of the video image; the second deals with the requirements for densitometric analysis. Fig. 16.1 illustrates these requirements. The angiographic procedure (*left panel*) is per-formed by injection of a known amount of contrast medium into the circulatory system. Preferably, the onset of injection should be trig-gered after a preset delay from the electrocardiographic R wave. An electric signal proportional to displacement of the syringe plunger on injection of the medium should also be recorded so that the time, rate, and duration of injected contrast medium can be related to the angiogram.

The image intensifier is connected to the video camera and also may be connected to a cinefilm camera, if simultaneous cineangio-cardiography is desired (Heintzen, 1971*a*, 1971*b*). The video X-ray images normally are stored on video tape. Up to 15 channels of analog data signals (e.g. intracardiac and circulatory pressures, blood flows, respiration, and electric events) can be recorded on the normally unused margins of the same video tape with the use of an amplitude modulation technique (Sturm *et al.*, 1973). This provides a per-manently synchronized recording of the temporal relationship of ana-tomic and hemodynamic events. For special densitometric procedures, a stop-action video disc recorder can be used.

X-ray exposure for videodensitometry is optimal when collimation of the X-ray beam restricts the image to the anatomic area of interest. In this way both the X-ray radiation scattering and the scattering within the intensifier are minimized. In addition the X-ray beam is filtered through a 0.5-mm sheet of copper placed between the X-ray tube and the patient. The range of X-ray tube operating conditions for a pulse duration of 2 milliseconds (60/second) may then be expected to be a voltage of 50 to 70 kv and a current of 200 to 250 ma.

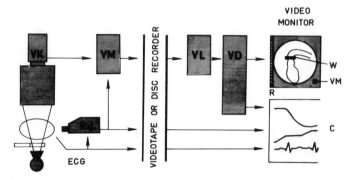

Figure 16.1 Schematic of the roentgen videodensitometer system. The *left panel* illustrates generation of X-ray video images. The video camera (*VK*) scans the intensified X-ray image and the video angiogram is stored on the tape recorder (or disc). The injector (*Inj.*) is triggered by the ECG, and the rate and volume of single or successive injections of contrast medium are recorded simultaneously on the video tape. A videomarker (*VM*) may be used to mark the video images in the sequence of the injector pulses. The onset of the injection pulses causes brightening of a selected region on the video monitor (*lower right*). In this way video images can be synchronized easily with associated analog data records. The *right panel* demonstrates the video signal processing for recording of the videodensitometric dilution curve (*C*) and other simultaneous events. Display of the video angiograms from the video magnetic tape recorder is followed by logarithmic conversion (*VL*) of the video signals. Finally, the photometric measurements from the sampling window (*W*) by the videodensitometer (*VD*) are performed. For this instant the sampling window is positioned over the ascending aorta. Synchronization of related simultaneous anatomic and hemodynamic events can be facilitated by amplitude modulation recording (*R*) of up to 15 analog signals in the rarely if ever used left margin of video raster.

A radiopaque calibration wedge can be used for quantitative concentration measurements. During the angiographic procedure, respiration should be stopped to avoid undesired changes in density and anatomic positions.

The *right panel* of Fig. 16.1 deals with the recording of dilution curves. The storage of the entire angiocardiogram on tape (or disc) permits selection of the optimal site for detection of contrast medium by real time, slow motion, or frame-by-frame display from the recorder. For selection of location of sampling field, the video signal is transmitted directly from the recorder to the videodensitometer. The densitometer allows selection of one or more sampling areas, which usually are rectangular. However, by use of a flying spot scanner assembly a sampling window of any shape, including automatic dynamic sampling only from the area of a silhouette of a moving structure such as the left ventricle, can be used (Ritman *et al.*, 1971). These sampling area positions and sizes are displayed to the investigator as electronically brightened regions (sampling windows) superimposed on the roentgenographic video image (monitor).

Dilution curves obtained from video angiograms after logarithmic conversion of the video image require a special video amplifier which is connected in series with the video tape recorder and the densitometer. Logarithmic conversion normally is performed only at the time of obtaining the curves because the conventional visual contrast of the angiocardiogram is considerably altered by the logarithmic amplifier.

When videodensitometry is applied to special problems, other operations may help to obtain accurate results. The three most important procedures are as follows. (1) Evaluation of radiation scattering in the image transfer chain. This can be performed by measuring videointensities in the video X-ray image when zero X-ray transmission is produced by a piece of lead shaped like the angiographically opacified structure (vessel) of interest (Bürsch *et al.*, 1971*b*). (2) Determination of the background X-ray absorption by the organ or tissue contiguous with the vascular structure of interest that is also opacified after the injection of dye. This is achieved by comparison of the videodensity measurements over the vessel and the adjacent tissue by means of two sampling fields (Smith *et al.*, 1971). (3) Compensation for cyclic nonspecific density signals caused by volume and position changes of the heart, blood vessels, or other structures under study. This can be handled readily by real-time computer processing (Sturm and Wood, 1971).

ADVANTAGES AND LIMITING FACTORS

Advantages of videodensitometry are essentially threefold. (1) The location and extent of the sampling site in the circulatory system can be accurately controlled. (2) A single injection may be used to obtain simultaneous dilution curves from many sites without further invasion and may be performed at leisure and repetitively. (3) Videodensitometric measurements have a high dynamic response and negligible delay limited only by the video field rate frequency of 60 cps and the image storage and sticking characteristics of the video camera (Ritman *et al.*, 1973).

The 30th-of-a-second image storage required for interlaced operation of conventional television systems does significantly degrade the capability of these systems to follow rapidly moving objects. Noninterlaced operation of these systems should be used to obtain adequate temporal resolution for measurement of left ventricular end-systolic and stroke volumes by roentgen videometry (Ritman *et al.*, 1973).

The physical and pharmacologic properties of the contrast media have significance as limiting factors for roentgenologic measurements. High iodine content leads to high viscosity and specific gravity of the solution. Consequently, blood flow alterations may become more significant if high concentrations, large volumes, or high rates of injection of contrast medium into the circulatory system are used (Bussmann *et al.*, 1971). These lead to immediate physical effects (e.g. altered flow) as well as delayed pharmacologic aftermaths. It is desirable to use the smallest quantities of contrast media that will provide an adequate signal-to-noise ratio on the video records. Although general rules with respect to various contrast media, weight of the patient, sites of injection, and sampling cannot be postulated, an indication of the order of magnitude may be taken from animal experiments where 0.2 to 0.5 ml/kg of body weight of contrast media (e.g. Renovist 69%, Urografin 76%) is injected into the cardiac chambers of the aorta without apparently significant changes in flow (Amorim, Tsakiris, and Wood, 1971; Bürsch *et al.*, 1971*b*). This result suggests that angiocardiographic procedures performed for clinical purposes (on injection of 0.5 to 1.5 ml/kg of body weight of contrast media) sometimes may not be appropriate for densitometric studies (A. A. Bove, H. C. Smith, R. E. Sturm, and E. H. Wood, *unpublished data*).

Superposition in the sampling window of simultaneously opacified portions of extraneous segments of the circulatory system may cause problems in recognition of the density changes in a specific segment of a blood vessel or heart chamber under study. These effects can be minimized if densitometric measurements are performed when the site of sampling is close to the place of injection. In this way significant opacification of other parts of the circulation frequently can be avoided or minimized during the period when changes in density are being recorded from the specific site in the circulation being studied.

FLOW MEASUREMENTS IN SINGLE VESSELS

Blood flow in a single vessel can be calculated from the measured velocity of the contrast medium in the vessel. For this purpose two detecting windows are positioned along the vessel at a measured distance apart. As shown in Fig. 16.2, two density curves will be obtained

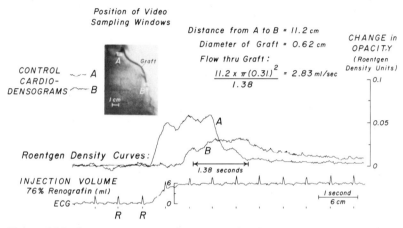

Figure 16.2 Computer-generated roentgen density curves recorded simultaneously over proximal (*A*) and distal (*B*) segments of an aorta to coronary artery saphenous vein graft in a patient with severe coronary artery disease. Determination of the difference in the mean transit times of the contrast medium at the two sites of 1.38 seconds, and the measured length of the graft between the two sites of 11.2 cm provides the basis for an estimation of the velocity of blood flow between these sites. The approximate volume rate of blood flow through the graft of 2.38 ml/ second can then be estimated as the product of blood velocity and vessel cross-sectional area. Measurements of vessel dimensions (length and diameter) were achieved by three-dimensional analysis of the vessel from biplane angiogram X-ray corrected for X-ray magnification. Blood flow of 160 ml/minute through this graft was determined by electromagnetic flowmeter after completion of extracorporal circulation and implantation of the graft at the time of surgery.

from the same indicator bolus. The measured distance between the sampling sites along the vessel divided by the time delay between the first and the second curve determines the velocity of the dye bolus. If the mean cross-section of the vessel can be determined from the angiogram, the flow can be calculated. The time delay between curves usually is determined from the difference of the appearance times, but for more accurate calculations of mean flow, the difference of the mean transit times should be used. The technique does not require knowledge of the actual indicator concentrations and hence circumvents several special problems caused by the image transfer system.

Videodensitometry has been employed for determination of flow measurements in such diverse vessels as the descending aorta (Bürsch, 1971), renal arteries (Riemann, 1970), coronary arteries (Smith *et al.*, 1971), and femoral arteries (Silverman, 1970). Accuracy of this method has been demonstrated with simultaneous control measurements in model and animal experiments. In flow models, the densitometric values were compared with direct volumetric determinations of flow, whereas, in animal experiments, electromagnetic flowmeters on the descending aorta were used (Table 16.1). Special aspects of the method are demonstrated in the application of flow studies to coronary artery saphenous vein bypass grafts. Difficulties encountered in these radiographic measurements are caused by superposition of the vessel of interest and the concurrently opacified heart tissue.

Table 16.1 Comparison of roentgen videodensitometric flow measurements with values found simultaneously by direct volumetric measurements (model) and electromagnetic flowmeter (animal)

Experiments		Range of flow	Standard deviation	Correlation coefficient
Type	No.			
Steady flow		*liters/min*	*liters/min*	
model	47	1–6	0.16	0.98
Animal	63	1–3	0.16	0.90

Real-time subtraction of tissue opacification from the vessel opacification may be achieved by using a dual window technique. In addition to the sampling window over the vessel, a second window is placed over the heart tissue immediately adjacent to the vessel (Fig. 16.2). The window overlaying the vessel samples both the dye in the vessel and the superposed opacified heart tissue. The window just beside the vessel samples only the dye in the heart tissue. In addition to this technique, elimination of periodic background changes (e.g. heart volume changes) can be most readily achieved by the use of a digital computer. This method is being applied successfully for flow measurements in aorta to coronary artery saphenous vein bypass grafts in man in an attempt to assess the efficacy of these implants after recovery from operation (Smith *et al.*, 1971).

DETECTION OF SHUNTS

Anomalous pathways of the circulation in and near the heart can be quantitated by videodensitometry. Records from various sites of X-ray images permit exploration of abnormalities in appearance times and amount of indicator appearing in different sections of the heart or the great vessels. The presence of left-to-right shunts in dogs with surgically created atrial septal defects and ventricular septal defects can be demonstrated by videodensitometry (Amorim *et al.*, 1971; K. Miyazawa *et al.*, 1973). All injections of contrast medium into the vena cava, right atrium, or left atrium resulted in premature appearance of dye in the right heart.

The method of choice for applying videodensitometry in quantitation of left-to-right shunts is by sampling over the right ventricular outflow tract in lateral plane after a single injection of contrast medium into the right atrium (or better, into one of the venae cavae). Early recirculation of the indicator (Fig. 16.3) occurs in the presence of a shunt. In accordance with conventional techniques of other dye methods, the areas under the curves may be used for calculating the ratio of the shunt flow to the pulmonary blood flow.

It must be mentioned that proper positioning of the sampling window is required to minimize superpositions by the aorta or the heart. Calibration procedures or morphologic measurements are not required. It is believed that this method is of value for the detection and measurement of the magnitude of intracardiac shunts in clinical diagnostic studies.

MEASUREMENTS OF CARDIAC OUTPUT
AND VENTRICULAR VOLUME CHANGES

Use of large videodensitometric sampling fields permits measurements of amounts of contrast medium in anatomically defined parts of the circulatory system. Advantages of this technique are illustrated by the following examples. In conventional cardiac measurements using the Stewart-Hamilton principle, the amount of injected dye is known

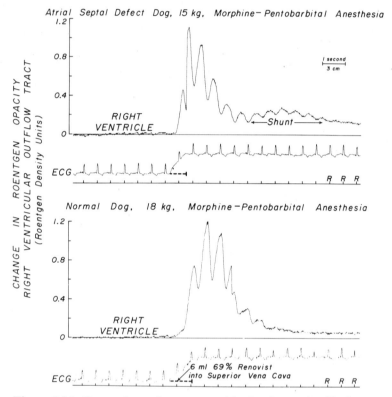

Figure 16.3 Comparison of roentgen videodensitometric dilution curves recorded over the right ventricular outflow tract after injections of 69% Renovist into the superior vena cava of a dog with atrial septal defect (*upper panel*) and a normal dog (*lower panel*). As expected, contrast medium was detected in the right ventricles of both dogs during the first systole after the injections into the cava. However, in the dog with atrial septal defect, a secondary deflection occurred after the major deflection caused by the initial passage of the contrast medium-blood mixture about 3.5 seconds after the onset of effects of the injection. It was absent in the curve recorded from the normal dog because of the recircula-tion of the portion of contrast medium-blood mixture which, after travers-ing the lungs, was shunted from left to right via the atrial septal defect. (From Amorim, D. deS., A. G. Tsakiris, and E. H. Wood. 1971. Use of roentgen videodensitometry for detection of left-to-right shunts in dogs with experimental atrial septal defect, p. 99–107. *In* P. H. Heintzen (ed.), Roentgen-, Cine- and Videodensitometry: Fundamentals and Applications for Blood Flow and Heart Volume Determination. Georg Thieme Verlag, Stuttgart. By permission.)

and a calibration of the dilution curve is performed by a separate procedure. In the case of videodensitometry, however, the amount of dye injected can be obtained directly from a dilution curve sampled from the densitometric window that covers the entire cardiac chamber into which the dye was injected (ventricle or atrium). Under appropriate conditions, the peak deflection of this dilution curve represents the amount of contrast medium injected. The cardiac output may then be calculated from the ratio of the peak deflection obtained from a window over the cardiac chamber to the area under the dilution curve obtained from a window overlying the aorta or any other great vessel. This method eliminates errors introduced by separate densitometric calibration procedures and due to uncertainty of the amount of indicator injected.

Accuracy of this method was tested in animal experiments, which showed results comparable with the conventional Cardio-green dye-dilution technique (J. Bürsch, R. Simon, and P. H. Heintzen, *unpublished data*). The probable accuracy of these measurements, including different injection techniques and sampling sites over the aorta, was equivalent to a standard deviation in the range of 0.25 to 0.30 liter/minute. The same principle of amount measurements (as distinct from concentration) also was successfully employed in detecting the washout of the amount of indicator from the left ventricle after an injection into this heart chamber.

Fig. 16.4 (*lower panel*) demonstrates how the large sampling field can be positioned over the left ventrical. The difference between "amount signals" from the ventricle and "concentration signals" from the ascending aorta is obvious. The stepwise decrease of the amount of contrast medium in the ventricle represents the ratio of the end-systolic volume to the end-diastolic volume of the ventricle provided homogeneous mixing of the indicator exists at the time of injection. The validity of this last assumption directly affects the applicability of indicator-dilution techniques for accurate measurements of heart volume.

Videodensitometry shows many of the problems inherent in the indicator-dilution techniques. Nonetheless, detection of the amount of indicator in the ventricle by videodensitometry may help to reduce these errors. The ascending aorta itself has to be considered as a mixing chamber, and all concentration measurements at this site may result in underestimation of values with respect to those in the ventricle, especially if the origin of the aorta is abnormally enlarged.

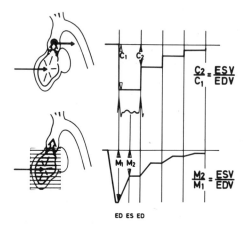

Figure 16.4 Comparison of the principle of concentration measurements from the ascending aorta and amount measurements of the indicator in the left ventricle by videodensitometry. *C,* density signals interpreted as concentration values, *M,* density signals interpreted as amount values, *ESV,* end-systolic volume of the heart chamber, and *EDV,* end-diastolic volume of the heart chamber.

If the amounts measured upstream to the aortic valves exclude this source of error, theoretically, the results will be more accurate. Results of animal experiments appear to confirm these considerations. Determinations using left ventricular sampling yielded end-diastolic volumes about 5% lower by "amount measurements" than by conventional "concentration measurements" of the indicator in the aorta (Bürsch, 1971). End-systolic and end-diastolic volumes of the ventricle (left or right) can be calculated from combined densitometric measurements of cardiac output and the residual fraction using the principle described by Holt (1956).

STUDIES OF VALVULAR FUNCTION AND QUANTITATION OF VALVULAR REGURGITATION

Effective use of videodensitometry has been made in valve function studies. Videodensitometry provides a sensitive method appropriate for detecting small indicator quantities regurgitating through incompletely closed valves. The high dynamic response with respect to density changes makes it possible to study the relationships of these changes to the phases of individual cardiac cycles. Consequently, the phasic relationships of atrial and ventricular contractions to valvular regurgitation have been investigated.

Initial studies dealt with exploration of the mechanisms for closure of the mitral valves (Williams et al., 1968, a and b; Williams and Wood, 1971). In these investigations the contrast medium was injected into the left ventricle. Measurements of the time and the degree of reflux of the indicator under different conditions of atrioventricular pacing of a heart-blocked dog have demonstrated that ventricular contraction is the essential element of complete mitral valve closure. Left ventricular filling from atrial contraction alone does not produce a sustained closure of the valves and normally is followed by a reflux of blood (functional insufficiency of mitral valve closure). This specific example shows that videodensitometry may be helpful in elucidation of hemodynamic mechanisms. Another important application of videodensitometry is directed toward quantification of regurgitant fraction of flow in the presence of anatomically incompetent valve closure.

One principle being applied with the use of these measurements is based on calculations of the ratio of forward flow to backward flow as determined from concentration measurements. The capability of sampling concentrations at multiple sites in the circulation without further invasion is of great advantage in application of roentgendensitometry, because conventional methods require at least two sampling catheters (upstream and downstream to the incompetent valve). Videodensitometry has proved useful in animal experiments for studies of mitral and aortic insufficiency (von Bernuth, Tsakiris, and Wood, 1970; Sturm and Wood, 1971).

Another principle of quantitation of regurgitant indicator utilizes amount measurements. The integrated density measurements over the left ventricle generate a washout curve of the contrast medium injected into the ventricle. From this curve the ratio of reflux to total stroke volume of the ventricle can be obtained. This result is illustrated in Fig. 16.5 where two density curves are shown, one without and the other with an aortic regurgitation obtained from the same animal. The regurgitant fraction can be determined from the amplitudes of the curves. These curves also indicate the ratio of the end-systolic to the end-diastolic left ventricular volume. Consequently, a single videodensogram sampled over a heart chamber allows quantification of the regurgitant fraction and the residual fraction of the ventricle.

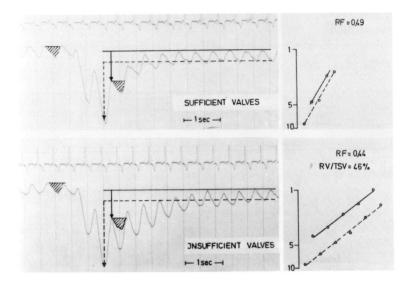

Figure 16.5 Example of two video densograms from the left ventricle of a pig with competent valve closure (*upper curve*) and after producing aortic insufficiency (*lower curve*). In both angiocardiographic procedures the contrast medium (Urografin) was injected during two consecutive diastolic phases of the left ventricle. In the lateral X-ray plane a single sampling window was positioned over the heart, covering the entire opacified left ventricle. Each dilution curve shows periodic density changes due to the volume and motion changes of the heart (*cross-hatched areas*). Relative amount measurements of dye were performed at end diastole (*broken lines*) and end systole (*continuous lines*). From those parts of the curves representing the indicator dilution, the amplitudes were measured considering the two different base lines in accordance with the different density values of the heart at the time of end diastole and end systole. The amplitudes were plotted sequentially on semilogarithmic paper as demonstrated in the two graphs positioned to the right of the curves. The residual fraction (*RF*) was determined by the ratio of the values of end-systolic to end-diastolic amplitudes. The regurgitant fraction (*RV/TSV*) was calculated by the ratio of the amplitudes representing the reflux of dye (values of end diastole minus preceding end systole) to the amplitude caused by the forward flow (difference of values between end diastole and end systole). (From Bürsch, J., R. Simon, and P. H. Heintzen. A new method of quantitating the amount of contrast medium using videodensitometry in circulatory indicator dilution studies. *Unpublished data.*)

Accuracy of this method was studied in animal experiments by comparison with electromagnetic flow meter values. Comparison of regurgitant fractions indicates a 1:1 ratio with the standard deviation of ±5.2% of the regurgitant fraction (Bürsch *et al., unpublished data*). If, in addition, an independent parameter is determined (e.g. the effective stroke volume), absolute values of the regurgitant flow and ventricular volumes can be calculated.

SUMMARY

Roentgen videodensitometry can be used to record dynamic variations in the roentgen density of transradiated anatomic structures with and without injection of roentgen contrast medium. This method has valuable applications in a large variety of hemodynamic investigations. When used in conjunction with injection of contrast medium, it affords advantages over conventional indicator-dilution methods because it measures simultaneously, and with high temporal resolution, the time course of the total amount of indicator in selected regions of the circulatory system and also provides simultaneous anatomic information from the angiogram. The total amount of indicator and volume of fluid being transradiated can be determined separately, whereas conventional indicators (hypertonic solutions, Cardio-green, and others) only allow measurement of indicator concentrations (the ratio of amount to volume). Radioactive indicator methods also supply values for the total amount of indicator in the region under study but have the disadvantage of relatively poor anatomic resolution. The examples given in this survey of applications of videodensitometry deal mainly with techniques using comparative intensity measurements obtained from angiocardiographic images, so that problems caused by the X-ray transfer system are minimized. It is believed that further improvements in the techniques of image transformation will lead to a still wider range of applicability of dynamic roentgen density and roentgen contrast medium indicator-dilution techniques. The use of orthogonal videodensitometry and dimension measurements (videometry) provides simultaneous (60/second) measurements of the depth of a structure (e.g. a blood vessel) being transradiated and the amount of contrast medium in the structure. Exploration of the application of simultaneous videometry and videodensitometric techniques is in the initial stage.

ACKNOWLEDGMENT

The techniques illustrated in Figs. 16.4 and 16.5 were developed and studied in the laboratory of Prof. P. H. Heintzen in the Department of Pediatric Cardiology of the University of Kiel, W. Germany. The authors are indebted to Professor Heintzen for these and his many other contributions to the development of roentgen densitometric techniques, which have contributed greatly to the contents of this chapter.

This investigation was supported in part by Research Grants HL-4664, HL-3532, and RR-7 from the National Institutes of Health, AHA-CI-10 from the American Heart Association, and NGR-24-003 from the National Aeronautics and Space Administration.

REFERENCES

Amorim, D. deS., A. G. Tsakiris, and E. H. Wood. 1971. Use of roentgen videodensitometry for detection of left-to-right shunts in dogs with experimental atrial septal defect, p. 99–107. *In* P. H. Heintzen (ed.), Roentgen-, Cine- and Videodensitometry: Fundamentals and Applications for Blood Flow and Heart Volume Determination. Georg Thieme Verlag, Stuttgart.

Bassingthwaighte, J. B., and R. E. Sturm. 1971. Calibration of the roentgen videodensitometer, p. 45–46. *In* P. H. Heintzen (ed.), Roentgen-, Cine- and Videodensitometry: Fundamentals and Applications for Blood Flow and Heart Volume Determination. Georg Thieme Verlag, Stuttgart.

Bürsch, J. 1971. Quantitative Videodensitometrie: Grundlagen und Ergebnisse einer röntgenologischen Indikatormethode. Habilitationsschrift, Kiel.

Bürsch, J., R. Johs, and P. Heintzen. 1971*a*. Validity of Lambert-Beer's law in roentgendensitometry of contrast material (Urografin) using continuous radiation, p. 81–84. *In* P. H. Heintzen (ed.), Roentgen-, Cine- and Videodensitometry: Fundamentals and Applications for Blood Flow and Heart Volume Determination. Georg Thieme Verlag, Stuttgart.

Bürsch, J., R. Johs, H. Kirbach, C. Schnürer, and P. Heintzen. 1971*b*. Accuracy of videodensitometric flow measurement, p. 119–132. *In* P. H. Heintzen (ed.), Roentgen-, Cine- and Videodensitometry: Fundamentals and Applications for Blood Flow and Heart Volume Determination. Georg Thieme Verlag, Stuttgart.

Bussmann, W.-D., W. Rutishauser, G. Noseda, B. Preter, and W. Meier. 1971. Influence of a new contrast medium (Metrizoate) on coronary blood flow, p. 133–139. *In* P. H. Heintzen (ed.), Roentgen-, Cine- and Videodensitometry: Fundamentals and Applications for Blood Flow and Heart Volume Determination. Georg Thieme Verlag, Stuttgart.

Heintzen, P. H. 1971*a*. Usefulness and limitation of conventional x-ray equipment for roentgendensitometric studies, p. 1–22. *In* P. H. Heintzen (ed.), Roentgen-, Cine- and Videodensitometry: Fundamentals and Applications for Blood Flow and Heart Volume Determination. Georg Thieme Verlag, Stuttgart.

Heintzen, P. H. 1971*b*. Videodensitometry with pulsed radiation, p. 46–56. *In* P. H. Heintzen (ed.) Roentgen-, Cine- and Videodensitometry: Fundamentals and Applications for Blood Flow and Heart Volume Determination. Georg Thieme Verlag, Stuttgart.

Heintzen, P., and J. Bürsch. 1966. Methods for the recording of radio-opaque dilution curves during angiocardiography. Proc. Ass. Eur. Pediatr. Cardiol. 2:39–43.

Heintzen, P., J. Bürsch, P. Osypka, and K. Moldenhauer. 1967. Röntgenologische Kontrastmitteldichtemessungen zur Untersuchung der Herz- und Kreislauffunktion. (Schluss). IV. Die Kalibrierung von Kontrastmitteldensitogrammen. Elektromedizin 12:145–157.

Heintzen, P., and H. Halsband. 1965. Fortlaufende photometrische Messung der Kontrastmitteldichte bei der Angiokardiographie: ein spezielles Indikatorverdünnungsverfahren. Z. Kreislaufforsch. 54:353–361.

Heintzen, P., and K. Moldenhauer. 1971*a*. The x-ray absorption by contrast material: theoretical considerations, p. 73–81. *In* P. H. Heintzen (ed.), Roentgen-, Cine- and Videodensitometry: Fundamentals and Applications for Blood Flow and Heart Volume Determination. Georg Thieme Verlag, Stuttgart.

Heintzen, P., and K. Moldenhauer. 1971*b*. X-ray absorption by contrast material using pulsed radiation, p. 85–88. *In* P. H. Heintzen (ed.), Roentgen-, Cine- and Videodensitometry: Fundamentals and Applications for Blood Flow and Heart Volume Determination. Georg Thieme Verlag, Stuttgart.

Heintzen, P., and J. Pilarczyk. 1971. Videodensitometry with contoured and controlled windows, p. 56–61. *In* P. H. Heintzen (ed.), Roentgen-, Cine- and Videodensitometry: Fundamentals and Applications for Blood Flow and Heart Volume Determination. Georg Thieme Verlag, Stuttgart.

Holt, J. P. 1956. Estimation of the residual volume of the ventricle of the dog's heart by two indicator dilution technics. Circ. Res. 4:187–195.

Marchal, M., M. T. Marchal, and R. Kourilsky. 1963. Aufzeichnung von Lungenkreislauf und Lungenventilation mittels photoelektrischer Zellen während der Röntgendurchleuchtung. Munch. Med. Wochenschr. 105: 1539–1546.

Miyazawa, K., H. C. Smith, E. H. Wood, and A. A. Bove. 1973. Roentgen videodensitometric determination of left to right shunts in experimental ventricular septal defect. Amer. J. Cardiol. 31:627–634.

Moros, G., R. J. Neri, and G. Villagordoa. 1953. Fluorodensography with radiopaque substance: a new method for hemodynamic investigation. Amer. Heart J. 45:495–499.

Riemann, H. 1970. Funktionsstudien mittels Videodensitometrie. Habilitationsschrift, Frankfurt.

Ritman, E. L., R. E. Sturm, and E. H. Wood. 1971. A biplane roentgen videometry system for dynamic (60/second) studies of the shape and size of circulatory structures, particularly the left ventricle, p. 179–211. *In* P. H. Heintzen (ed.), Roentgen-, Cine- and Videodensitometry: Fundamentals and Applications for Blood Flow and Heart Volume Determination. Georg Thieme Verlag, Stuttgart.

Ritman, E. L., S. A. Johnson, R. E. Sturm, and E. H. Wood. 1973. The television camera in dynamic videoangiography. Radiology 107:417–427.

Rutishauser, W. 1969. Kreislaufanalyse mittels Rontgendensitometrie. Huber, Bern.

Rutishauser, W., W.-D. Bussmann, G. Noseda, W. Meier, and J. Wellauer. 1970. Blood flow measurement through single coronary arteries by roentgen densitometry. I. A comparison of flow measured by a radiologic technique applicable in the intact organism and by electromagnetic flowmeter. Amer. J. Roentgenol. Radium Ther. Nucl. Med. 109:12–20.

Rutishauser, W., H. Simon, J. P. Stucky, N. Schad, G. Noseda, and J. Wellauer. 1967. Evaluation of roentgen cinedensitometry for flow measurement in models and in the intact circulation. Circulation 36:951–963.

Silverman, N. R. 1970. Television fluorodensitometry: technical considerations and some clinical applications. Invest. Radiol. 5:35–45.

Sinclair, J. D., W. F. Sutterer, J. L. Wolford, E. Armelin, and E. H. Wood. 1960. Problems in comparison of dye-dilution curves with densitometric variations at the same site in the circulation measured from simultaneous cineangiograms. Mayo Clin. Proc. 35:764–773.

Smith, H. C., R. L. Frye, G. D. Davis, J. R. Pluth, R. E. Sturm, and E. H. Wood. 1971. Simultaneous indicator dilution curves at selected sites in the coronary circulation and determination of blood flow in coronary artery-saphenous vein grafts by roentgen videodensitometry, p. 152–157. *In* P. H. Heintzen (ed.), Roentgen-, Cine- and Videodensitometry: Fundamentals and Applications for Blood Flow and Heart Volume Determination. Georg Thieme Verlag, Stuttgart.

Stauffer, H. M., M. J. Oppenheimer, L. A. Soloff, and G. H. Stewart III. 1957. Cardiac physiology revealed by the roentgen ray: applications of electrokymography, biplane angiocardiography and cinefluorography. Amer. J. Roentgenol. Radium Ther. Nucl. Med. 77:195–206.

Sturm, R. E., E. L. Ritman, R. J. Hansen, and E. H. Wood. Simultaneous recording of multichannel analog data and video images on the same video tape or disc. J. Appl. Physiol. (in press)

Sturm, R. E., J. J. Sanders, and E. H. Wood. 1964. A roentgen video-densitometer for circulatory studies. Fed. Proc. 23:303.

Sturm, R. E., and E. H. Wood. 1971. Roentgen image-intensifier, television, recording system for dynamic measurements of roentgen density for circulatory studies, p. 23–44. *In* P. H. Heintzen (ed.), Roentgen-, Cine- and Videodensitometry: Fundamentals and Applications for Blood Flow and Heart Volume Determination. Georg Thieme Verlag, Stuttgart.

Von Bernuth, G., A. G. Tsakiris, and E. H. Wood. 1970. Quantitation of experimental aortic regurgitation by videodensitometry in intact dogs. Circulation 42 (Suppl 3):60.

Williams, J. C. P., R. A. Vandenberg, T. P. B. O'Donovan, R. E. Sturm, and E. H. Wood. 1968a. Roentgen video densitometer study of mitral valve closure during atrial fibrillation. J. Appl. Physiol. 24:217–224.

Williams, J. C. P., R. A. Vandenberg, R. E. Sturm, and E. H. Wood. 1968b. Presystolic atriogenic mitral reflux developed at abnormally long PR intervals. Cardiovasc. Res. 2:271–277.

Williams, J. C. P., and E. H. Wood. 1971. Application of roentgen video-densitometry to the study of mitral valve function, p. 89–98. *In* P. H. Heintzen (ed.), Roentgen-, Cine- and Videodensitometry: Fundamentals and Applications for Blood Flow and Heart Volume Determination. Georg Thieme Verlag, Stuttgart.

Wood, E. H., R. E. Sturm, and J. J. Sanders. 1964. Data processing in cardiovascular physiology with particular reference to roentgen video-densitometry. Mayo Clin. Proc. 39:849–865.

chapter 17/ Regional Blood Flow Measurement

Norman F. Paradise/I. J. Fox
University of Minnesota Department of Physiology/
Minneapolis, Minnesota

The blood flow to a regional volume of tissue can be measured if all of the arterial inflow and/or venous outflow vessels of the region are accessible. Blood flow to the region can then be determined directly by measuring the time it takes to accumulate in a graduated cylinder the volume of blood flowing either to or from the region. Indicator-dilution techniques, techniques in which the arterial or venous blood concentration of indicator must be known, are used to obtain a quantitative estimate of the blood flow per unit volume of tissue in which the tracer is distributed. However, even less direct methods for estimating regional blood flow must be used if the venous vessels are inaccessible and if the arteries supplying the region are sufficiently small to prevent collection of blood from them. Thus, for example, a direct measurement of the regional blood flow to a 0.1-ml volume of tissue of a 100-g organ would seem to be quite impossible by either of the techniques mentioned above. The purpose of this selective review will be to examine briefly some of the methods available for measuring flows to regions sufficiently small so as to prevent either the collection of the total arterial inflow or venous outflow or the collection of venous blood samples pertaining only to that region whose blood flow it is desired to measure. The indirect methods of measuring regional blood flow currently in use have been developed, in part, to circumvent the inability to sample the arterial inflow or venous outflow from these small regions.

OBSERVATIONAL METHOD OF BLOOD FLOW MEASUREMENT

The quartz-rod transillumination technique permits direct observation of regional blood flows. Wakim and Mann (1942) placed the tip of an illuminator under the free edge of amphibian and mammalian livers and made microscopical observations in various regions. When the observations were extended over several hours, considerable intermittency in the flows became apparent. Differences in intra- and interlobular flow were noted. These investigators estimated that more than three-fourths of the hepatic sinusoids are in an inactive state, whereas only about one-fourth have flow passing through them under conditions in which neither excitatory nor inhibitory influences were acting on the intact amphibian liver. Sinusoidal blood flow was also found to be intermittent in the mammalian liver. However, shifts in circulatory activity were found to be more frequent in the mammalian than in the amphibian liver. Waxing and waning of flow occurred over a period of several minutes with changes in the rate of sinusoidal flow occurring asynchronously, there being no regularity in the sequence in which the changes in the flow rate occurred in the different groups of sinusoids.

An excellent qualitative assessment of regional flows is possible with this technique. Quantitative estimates of single vessel blood flow can also be made by measuring red cell velocities and vessel diameters. The method, however, is inapplicable to measurement of blood flow in deeper tissues. For example, the inner layers of the myocardium are inaccessible to observation by this method. Thus, although the utility of the technique is limited to superficial or thin tissue volumes, it is undoubtedly very useful and informative within the limits of its applicability.

BLOOD FLOW DETERMINATION USING NONTRAVERSING INDICATORS

Radioactively labeled microspheres have been utilized for estimating regional blood flow to various organs. The principle of the method has been recently discussed by Domenech and co-workers (1969): (1) if the microspheres are well mixed at the root of the aorta, (2) if they are distributed in proportion to the blood flow, and (3) if they do not escape from the organs, e.g. via the venous blood, then the fractional distribution of the microspheres in the organs will be in proportion to the fraction of the cardiac output going to the various regions of the different organs. Thus, regional myocardial blood flow is determined by injecting labeled microspheres into the left atrium while simultaneously collecting and timing the volume flow through a cannula inserted into an artery, e.g. the femoral or internal mammary artery, the entire blood flow through which is collected in a graduated cylinder (Domenech et al., 1969). The total amount of radioactivity in the timed blood sample is determined by counting the sample in a well scintillation detector. The flow to any volume of tissue may then be determined by excising a piece of tissue of the desired volume and similarly measuring the radioactivity, in counts per minute (CPM), present in that volume by counting the tissue in a well scintillation detector. The regional flow ($F_{regional}$) is then calculated according to the following equation, in which the brachial artery was the arterial vessel cannulated:

$$F_{regional} = \frac{F_{brachial\ artery}}{CPM_{in\ the\ collected\ arterial\ blood\ volume}} \cdot CPM_{in\ tissue\ sample}. \quad (17.1)$$

Domenech and co-workers (1969) and Sugishita and co-workers (1971) have recently utilized this method to estimate the regional myocardial blood flow to the left ventricular subendocardial and subepicardial tissue. The ratio of endocardial:epicardial flow was found to vary with the diameter of the injected microspheres (Domenech et al., 1969). The mean endocardial blood flow was found in nine experiments to be 2.25 ± 0.82 times greater than the simultaneous epicardial flow when the injected microspheres were 51 to 61 μm in diameter. When microspheres of 21-μm diameter were used the mean endocardial to epicardial flow ratio was found to be reduced to 1.43 ± 0.91, and with microspheres of 14.1-μm diameter this ratio was reduced even further to 1.28 ± 0.37. In close correspondence with these results, Sugishita and co-workers (1971) found the mean endocardial to epicardial blood flow ratio of the left ventricular free wall to be 1.26 ± 0.28 when microspheres of 15 μm were used.

These results indicate that one or more of the assumptions of this method for estimating regional blood flow is violated by at least two of the three sizes of microspheres used by Domenech and co-workers (1969). The microsphere diameter which is optimal for calculating regional flow, if any, cannot be determined from the above experiments. The reasons for the differences in the calculated regional flows can, however, be postulated, and the optimal microsphere diameter for these studies can then be inferred. One reason suggested for the discrepancy in the results is that the smaller microspheres are probably more evenly dispersed across the lumen of the larger coronary arteries and thus more evenly distributed to the subepicardial and subendocardial arteries (Domenech *et al.*, 1969). The tendency for the larger microspheres to be concentrated near the axis of the stream, then, would prevent these microspheres from distributing in proportion to the flow.

The above observations suggest that the smaller microspheres might be distributed more nearly in proportion to the blood flow and that they are more likely to measure flow to the myocardial capillaries. However, it should be remembered that the use of microspheres, even those of small diameter, to measure regional blood flow is neither a necessarily accurate nor valid method, which is, of course, also true of the other techniques to be described. Further experimental verification is required for the measurements of regional flows to tissues no thicker than the subendocardial or subepicardial layers of the ventricle, obtained by using microspheres. Thus, Domenech and co-workers concluded that it is unlikely that any of the different sizes of microspheres used by them measure blood flow to small portions of the ventricle.

Recently Wallin and co-workers (1972) described a technique for measuring intrarenal plasma flow using antiglomerular basement membrane (anti-GBM) antibody which, according to the authors, meets the criteria for an ideal indicator in that (1) it has the same rheologic properties as plasma, (2) it is totally extracted in one passage through the kidneys, (3) it is specifically bound to the glomerular basement membrane and thus provides an accurate index of absolute glomerular blood flow, and (4) it may be administered intravenously and measures the plasma flow distribution over an integrated time period rather than in an instantaneous period. Obviously this method is limited to measurement of regional flow in the kidney, actually, in the glomerulus containing portion, i.e. the nonmedullary portion of the kidney. Simultaneous regional flows calculated using anti-GBM antibody and 15-μm radioactive microspheres revealed a 1.5- to 2-fold greater peak superficial cortical blood flow using the microspheres, which the authors attributed to selective streaming of the microspheres.

BLOOD FLOW DETERMINATION USING DIFFUSIBLE INDICATORS

Method of Internal Calorimetry

In internal calorimetry, heat is used as a diffusible indicator (Grayson *et al.,* 1971). A heated thermocouple is inserted into the tissue of an organ (in these experiments, the probe was inserted 1 mm below the epicardial surface of the dog heart), and the apparent thermal conductivity of the tissue in the vicinity of the probe is determined. In living tissue there is an apparent increase in the conductivity because heat is removed from the region by the flowing blood. The apparent increase in conductivity has been found to be related to blood flow in the tissue volume surrounding the probe, thus the technique allows an estimation of regional blood flow differences.

The heating circuit supplies the heater of the probe with currents of different magnitude. The increment in tissue temperature which is produced by an alteration in the heating current is measured. Since apparent tissue conductivity is linearly related to the inverse of the temperature increment, the constant of proportionality can be independently determined, and thus the apparent conductivity of the tissue can be calculated. The contribution of blood flow to the apparent conductivity of the living, perfused tissue is determined by subtracting the value found for the conductivity of dead, unperfused tissue. It was found that these differences in apparent conductivity between perfused, living and unperfused, dead tissue were linearly related to blood flow. The dimension of volume, however, is absent from the measurement, and the technique, therefore, does not permit quantitation of absolute regional flows. Differences in regional blood flow can be determined, however.

Several features of the method might be disadvantageous. That the probe can be inserted only a few millimeters at most from the organ surface prevents investigation of the blood flow to deeper tissues. In addition, tissue trauma from the inserted probe and elevated tissue temperatures from the heater of the probe (approximately 0.5°C) might markedly affect flow in the vicinity of the probe. Thus, although this method may appear to be adequate for determining differences in regional blood flows, the measured normal regional flows themselves may be altered by the experimental procedure.

Method of Tissue Temperature Detection with Thermistor Probes

Cameron (1970) estimated regional cerebral blood flow by measuring the rate of temperature change of the cortical surface after the cerebral tissues had been artificially cooled by 0.2 to 0.3°C. The brain was cooled by the injection of cool saline into the carotid artery for a period of 10 seconds or 1 minute. The return of the cortical tissue temperature toward normal is recorded, and the flow per gram of tissue is calculated from the half-time of the exponential rise of the lowered tissue temperature. The assumptions of the calculation are essentially the same as for equation (17.3) to be described below, namely that (1) the arterial inflow and venous outflow be constant and equal during the period of measurement, (2) the tissue temperature be equal to the temperature of the venous blood leaving that tissue (i.e. diffusion equilibrium is assumed to exist), (3) the region under investigation be homogeneously perfused, (4) the only means by which the indicator can flow into or out of this region be via the blood, and (5) the indicator does not recirculate. In addition, it is assumed that the specific heat of the tissue and blood are the same. The results indicate that the technique will measure regional cerebral cortical blood flow with approximately the same degree of accuracy as those techniques using ^{85}Kr and ^{133}Xe.

An important difference between this technique and other methods using diffusible indicators is that heat diffuses much more rapidly through tissue than do inert gases, water, alkali metals, or other molecules. The advantage of this method, therefore, is that the assumption of diffusion equilibrium is more likely to be fulfilled when heat is the indicator. This greater diffusibility would, however, be disadvantageous if countercurrent effects were present (see below).

The assumption that heat is removed from the tissue only by the blood is obviously invalid in these experiments (Cameron, 1970). The thermistor with which temperature changes were monitored was placed on the cortical surface just inside the dura. Thus heat could readily be transferred from the surrounding tissue to the cortex when the cortical temperature was lowered. Consequently, the return of cortical surface temperature to normal after cooling is in part due to diffusion of heat from the surrounding tissue. The extent to which assumption 4 was violated cannot be assessed without a knowledge of the temperature and of the proximity of the dura and surrounding tissue to the cortex during the experimental procedure. This problem could be circumvented by implanting the thermistor probe beneath the surface of the brain which would, however, introduce other problems. The thermistor probes used in these experiments were 1.5 mm in diameter, and it is unlikely that any such implantation could be performed atraumatically and without alteration of the normal blood flow.

The thermistor has been estimated to detect flow in a region no larger than 0.5 ml. Such tissue volumes are of the size utilized by Thompson and co-workers (1959) and by Griffen and co-workers (1970) in other procedures, and, as will be discussed, it is possible that a tissue volume of this size may not be homogeneously perfused. A thermistor probe that monitored the temperature of a tissue volume of the order of a few microliters would be more likely to avoid the problem of intraregional perfusion heterogeneity.

The injected saline and the temperature change which it produces in the arterial blood might also be expected to alter cerebral blood flow per se. Betz (1965) found that regional cerebral blood flows were altered by injection of saline at physiological temperatures into the carotid artery. The variations in flow were dependent on the amount of saline injected and on the duration of the injection. Reduction of the temperature of the injected saline would be expected to affect arteriolar and precapillary smooth muscle tone. To produce a decrease in cerebral tissue temperature of 0.2 to 0.3°C by means of a 10-second or 1-minute saline infusion, the temperature of the arterial blood must be lowered considerably below this amount. When the cooled saline infusion is stopped, the inflow of blood of normal temperature might be followed by reactive hyperemia and excessive blood flows at the very time at which the flow measurement is made.

Method of Hydrogen Detection with Platinum Electrodes

The oxidation of molecular hydrogen to hydrogen ions at the surface of a platinum catalyst is the basis of another method for measuring regional blood flow. The current generation by this reaction at the surface of a platinum electrode has been shown to be proportional to the hydrogen concentration when the external resistance in the circuit is low (Hyman, 1961). The circuit employed consists of a platinum-platinum black electrode and a calomel half cell. When platinum electrodes are inserted into a tissue, the changes in current are assumed to follow in a linear fashion the changes in the concentration or the partial pressure of hydrogen gas in a tissue cylinder around the electrode tip. Changes in the tissue concentration of H_2 in the vicinity of the electrode are then used in the calculation of the regional blood flow. Simultaneous regional flows can be estimated by recording from several electrodes inserted into the organ at different positions.

Aukland et al. (1964) employed hydrogen desaturation and equation (17.3) to calculate regional flows. Good agreement was found between calculated regional ventricular myocardial blood flow and directly measured total myocardial venous outflow. These investigators concluded that the myocardium was homogeneously perfused; local blood flow variations were not found to be large. The method was also felt to predict adequately regional flows in the renal cortex, but conclusions as to its adequacy for flow determinations in skeletal muscle could not be drawn.

Recently, Fieschi and co-workers (1969) compared this technique in subcortical structures of the brain with the autoradiographic technique for measuring local blood flow (see below). The blood flow in the caudate nucleus of 11 cats measured in the awake state, using hydrogen desaturation, was 103 ml/100 g/minute. Reivich and co-workers (1969) obtained a value of 102 ml/100 g/minute for flow in the same nucleus, using the autoradiographic technique. When local blood flow in the caudate nucleus was measured independently by means of both the H_2 clearance method and the ^{14}C-antipyrine autoradiographic method within a few minutes of each other in the same seven anesthetized cats (Fieschi et al., 1969), the average value was 56.4 ml/100 g/minute by the H_2 method and 50.7 ml/100 g/minute by the ^{14}C-antipyrine method. The difference of 5.7 ml/100 g/minute was not statistically significant. In addition, the concentration of ^{14}C-antipyrine, as judged from the autoradiograms, was found to be uniform throughout the caudate nucleus, including the area in which the platinum electrode had been inserted.

Fieschi and co-workers (1969) concluded that the H_2 method reliably measures blood flow in the caudate nucleus. The results were also taken to indicate that neither the acute implantation of the 0.4-mm diameter electrodes nor their presence in the tissues for a period of 3 days influenced the local circulation "to any great extent." However, this technique may not be applicable to regional blood flow determination in all organs. Using microelectrodes with a tip diameter of 2 to 5 μm, Stosseck (1970) found that the clearance of inhaled H_2 from the cat cerebral cortex and subarachnoid space was considerably influenced by the location of arteries in the tissue adjacent to the microelectrode. The half-desaturation times of the clearance curves varied between 44 seconds when a microelectrode was located in the center of a pial artery and 265 seconds when the microelectrode was about 500 μm from the artery. Shunt diffusion from pial veins to pial arteries was found to contribute significantly to the venous H_2 clearance curves when an artery and vein were in close proximity; when the artery and vein were 200 μm apart and were each about 3 mm in length, the half-desaturation time was nearly identical in both, being 40 seconds in the artery and 45 seconds in the vein. The clearance curves of veins situated a normal distance from pial arteries showed a half-desaturation time of about 130 seconds. It is evident, therefore, that the location of the platinum electrode is important to the accurate estimation of regional blood flow. If the electrode is in close proximity to an artery, the tissue clearance of H_2 will be markedly affected and, consequently, capillary blood flow will not be accurately estimated.

Method of External Monitoring of Locally Labeled Depots

The measurement of the washout rate of local depots of inert radioactive gases (e.g. xenon or krypton) or of other diffusible indicators (e.g. sodium or iodine) is another technique by which regional capillary blood flows have been estimated. In these techniques the tissue concentration is assumed to be equal to some fraction of the venous blood concentration (i.e. $C_t = mC_v$, where m varies between 0 and 1). In principle, a volume of tissue is labeled with isotope either by introducing the label directly into the tissue via a hypodermic needle or via diffusion from the tissue's surface. The rate of disappearance of the label is then measured by a collimated scintillation detector. From the recorded disappearance rate data, the blood flow to the region in which label had been deposited is then calculated.

Kety (1951, 1960) has obtained an expression for the calculation of capillary blood flow from the clearance of a nonmetabolizable substance introduced directly into the tissues. It is implicitly assumed in his approach that the blood flow in the region of tissue from which the tracer is being cleared is uniform, i.e. each unit volume of tissue receives the same blood flow. It is also assumed that arterial and venous blood flows are constant and equal during the period of measurement, i.e. that a steady state exists. The disappearance of isotope is then described by the following equation:

$$\frac{dC_r}{dt} = -\frac{m_r F_r C_r}{z_r V_r},$$ (17.2)

where the subscript r refers to the region containing the isotope, C is the concentration remaining in the labeled region, F is the blood flow, V is the tissue volume in which the tracer is distributed, z is the blood-tissue partition coefficient, m denotes the extent to which diffusion equilibrium between blood and tissue for a particular indicator is achieved during a single passage of blood from the arterial to the venous end of the capillary, and t is time.

Diffusion equilibrium is commonly assumed to exist, i.e. $m = 1$, so that the following equations describe the washout of isotope from a local depot:

$$\frac{dC_r}{dt} = -\frac{F_r C_r}{z_r V_r},$$ (17.3)

$$C_r(t) = C_r(t = 0)e^{-(F_r t/z_r V_r)}.$$ (17.3a)

From the observed tissue clearance, using these equations, the regional blood flow per unit volume of tissue can be estimated.

Lassen (1967) and Sejrsen and Tonnesen (1968) have assumed that diffusion equilibrium exists only during the initial period of clearance, and, consequently, they calculated the flow-volume relations only from the initial slope of the clearance curve using the equation:

$$\frac{F_r}{V_r} = z_r \left(\frac{-dC_r/dt}{C_r}\right)_{t=0}.$$ (17.4)

This equation is equivalent to equation (17.3), except that it is assumed to be valid for only a short period of time following the labeling. One necessary consequence of the assumption that, for the tracer, equilibrium between venous blood and tissue exists only during the initial stages of the washout is that the degree of equilibration, m, is a function of time. It is not clear how the tissue and venous blood concentrations can be equal during the early moments of the washout when, at later times, the venous blood concentration at any one moment is less than the tissue concentration. It would appear from the assumptions of the method that the degree of diffusion equilibrium would have to remain constant so long as the blood flow remained constant, and it is not apparent how equation (17.4) can correct for the violation of this assumption. In other words, it is not clear how there could exist a variation with time in the degree of equilibration between the tissue and its draining venous blood.

Kirk and Honig (1964) estimated the regional blood flow to the inner and outer layers of the left ventricle of dogs while the whole heart was being perfused at a constant rate. Na-^{131}I in saline, a diffusion-limited indicator (in violation of the assumptions of the method), was injected either 2.5 or 7 mm below the epicardial surface of the left ventricle to determine flow to the epi- or endocardial regions, respectively. The half-time of the washout was determined according to equation (17.2), and the effective blood flow was calculated assuming that ^{131}I was distributed in 30% of the tissue volume. Values for blood flow of 57 and 45 ml/100 g/minute were calculated for the superficial and deep myocardial depots, respectively. If, however, the 0.01-ml injection volume significantly increased the interstitial fluid volume, it was estimated that the actual flows might have been twice the calculated flows.

Andersen and co-workers (1969) repeated the work of Kirk and Honig (1964) using ^{133}Xe. ^{133}Xe in 0.1-ml volumes was injected at depths of 2 and 8 mm from the epicardial surface of the left ventricle of dogs. The isotope was assumed to wash out from two compartments, and, thus, the washout curves were considered to be double exponentials. The curves were graphically resolved into two components. Blood flow was calculated from the more rapidly decreasing exponential according to equation (17.3a). The blood-tissue partition coefficient was set equal to 0.77 in the equation. The second compartment was considered to be due to the removal of ^{133}Xe from myocardial fat, to scatter from extramyocardial tissue, and to recirculation. The calculated values for blood flow were 108 ± 26 ml/100 g/minute for the superficial layer of myocardium and 71 ± 24 ml/100 g/minute for the deeper layer. Thus, the flow to the superficial layers of the myocardium is seen to be approximately 30% greater than the flow to the deeper layers, as in the case of the study by Kirk and Honig, but the absolute values for the flows are considerably larger than in the latter study.

Sullivan and co-workers (1967) determined clearance constants after intramyocardial injection of ^{85}Kr in isotonic saline solution. ^{85}Kr deposits were made in the left ventricle of dogs and human subjects at a uniform depth of 3 mm for all measurements. As might be expected, these investigators found that myocardial blood flow is unevenly distributed in the presence of significant coronary artery disease. For example, in a patient with diffuse narrowing of the left anterior descending artery there was a fall in clearance constants along the vessel, the clearance constant having a value of 3.06 proximally, i.e. upstream, a constant of 0.96 near midvessel, and of 0 distally, i.e. downstream. Calculation of myocardial blood flow from the clearance constant requires a knowledge of the partition coefficient of the tracer. Sullivan and co-workers felt that the partition coefficient might differ in tissues perfused by diseased as compared to normal vessels. Thus, a difference in clearance constant between any two regions could reflect either a difference in nutrient flow or a difference in partition coefficient or both. There was no way to distinguish between these alternatives in these experiments.

Sejrsen and Tonnesen (1968) labeled a local region of cat gastrocnemius with ^{133}Xe by allowing the gas to diffuse into the tissue from the surface of the muscle. For this purpose, a Plexiglas cup containing ^{133}Xe was applied to the surface of the muscle for 30 seconds. After the cup was removed and the excess xenon was blown away, the clearance of the isotope from the tissue was measured by a collimated scintillation detector. Blood flow per unit weight of tissue was calculated from the initial slope of the disappearance curve of the tracer (equation 17.4). The calculated flows to the locally labeled regions were found to be similar, on a per volume basis, to the total venous blood outflow from the whole gastrocnemius. The venous outflow varied between 3.9 and 41.9 ml/100 g/minute and the calculated local blood flow varied between 3.0 and 41.5 ml/100 g/minute. These investigators concluded that the results of the gas-labeling experiments indicated that the blood flow to the local area was representative of the blood flow to the whole muscle; no evidence for a systematic "higher-than-average" or "lower-than-average" blood flow to the superficial layer of the muscle was found in their experiments.

An obvious criticism of the initial slope method is that no objective criteria are established for determining the initial slope. The initial slope was defined as those points for which the clearance rate was constant. It appears from their figures (Sejrsen and Tonnesen, 1968) that the selection of the initial slope must, of necessity, be arbitrary to a large degree since a straight line can be fitted to any two or more of the large number of points involved. Another criticism of the method is that it can be applied only to superficial tissue volumes. Labeling of the deeper tissues is impossible and, therefore, the method has limited applicability. An obvious advantage of this method, however, is the relatively atraumatic labeling procedure as compared to the needle injection technique, but this advantage is again counterbalanced by its limited applicability in the intact, skin-covered normal animal and man.

Hyman (1960) has discussed some of the general experimental difficulties of the local labeling technique. Introduction of the label into the tissue and the detection method both may influence the results. It is assumed that the introduction of label does not significantly alter the conditions of the flow under study. Thus, to minimize the disturbance of the tissue under study, the injection of label is usually made in a solution of isotonic saline. However, the volume of injected fluid has been found to influence the clearance rate (Warner *et al.,* 1953). The clearances were greatest for injected volumes of less than 0.1 ml. Hyman (1960) has suggested that large injection volumes might decrease the apparent clearance rates because of distortion of the tissue and compression of neighboring blood vessels. To minimize the possible error which could be attributed to the volume of the injected fluid, volumes as small as 0.01 ml have been used (Kirk and Honig, 1964). Yet a possible error in the flow measurement of 100% was postulated by these authors when such a volume of only 0.01 ml was used. As noted previously, labeling the surface tissue by diffusion avoids errors due to tissue trauma from the injection needle and also perhaps from leakage along the needle track.

The detection system used in the local labeling technique might also contribute to the errors of the method, as noted above. Ideally, the detectors should be insensitive to minor redistribution effects and yet be sufficiently collimated to be insensitive to radioactivity in any area other than the injected depot (Hyman, 1960). Thus, Sejrsen and Tonneson (1968) situated the scintillation detector at a distance of 14 cm from the muscle surface and had the collimation sufficiently wide so that impulses could be detected from the entire depot (1 cm in diameter in these experiments). In local labeling experiments, the local tissue clearance of isotope can be accurately related to local blood flow only if the blood flow removes the isotope rapidly from the angle through which impulses can be detected, i.e. the acceptance angle of the detector. Ideally, from a geometrical standpoint, this would occur if blood flowed only in a direction perpendicular to the axis of the collimator. If, however, blood flowed in a direction parallel to the axis of the collimator, that is, from the superficial to the deeper tissues, the rate of isotope detection would be changed only slightly until flow changed direction and removed the isotope perpendicularly out of the detector's acceptance angle. In general, all nondepot activity detected, that is, activity in the deeper tissues in the case of flow parallel to the axis of the collimator, would reduce the apparent rate of clearance from the local depot. Thus, the rate of clearance from a local tissue depot would be underestimated, and, consequently, local blood flow would be underestimated.

Andersen and co-workers (1969), in their study of myocardial blood flow, used a wide-angle collimator which allowed detection of activity from the whole heart. Thus, the rate of clearance of activity solely from the local depot in the myocardium, as theory requires, was not accurately measured in their technique. The clearance measurements in these experiments were unavoidably influenced by the location and amount of radioactivity in the venous blood draining the local depot. The observed clearance under this circumstance will definitely be lower than the actual clearance from the local depot. Although it is difficult to assess the degree of error which such wide collimation contributes to the flow calculation, the results from these experiments must be accepted with caution. Other investigators (e.g. Sullivan *et al.,* 1967; Kirk and Honig, 1964) neglect to report both the location and the degree of collimation of their detection apparatus. Of course, no assessment of potential errors in the counting technique is possible in such instances.

In summary, the clearance of radioactive isotopes from local depots is used quantitatively to measure regional blood flow. The calculations are based on the assumptions of the presence of regional perfusion homogeneity and of diffusion equilibrium for the tracer between blood and tissue. A violation of either of these assumptions could lead to errors in the flow estimation. In this regard, it is interesting to note that Lassen (1964) found the clearance of Na and Xe from skeletal muscle to be the same at low blood flows, but that the clearance of Na was considerably lower than that of Xe at high blood flows. This was taken as evidence that the transcapillary movement of Na is limited by diffusion at the higher flows and therefore that the assumption of diffusion equilibrium was not valid for this tracer.

In addition to possible violations of the underlying assumptions, labeling procedures might substantially alter the regional flow characteristics, and inappropriate counting geometry might significantly influence clearance measurements so that local blood flows calculated from them may not be accurate. However, it seems likely at least that differences in regional clearances of tracer are related to differences in regional blood flows.

Autoradiographic Method Using ^{14}C-Antipyrine

In the autoradiographic method using ^{14}C-antipyrine an organ is perfused for a short period of time with blood containing this diffusible radioactive indicator. The perfusion is stopped well before blood-tissue equilibration for the tracer is reached, and tissue samples are removed for analysis of tracer concentration (Landau *et al.*, 1955; Reivich *et al.*, 1969). The theoretical basis of this approach has been discussed (Kety, 1960). The type of perfusion which has been used is one in which the arterial tracer concentration is variable over time. For tissue volumes perfused with blood in which the arterial concentration is variable but zero at time zero, the regional tissue concentration, in theory, varies with time according to the following equation (Kety, 1960):

$$C_r(T) = \frac{F_r}{V_r} \int_0^T C_a(t) e^{-(F_r/z_r V_r)(T-t)} dt, \qquad (17.5)$$

where $C_r(T)$ is the concentration of tracer material in the tissue at time T, and $C_a(t)$ is the arterial blood concentration of tracer. The other symbols have been previously defined. Knowledge of the tissue concentration of tracer, the blood tissue partition coefficient, and the time course of the arterial tracer concentration changes allows the F/V ratio to be calculated for any tissue sample. The assumptions of this equation are the same as those of equations (17.3) and (17.3a).

Autoradiography is used to determine the regional tissue concentration of the radioisotope. Reivich and co-workers (1969) administered ^{14}C-antipyrine intravenously to awake cats for 1 minute. During the 1-minute perfusion period, arterial blood samples were obtained via a polythene catheter advanced into the descending aorta. Arterial blood samples (0.04 ml) were collected every 6 seconds. At the end of the 1-minute perfusion period, the animal was rapidly killed by injection of a saturated KCl solution, and the brain was rapidly removed and frozen. Sections (20 μm) were cut in a cryostat and prepared for exposure to X-ray film. Standards containing known concentrations of ^{14}C-antipyrine were also included on each film, thus for each film a calibration curve could be obtained which related optical density to the ^{14}C-antipyrine concentration. The blood-tissue partition coefficient for antipyrine was found to be 0.99. After the observed time course of the arterial tracer concentration was corrected for catheter smearing (Fox, Sutterer, and Wood, 1957), the regional blood flows were calculated according to equation (17.5). The calculated regional blood flows varied between 1.64 ± 0.14 ml/g/minute for the lateral geniculate body and 0.24 ± 0.01 ml/g/minute for the white matter of the cerebellum. The regional flow values obtained by these investigators agreed closely with values obtained earlier by similar but more cumbersome procedures (Landau et al., 1955).

Differences in tissue concentration of tracer observed in these experiments are, however, not necessarily attributable only to differences in the flow-volume ratios. In addition to F/V differences, the observed heterogeneity in tissue concentration of tracer could be due, in part, to one or more of the following factors: (1) regional differences in capillary permeability to the tracer, (2) differences in transit time from the injection site to the regional capillary beds (i.e. differences in precapillary transit times) (Griffen et al., 1970), (3) regional differences in blood-tissue partition coefficients, (4) regional differences in arterial blood content per unit volume of tissue (Griffen et al., 1970). It is unlikely, however, that any one of these factors alone could create the large differences in tissue concentration observed, but these factors cannot be completely neglected.

The extent to which differences in capillary permeability might have contributed to the observed differences in tissue concentration is difficult to assess. It seems unlikely that a diffusible molecule such as ^{14}C-antipyrine would be characterized by differences in its regional capillary permeability coefficient of such a magnitude as to result in the differences in tissue concentration of the magnitude observed. Thus, although this factor should be considered, it probably plays a negligible role, if any, in the etiology of the observed tissue concentration differences.[1]

A factor that might contribute significantly to tissue concentration differences is differences in arterial (precapillary) transit times. If, for example, the delivery of isotope to one region of the brain as compared to another is delayed by 5 or 10 seconds, the final tissue concentration

---

[1]Eckman and co-workers recently reported the existence of such regional variations in the capillary permeability coefficient of ^{14}C-antipyrine in the brain and concluded that regional variations in ^{14}C-antipyrine uptake by cerebral tissue were a function of both variations in capillary permeability and in blood flow (Eckman, W. W., R. D. Phair, J. D. Fenstermacher, and L. Sokoloff. Fed. Proc. 32:324Abs., 1973).

in the region after a 1-minute perfusion period might be considerably less than the concentration in a region with no delay. Consequently, the F/V ratio in the region with the transit time delay will be underestimated. To illustrate the extent to which F/V ratios might be underestimated, various arterial concentration functions were substituted into equation 17.5, and the F/V ratios were calculated. Three functions have been chosen: (1) $C_a = C_0(1 - e^{-at})$, where $C_0 = 1$ and $a = 2$, thus $C_a = 1/2$ when t is approximately 30 seconds (i.e. $T_{1/2} = 30$ seconds); (2) $C_a = k$, where k is a constant; (3) $C_a = Kt$, where $k = 1$. For each of these functions the extent to which the actual F/V ratios might be underestimated after a 1-minute perfusion because the delivery of labeled blood to a particular region was delayed by 5 or 10 seconds can be calculated. With respect to the work of Reivich and co-workers (1969), this is equivalent to saying that the arrival of tracer in two regions of the brain occurred 5 and 10 seconds, respectively, after the tracer was detected in the descending aorta. It is apparent from these sample calculations (Table 17.1) that transit time differences can substantially influence the calculated regional F/V ratios.

The extent to which possible transit time differences would contribute to the error in the flow-volume ratio calculations in experiments of the type of Reivich and co-workers (1969) can be reduced by perfusing for a longer period of time (Griffen et al., 1970). A 5- or 10-second delay is a considerable fraction of the total duration of a 1-minute perfusion. A longer perfusion period (e.g. 3 minutes), which

Table 17.1 Underestimation of F/V ratios when the precapillary transit of tracer is delayed 5 or 10 seconds during a 1-minute perfusion[a]

Arterial concentration input function	Underestimation with 5-second transit time delay	Underestimation with 10-second transit time delay
	%	%
$C_a = C_o(1 - e^{-2t})$	13	27
$C_a = k$	8	17
$C_a = t$	17	34

[a]Three arterial concentration input functions were substituted into equation (17.5), and the F/V ratios were calculated. The actual F/V ratio was assumed to be unity. Equation (17.5): $C_r(T) = \dfrac{F_r}{V_r} \displaystyle\int_0^T C_a(t) e^{-(F_r/z_r V_r)(T-t)} dt$.

would considerably reduce the influence which a 10-second delay would have on the differences in the observed tissue concentrations of tracer, has the disadvantage of obscuring true differences in F/V ratios between regions because such differences are less detectable the closer one approaches blood-tissue equilibration for the tracer.

Since antipyrine distributes in organ tissue water, variations in the water volume per unit weight of tissue might contribute to the observed differences in tissue tracer concentration. For example, if the flow per unit water volume were the same to two regions but one region contained 80% water by weight and the other 70%, autoradiographic methods would probably show a greater concentration of antipyrine per unit weight of tissue in the sample containing 80% water. However, on the basis of rough calculations, regional water content differences could not contribute greatly to the observed tissue concentration differences found by Reivich and co-workers (1969). The extent of such a contribution, if any, could easily be determined by drying regional tissue volumes to constant weight and calculating the regional water contents. In this regard, Reivich and co-workers (1969) found that the regional blood-tissue partition coefficient for antipyrine which would be sensitive to differences in water content varied only between 0.99 and 0.97.

During the early moments of perfusion, the concentration of tracer in the arterial blood is, of course, much higher than in the other tissues. Therefore, in tissue specimens containing a higher than average content of arterial blood per unit weight of tissue, the concentration of tracer in the nonvascular tissue would be overestimated. To illustrate this possibility, assume that an organ is perfused by arterial blood containing a unit concentration of tracer. Further assume that the perfusion is stopped at a time when the nonarterial tissue concentration is 0.1 times the arterial blood concentration. Then, if one tissue region has an arterial blood content of 4% and another has an arterial blood content of only 2%, the observed concentration of tracer in tissue from the region with the higher content of arterial blood will be 17% greater than that in the region with the lower content of arterial blood. It would appear that the degree to which variations in the arterial blood volume of tissue samples may contribute to the observed heterogeneity of tissue concentrations of tracer may be determined by employing an autoradiographic technique using indicators which are confined to the vascular compartment. By such a technique it may not be possible, however, to distinguish whether the differences in blood volume are due to differences in arterial or in venous blood content, and, of course, the important parameter is the volume of arterial blood contained in the tissue sample.

In this regard, it is worth noting that the concentration of a diffusible indicator in aortic blood is not necessarily indicative of the precapillary indicator concentration. Stosseck (1970) found that, when H_2 was inhaled by cats, the simultaneous concentrations of H_2 in the aorta and in a pial artery were quite different. The calculated regional cortical flow was 25% greater when the arterial concentration changes were recorded in the pial artery instead of in the aorta. When the mean organ blood flow was reduced, the calculated flows using either the aortic or the pial artery H_2 concentration differed by 100%. The difference between the pial artery and aortic blood indicator concentration time curves has been attributed to diffusion of hydrogen from the smaller arterial vessels (Stosseck, 1970). Consequently, the concentration of diffusible indicators in large arteries cannot be assumed to be the same as that arriving at the capillaries. There is apparently a degree of precapillary exchange of H_2 sufficient to affect the calculation of regional blood flow. This means that the time course of the arterial concentration changes of antipyrine found by Reivich and co-workers by sampling from the aorta may not represent the actual arterial concentrations arriving at the cerebral capillaries, although, on the basis of differences in molecular weight and lipid solubility, the error stemming from precapillary loss of indicator would be expected to be less for antipyrine or tritiated water (see below) than for H_2.

Other factors which might contribute to errors in the determination of the tissue concentration of ^{14}C-antipyrine from autoradiograms have been discussed (Reivich et al., 1969). Variations in the thickness of the 20-μm sections would produce apparent changes in the antipyrine concentration. In addition, there is variability in making replicate readings of the optical density of the autoradiograms as well as of the standards. The variability of replicate determinations of the concentrations within single tissue samples was found to be 1.9% of their mean value. The variability within sections is thought to be due mainly to errors in making duplicate optical density measurements. The variability between sections is thought to include this error plus the error due to variations in section thickness.

Regional Measurement of Tissue Isotopic Water Content

Regional blood flow has been estimated in a fashion similar to that using ^{14}C-antipyrine with the use of isotopic water (Griffen et al., 1970; Thompson et al., 1959). After a volume of blood far short of that required for equilibration of the tissue water with the isotopic water in the perfusing blood had flowed into the organ, the perfusion was stopped, the organ was removed, and approximately 0.5- to 1.0-g samples were taken for isotopic analysis. The water was quantitatively distilled under vacuum from the tissue and blood samples. Deuterium oxide (D_2O) was analyzed in duplicate by mass spectrometry, and tritiated water was analyzed by liquid scintillation spectrometry. The average difference in D_2O concentration between duplicate samples was found to be less than 1% (Thompson et al.,1959).

If the arterial tracer concentration is kept constant during the perfusion, as in the experiments of Griffen et al. (1970) and Thompson et al. (1959), the regional flow-volume ratios can be calculated from the following equation, the assumptions of which are the same as those of equations (17.3) and (17.3a) (Kety, 1951; Johnson et al., 1952):

$$C_r(T) = Z_r C_a (1 - e^{-(F_r T/z_r V_r)}).$$ (17.6)

A frequency distribution of normalized F/V ratios for the dog liver was found to appear nearly Gaussian and to have a coefficient of variation equal to 0.53 (Griffen et al., 1970). Thompson and co-workers (1959) demonstrated the presence of a relatively small degree of regional perfusion heterogeneity (small differences in regional F/V ratios) in the dog gastrocnemius as compared to that in rat liver. The degree of perfusion heterogeneity which they found, which was determined from the D_2O concentrations of tissue samples from the gastrocnemius, was too small to permit accurate prediction of the D_2O concentration actually found in the venous outflow of the muscles, thus they postulated the presence of additional, heterogeneously perfused regions which were missed by their sampling technique. Thus, Thompson and co-workers raised the possibility (among others) that either: (1) the degree of heterogeneity, that is, the variability in the D_2O concentrations between tissue samples of the size taken by them, may not have been representative of a greater degree of heterogeneity actually present between the tissue regions of this size in their muscles, or (2) the variability in the tissue D_2O concentrations was minimized by the large size of their samples and that, therefore, their tissue samples did not reveal the greater degree of heterogeneity actually present within them.

In a recent investigation (Paradise and co-workers, 1971), the degree of perfusion heterogeneity in both unisolated and in completely isolated skeletal muscles, *in situ* and contracting in both instances, was compared by measuring regional variations in the concentration of tritiated water in these muscles after perfusing them with blood labeled with this isotope. The tissue tritium concentration was determined in tissue samples having an average weight of 80 mg. Thus, these samples were considerably smaller than those taken by Griffen and co-workers (1970) and by Thompson and co-workers (1959).

The primary purpose of measuring regional blood flows in this study was to determine the degree of perfusion heterogeneity demonstrable in skeletal muscle when tissue samples of this size are taken for isotopic analysis. A considerable degree of perfusion heterogeneity was demonstrated in the unisolated muscles, and this increased more than twofold in the surgically isolated muscles. The degree of perfusion heterogeneity was shown to be correlated with the degree of edema formation, the latter being much more marked in the surgically isolated muscles. Thus, surgical dissection and edema formation were shown to seriously alter tissue perfusion. Blood flow to the myocardium has also been measured using tritiated water and tissue samples up to 1 to 2 grams in size by Palmer and co-workers (1966) and by Bloor and Roberts (1965).[2]

The previous discussion describing alternative explanations for the observed heterogeneity in tissue tracer concentration is also applicable to these tissue sampling techniques of Thompson *et al.*, Griffen *et al.*, and Paradise *et al.* In addition to the four factors listed above, regional dissimilarities in the rate at which non-water hydrogen exchanges with the water hydrogen could contribute somewhat to this heterogeneity in tissue concentration (Griffen *et al.*, 1970). This factor was thought to account for no more than a minor part of the observed heterogeneity of regional tissue concentrations in the liver, however. Similarly, regional variations in arterial blood content and in water content were thought to be relatively unimportant in contributing to the heterogeneity in the tissue concentrations found in the liver because of the uniformity of the histological structure of the liver (Griffen *et al.*, 1970). The fact that no topographical regularity was found in the tissue concentration of tracer was taken as significant evidence that the dispersion of precapillary transit times in the liver

[2]The technique of Paradise and co-workers, 1971, using tritiated water and small tissue samples has recently been applied also to the measurement of regional myocardial blood flow (Paradise, N.F., H. B. Burchell, D. A. Gerasch, and I. J. Fox. Fed. Proc. 32:403Abs., 1973).

was not the chief factor causing differing local isotope concentrations (Griffen *et al.*, 1970).

Measurement of Regional Tissue ⁸⁶Rb Content: Plasma Clearance Method

^{86}Rb is a diffusible indicator that has been used in the estimation of regional blood flow. Moir and DeBra (1967) used rubidium for a qualitative estimation of regional blood flow to the inner and outer layers of the left ventricle of the dog. These investigators perfused the common left coronary artery from a reservoir for exactly 2 minutes. ^{86}Rb was infused into the perfusion system at a constant rate. At the end of the perfusion period, the heart was electrically fibrillated and excised. Tissue samples were removed from the left ventricle, divided into inner and outer halves, and then assayed for radioactivity in a scintillation counter. The uptake of ^{86}Rb was expressed as counts per minute per gram of myocardium and used as a "directional index" of the distribution of the metered common left coronary flow to both the endocardium and epicardium. No attempt was made to use the ^{86}Rb tissue concentrations to calculate regional blood flows. Thus, differences in the ^{86}Rb tissue concentrations can only qualitatively reflect blood flow differences, since no theoretical expression relating tissue concentration and flow was derived or utilized by these investigators. The endocardial and epicardial uptake of ^{86}Rb was found to be similar under conditions of normal perfusion, but the endocardial uptake was considerably reduced during ischemia.

Love and Burch (1957, *a* and *b*) injected ^{86}Rb in saline into the femoral vein of dogs by means of a variable speed syringe to keep the arterial concentration of ^{86}Rb constant. The dogs were killed at different times after the onset of the infusion, and 1-g tissue samples were collected from different parts of the heart. These were digested in HNO_3 for determination of their ^{86}Rb and potassium content. The initial myocardial ^{86}Rb clearance was calculated for various tissue samples and used as a minimal value for the rate of plasma flow. Clearance is here defined as in renal physiology. The clearance of a substance is the amount of plasma completely cleared of the substance per minute. Mathematically, $C = F(C_a - C_v)/C_a$, where C is the clearance of a substance from plasma, C_a and C_v are the arterial and venous concentrations of the substance, respectively, and F is the plasma flow. Small differences in ^{86}Rb clearance were found between the outer, middle, and inner portions of the left ventricle (Love and

Burch, 1957*b*). Clearance of plasma ^{86}Rb by the middle portion of the wall averaged 8.0% more than that of the outer third and 5.8% more than that of the inner third. Again in this study, the calculated ^{86}Rb clearance values were not explicitly related to regional myocardial blood or plasma flows. The clearances were, however, assumed to be related to plasma flows.

Moir (1966), however, has shown that the relationship between blood flow and ^{86}Rb clearance is nonlinear and that flow is underestimated over a wide range, but particularly at the high flow rates. Furthermore, the ^{86}Rb extraction ($E = C_a - C_v/C_a$) was found to vary inversely with flow. Therefore, for the clearance of ^{86}Rb to be accurately related to blood flow, it is necessary to know the ^{86}Rb extraction as a function of the flow. For these reasons, the ^{86}Rb clearance method is not adequate for accurate determination of regional blood flows because the necessary quantitative relationship between flow and clearance has not been established. The clearance of ^{86}Rb in ventricular tissue is neither equal to nor directly proportional to the myocardial blood flow.

This clearance method is thought to underestimate coronary flow rates because the time that the isotope spends in the capillary is felt to be a determinant of the myocardial extraction of [86]Rb (Moir, 1966). The data of Renkin (1959) support this view. The transport of [42]K [an element assumed to have properties quite similar to those of [86]Rb (Love and Burch, 1957*a*)] from blood to tissue in skeletal muscle was found to be a function of the blood flow, but the nature of the relation between these two quantities was not immediately clear. The [42]K transport also appeared to be related to the time spent by the blood in the capillaries. At low blood flows, this time was obviously long, and the extraction approached completion. At high blood flows, this time was short, and the extraction was reduced. Therefore, the clearance of molecules such as ^{86}Rb and [42]K from plasma to any homogeneously perfused volume of tissue seems to depend on the rate of blood flow and on the capillary permeability to the tracer. Renkin (1959) has derived the following equation:

$$C = F(1 - e^{-PS/F}),\tag{17.7}$$

where C is the clearance, F is blood flow, P is the permeability coefficient of the diffusion barrier, and S is its surface area. This model indicates that, in a homogeneously perfused region, clearance and flow are related in a well defined manner if the PS product is known.

Thus, since the calculated regional clearances are not simply related to regional blood flow when nonflow-limited tracers such as ^{86}Rb are used, the ^{86}Rb clearance method is not adequate for accurate determination of regional flows. The clearance method might perhaps be applicable for tracer molecules that are rapidly transported across capillary membranes in comparison to blood flow (e.g. tritiated water or ^{14}C-antipyrine).

COMPARISON OF TECHNIQUES

Each of the above mentioned methods utilizing diffusible indicators (i.e. local labeling, autoradiography, H_2 polarography, heat clearance, quantitative water distillation, the tritiated water technique, and the plasma clearance method) in the determination of regional tissue indicator concentrations employs similar equations for calculating regional blood flows (i.e. equations (17.3), (17.3a), (17.4), (17.5), and (17.6)). The underlying assumptions are the same for all techniques. Consequently, the technique which (1) most accurately measures tissue concentration, which (2) does not cause a violation of the assumptions, and which (3) least disturbs the perfusion of the region would seem to be best suited for the estimation of regional blood flow.

The external monitoring (local labeling), H_2 clearance, and autoradiographic techniques, in principle, determine flows to much smaller regional volumes than do the heat clearance and water distillation techniques. In the local labeling experiments, the tissue volume to which the perfusion is determined is not much greater than the volume of the injected label, a volume as small as 0.01 ml. The autoradiographic technique enables the tissue concentration to be determined from 20-μm sections. The area of the section in which the tracer concentration is to be determined can probably be made as small as desired (e.g. less than 1 mm^2). Consequently, the regional volumes in which tissue concentrations may be determined are much smaller in the local labeling, autoradiographic, and H_2 clearance techniques. As a result, the assumption of perfusion homogeneity in the region sampled is less likely to be violated with these methods than with the heat clearance and water distillation techniques.

Nevertheless, it is not certain that the autoradiographic and local labeling techniques measure regional concentrations of tracer as accurately as the methods measuring regional isotopic water content. As has already been noted, there is the possibility that radioisotope detection occurs from regions away from the site of the injection depot in the local labeling experiments. However, local labeling techniques, since they are nondestructive, permit repetition of the flow measurement under different conditions. The autoradiographic method is subject to calibration errors as well as to errors stemming from variations in the thickness of the sections as already noted. Even the H_2 clearance technique which would seem to be the most appropriate of these methods for measuring regional flows since the H_2 concentration is monitored from a small tissue volume, suffers from possible inaccurate determination of the H_2 concentration under certain circumstances previously described. It seems that, under the most favorable conditions, the H_2 clearance method may have the least number of objectionable factors that could cause inaccurate blood flow estimations. The favorable comparison of this method with the autoradiographic method is evidence for the validity of the H_2 clearance method, but one would wish to see further comparisons before rendering a final judgment.

The tissue sampling technique of Thompson and co-workers (1959) would be a simple and accurate method if it were possible to obtain an accurate determination of the isotopic water content in much smaller tissue volumes. Small tissue volumes are necessary to circumvent the suspected intraregional heterogeneity of the large tissue volumes. The study of Paradise et al. with tritiated water and 80-mg samples represents a step in this direction. Thus, if the accuracy of the determination of the tracer concentration in small tissue volumes were assured, the method would be superior to the other methods for the following reasons: (1) the tissue concentration is accurately determined, (2) the volume of the sampled region of tissue is easily determined so that actual flows can be calculated, (3) the geometrical problems of external monitoring are avoided, (4) the method is considerably simpler than measuring tissue concentrations by autoradiography, (5) the tissue is not subjected to probes, needles, or electrodes —procedures which may influence the blood flow to the region of tissue under study, and (6) simultaneous measurement of the flows to many regions is possible.

ACKNOWLEDGMENT

We thank Dr. H. B. Burchell for encouraging us to undertake this task and for kindly reading the manuscript. This work was supported in part by Research Grant HE08068 and Training Grant HE05222 from the United States Public Health Service.

REFERENCES

Andersen, H., H. Bagger, and H. Gotzsche. 1969. Non-uniform blood flow in the left ventricular wall of dogs measured by the Xe^{133} wash-out technique. Acta Physiol. Scand. 76:376–382.

Aukland, K., B. F. Bower, and R. W. Berliner. 1964. Measurements of local flow with hydrogen gas. Circ. Res. 14:164–187.

Betz, E. 1965. Local heat clearance from the brain as a measure of blood flow in acute and chronic experiments. Acta Neurol. Scand. Suppl. 14:29.

Bloor, C. M., and L. E. Roberts. 1965. Effect of intravascular isotope content on the isotopic determination of coronary collateral blood flow. Circ. Res. 16:537–544.

Cameron, B. D. 1970. Determination of regional cerebral cortical blood flow using a heat clearance technique. Phys. Med. Biol. 15:715.

Domenech, R. J., J. I. E. Hoffman, M. I. M. Noble, K. B. Saunders, J. R. Henson, and S. Subijanto. 1969. Total and regional coronary blood flow measured by radioactive microspheres in conscious and anesthetized dogs. Circ. Res. 25:581–596.

Fieschi, C., M. Nordini, and A. Bartolini. 1969. The hydrogen gas method to measure local blood flow in subcortical structures of the brain with a comparative study with the ^{14}C-antipyrine method. Exp. Brain Res. 7:111.

Fox, I. J., W. F. Sutterer, and E. H. Wood. 1957. Dynamic response characteristics of systems for continuous recording of concentration changes in a flowing liquid (for example, indicator-dilution curves). J. Applied Physiol. 11:390–405.

Grayson, J., R. L. Caulson, and B. Winchester. 1971. Internal calorimetry-assessment of myocardial blood flow and heat production. J. Applied Physiol. 30:251–257.

Griffen, W. O., Jr., D. G. Levitt, C. J. Ellis, and N. Lifson. 1970. Intrahepatic distribution of hepatic blood flow: single-input studies. Amer. J. Physiol. 218:1474–1479.

Hyman, C. 1960. Peripheral blood flow measurements: tissue clearance. Methods Med. Res. 8:236–242.

Hyman, E. I. 1961. Linear system for quantitating hydrogen at a platinum electrode. Circ. Res. 9:1093–1097.

Johnson, J. A., H. M. Cavert, and N. Lifson. 1952. Kinetics concerned with distribution of isotopic water in isolated perfused dog heart and skeletal muscle. Amer. J. Physiol. 171:687–693.

Kety, S. S. 1951. The theory and applications of the exchange of inert gas at the lungs and tissues. Pharmacol. Rev. 3:1–41.

Kety, S. S. 1960. Theory of blood-tissue exchange and its application to measurement of blood flow. Methods Med. Res. 8:223–227.

Kirk, E. S., and C. R. Honig. 1964. Nonuniform distribution of blood flow and gradients of oxygen tension within the heart. Amer. J. Physiol. 207:661–668.

Landau, W. M., W. H. Freygang, Jr., L. P. Rowland, L. Sokoloff, and S. S. Kety. 1955. The local circulation of the living brain; values in the unanesthetized and anesthetized cat. Trans. Amer. Neurol. Assoc. 80:125.

Lassen, N. A. 1964. Muscle blood flow in normal man and in patients with intermittent claudication evaluated by simultaneous Xe^{133} and Na^{24} clearances. J. Clin. Invest. 43:1805–1812.

Lassen, N. A. 1967. On the theory of the local clearance method for measurement of blood flow including a discussion of its application to various tissues. Acta Med. Scand. Suppl. 472:136–145.

Love, W. D., and G. E. Burch. 1957a. A study in dogs of methods suitable for estimating the rate of myocardial uptake of Rb^{86} in man, and the effect of 1-norepinephrine and Pitressin[R] on Rb^{86} uptake. J. Clin. Invest. 36:468–478.

Love, W. D., and G. E. Burch. 1957b. Differences in the rate of Rb^{86} uptake by several regions of the myocardium of control dogs and dogs receiving 1-norepinephrine or Pitressin[R]. J. Clin. Invest. 36:479–484.

Moir, T. W. 1966. Measurement of coronary blood flow in dogs with normal and abnormal myocardial oxygenation and function. Circ. Res. 19:695–699.

Moir, T. W., and D. W. DeBra. 1967. Effect of left ventricular hypertension ischemia and vasoactive drugs on the myocardial distribution of coronary flow. Circ. Res. 21:65–74.

Palmer, W. H., W. M. Fam, and M. McGregor. 1966. The effect of coronary vasodilatation (dipyridamole induced) on the myocardial distribution of tritiated water. Canad. J. Physiol. Pharmacol. 44:777–782.

Paradise, N. F., C. R. Swayze, D. H. Shin, and I. J. Fox. 1971. Perfusion heterogeneity in skeletal muscle using tritiated water. Amer. J. Physiol. 220:1107–1115.

Reivich, M., J. Jehle, L. Sokoloff, and S. S. Kety. 1969. Measurement of regional cerebral blood flow with an antipyrine-^{14}C in awake cats. J. Applied Physiol. 27:296–300.

Renkin, E. M. 1959. Transport of potassium[42] from blood to tissue in isolated mammalian skeletal muscles. Amer. J. Physiol. 197:1205–1210.

Sejrsen, P., and K. H. Tonnesen. 1968. Inert gas diffusion method for measurement of blood flow using saturation techniques. Circ. Res. 22: 679–693.

Stosseck, K. 1970. Hydrogen exchange through the pial vessel wall and its meaning for the determination of the local cerebral blood flow. Pflügers Arch. 320:111.

Sugishita, Y., S. Kaihara, H. Yasuda, M. Iio, S. Murao, and H. Ueda. 1971. Myocardial distribution of blood flow in the dog, studied by the labeled microsphere. Jap. Heart J. 12:50.

Sullivan, J. M., W. J. Taylor, W. C. Elliott, and R. Gorlin. 1967. Regional myocardial blood flow. J. Clin. Invest. 46:1402–1412.

Thompson, A. M., H. M. Cavert, N. Lifson, and R. L. Evans. 1959. Regional tissue uptake of D_2O in perfused organs: rat liver, dog heart, and gastrocnemius. Amer. J. Physiol. 197:897–902.

Wakim, K., and F. C. Mann. 1942. The intrahepatic circulation of blood. Anat. Rec. 82:233.

Wallin, J. D., A. H. Israelit, F. C. Rector, Jr., and D. W. Seldin. 1972. Intrarenal plasma flow distribution during micropuncture in the dog. Amer. J. Physiol. 222:649–652.

Warner, G. F., E. L. Dobson, N. Pace, M. E. Johnston, and C. R. Finney. 1953. Studies of human peripheral blood flow: effect of injection volume on the intramuscular radiosodium clearance rate. Circulation 8:732–734.

part 3/ EQUIPMENT AND MATERIALS

chapter 18/ Indocyanine Green

Michael R. Tripp/Claude R. Swayze/I. J. Fox
Department of Physiology/University of Minnesota/
Minneapolis, Minnesota

GENERAL PROPERTIES

Indicator-dilution curves using nondiffusible indicators are widely performed in man and animals for measurement of cardiac output and local blood flows, as well as for diagnostic purposes (Fox and Wood, 1960a; Fox, 1962; Fox, 1966). In 1956 a new indicator dye, Indocyanine green (ICG) (Fig. 18.1), was introduced which has properties especially suited for this purpose. (Fox et al., 1956; Fox et al., 1957b; Fox and Wood, 1960b; Fox et al., 1959).

The primary advantage of ICG in indicator-dilution techniques is related to its spectral properties. Its peak spectral absorption for dilute solutions in blood (800 nm) (Fig. 18.2) lies at an isosbestic point for reduced hemoglobin and oxyhemoglobin, which is located in a spectral region at which hemoglobin solutions themselves have a low optical density. This obviates interference in the continuous recording of dilution curves from the presence of fluctuations in the oxygen saturation of the blood of animals, normal man, or patients with congenital heart disease (Wood et al., 1955; Fox et al., 1957b). As the concentration of fresh aqueous solutions is increased, ICG aggregates to form dimers and higher order micellar aggregates, the

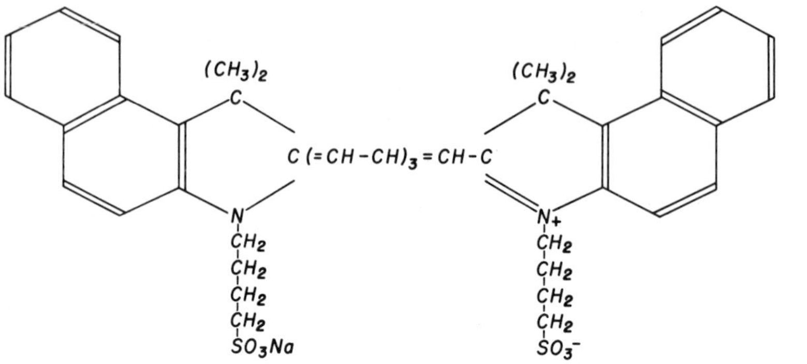

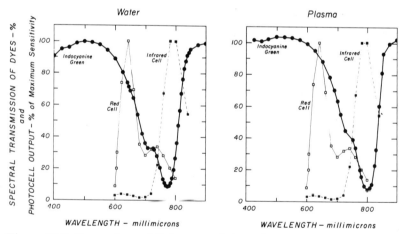

Figure 18.1 Structural formula of Indocyanine green: anhydro-3,3,3',3'-tetramethyl-1,1'-(4 sulfobutyl)-4,5,4',5'-dibenzoindotricarbocyanine hydroxide sodium salt. Molecular weight: 775. Note that ICG has the structure of a detergent.

"critical micellar concentration" being approximately 1.6×10^{-5} to 2.0×10^{-5} M or 12 to 15 mg/liter. Beginning at slightly lower concentrations, the absorption spectrum of ICG in aqueous solution

Figure 18.2 Comparison of light transmission of equal concentrations (5 mg/liter) of Indocyanine green in water and in plasma, and spectral sensitivities of oximeter photocells. These studies were performed immediately after preparation of fresh solutions of dye in water and in plasma. Note shift in absorption peak of dye from 775 nm in aqueous solution to 800 nm when dye is dissolved in plasma.

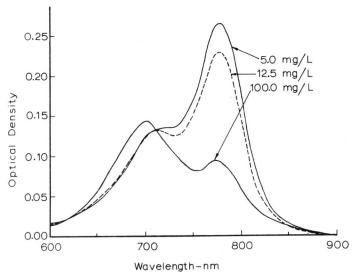

Figure 18.3 Absorption spectra of Indocyanine green in water at three different concentrations. With increasing concentration, the absorption peak at 775 nm progressively decreases whereas that at 700 nm progressively increases. These spectral changes with increasing concentration are evidence for the formation of dimers and higher aggregates. Optical densities were normalized by dividing measured values by concentrations in mg/liter and correcting to a path length of 1 cm. (Tripp and Fox, 1973).

changes; at higher concentrations, a second absorption peak appears in the region of 700 nm as the primary peak at 775 nm becomes attenuated (Baker, 1966; Saunders *et al.,* 1970; Tripp and Fox, 1973) (Fig. 18.3). Sutterer and co-workers (1966) noted a shift in the absorption peak from 780 to 900 nm when concentrated (1.25 mg/ ml) aqueous solutions of ICG were kept at room temperature over a period of 2 to 7 days. The probable aggregation causing this shift was found to be irreversible, in contrast to the rapid dissociation upon dilution of the aggregates present in freshly prepared concentrated solutions (*vide infra*). It was also noted by Sutterer and associates that, in dilute aqueous solutions (1.5 mg/liter), the absorbance of ICG at wave lengths between 650 and 1000 nm completely disappeared over this period. This is explainable on the basis of "oxidative fading" (Hamer, 1964).

ICG is bound to the plasma proteins and is therefore distributed in the intravascular space. Blood volumes calculated by using ICG (the blood levels of the dye at zero time having been estimated by extrapolation) were not significantly different from those calculated by means of radioiodinated serum albumin (Cherrick *et al.*, 1960; Wheeler *et al.*, 1958; Bradley and Barr, 1968). The dye was shown in earlier studies to be primarily bound to plasma albumin (Fox and Wood, 1960*b*; Cherrick *et al.*, 1960). However, a more recent study by Baker (1966) found ICG to be largely associated with the α_1-lipoproteins of the globulins. Abnormal patterns of protein binding have been observed in acute viral hepatitis, but dye concentrations were not reported(Janecki and Krawcynski, 1970). An explanation for these discrepancies is not available.

The dye was found to be rapidly and completely excreted by the liver. This permits the performance of large numbers of dilution curves in the same animal without undue dye accumulation, helps to minimize its toxicity, and immediately suggested its use in the measurement of liver blood flow and liver function. The following is an illustrative sampling of the extensive literature of these subjects. Although ICG has an excretion half-time of less than 10 minutes in normal humans and dogs (Fox *et al.*, 1957*a*; Wheeler *et al.*, 1958; Cherrick *et al.*, 1960; Hunton *et al.*, 1960; Rapaport *et al.*, 1959; Reemtsma *et al.*, 1960), the excretion rate is considerably prolonged in some cases of impaired liver function, for example in cirrhosis (Cherrick *et al.*, 1960; Hunton *et al.*, 1960; Reemtsma *et al.*, 1960), and after experimental myocardial infarction in the dog (Hood *et al.*, 1968). Additionally, bilirubin, ethanol, and BSP may reduce the dye's excretion rate (Wheeler *et al.*, 1958; Hargreaves, 1966; Kotelanski *et al.*, 1969; Groszmann *et al.*, 1969). Excretion has followed first or second order kinetics in different studies (Wheeler *et al.*, 1958; Jaross and Schentke, 1969). Although the use of ICG in the measurement of hepatic-splanchnic blood flow is the subject of another chapter (Chapter 12), its use as a test of liver function, which is the subject of widespread study, is not within the purview of this book.

The fact that ICG does not stain the skin and mucous membranes also helps to assure that there is no limitation on the number of curves which may be performed in a single person. Its high degree of absorbance at its spectral absorption peak in dilute solution (800 nm), as compared to Evans blue, permits the injection of small doses of dye in the performance of dilution curves.

ICG was shown to follow Beer's law in whole blood up to moderate concentrations (12 to 15 mg/liter), with only a slight change in the slope of the linear relationship between concentration and optical density, dictated by this law, at concentrations up to 24 mg/liter, which is convenient for calibration purposes. These data were obtained in whole blood using a modified oximeter; both of these circumstances violate assumptions on which the Beer's law relationship is based (Fox and Wood, 1960*b*). In recent studies using a spectrophotometer or filter photometer, a linear relation between concentration and optical density of ICG in plasma has been found up to a concentration of 25 mg/liter (Saunders *et al.*, 1970; Sekelj *et al.*, 1967), but others have reported deviations from the linear relation at low concentrations (Simmons and Shephard, 1971; Gathje *et al.*, 1970).

Finally, studies indicated the rapid stabilization of the optical density of the dye-blood mixture after its addition to whole blood. These studies are discussed further below.

Thus ICG was free of the disadvantages which complicated the use of other nondiffusible indicators, notably the blue dyes, for hemodynamic studies (Zijlstra and Mook, 1962; Fox and Wood, 1957*b;* Nicholson *et al.*, 1951; Lacy *et al.*, 1955). The basic chemistry of the cyanine dyes has been reviewed (Hamer, 1964; Brooker, 1966) and will be considered only in regard to specific questions raised in the present chapter.

Validation of the ICG dilution technique has been the subject of numerous investigations. Fox and co-workers (1957*b*) originally used an indirect approach, comparing flow measurements made with ICG and Evans blue in animals. No significant difference between flow values from successive ICG and Evans blue dilution curves was found in these studies. Earlier studies had validated the dilution technique by comparison of the results obtained using Evans blue with those determined by direct flow measurements (Kinsman *et al.*, 1929) and by the direct Fick procedure (Moore *et al.*, 1929; Hamilton *et al.*, 1948). A number of later studies have also shown close agreement between the flow values determined by the use of ICG and values from simultaneously performed direct Fick procedures (Richardson *et al.*, 1959; Miller *et al.*, 1962; Grenvik, 1966; Klocke *et al.*, 1968). The results from other indirect methods have also been compared with those from ICG dilution techniques with good agreement being reported. The study by Klocke and co-workers (1968), for example, found no significant difference between the values obtained by the use of ICG or by the direct Fick procedure and those from using dissolved hydrogen as the indicator. The hydrogen concentration was measured continuously with an intra-arterial platinum electrode (calibrated by gas chromatography) after constant-rate or sudden-single injections. They also found that flow values from these methods compared closely with direct volumetric determinations of flow using a right heart by-pass preparation and a volumetrically calibrated pump. Similarly good agreement between ICG dilution and direct volumetric measurements has been reported for a heart-lung preparation (Sekelj et al., 1967), and for isolated dog forelimb preparations (Baker and O'Brien, 1964; Wolthuis and co-workers, 1969). However, a number of studies have reported erroneous flow measurements with ICG. The probable source of these discrepancies, the spectral stabilization of ICG, will be considered in detail in the following section.

SPECTRAL STABILIZATION

Rapid stabilization of the optical density of the dye-blood mixture after addition of the dye to blood is essential to the continuous, quantitative, photometric recording of indicator-dilution curves. This is particularly true of the more recent techniques, partly made possible by ICG, in which the injection and sampling sites have been brought very closely together, thus shortening the time available for optical stabilization. Consider, for example, the following: right atrial or ventricular injection and pulmonary artery sampling (Fox and Wood, 1957a; Fritts et al., 1957); thoracic aorta flow measurement in man (Grace and co-workers, 1957); and especially, techniques using fiber-optic detecting devices (Polanyi and Hehir, 1960; Fox et al., 1964), in which there may be a minimum of delay before detection of the dye-blood mixture after dye injection, since sampling of the dye-blood mixture through hydraulic tubing is not required (Fox et al., 1957c). For these reasons, studies were originally performed by comparing the rate of stabilization of ICG and blood with that of Evans blue and blood; the spectral absorption of the latter mixture was previously assumed to stabilize almost instantaneously (Andres et al., 1954). These experiments employed a Waring Blendor from which blood was continuously sampled via a cuvette-needle assembly. The system had a very slow response time (Fox et al., 1957c) largely due to a slow mixing time, thereby making it impossible to draw conclusions concerning spectral stabilization in less than 5 seconds. It was nonetheless reassuring that the spectral absorption of the mixture of ICG and blood appeared to stabilize in at least that period of time. From this comparison with Evans blue, the inference was drawn that the spectral stabilization of the ICG dye-blood mixture was rapid enough to permit its use in techniques for continuous recording of indicator-dilution curves. Further evidence for this inference was the basic data showing no systematic difference between flows calculated from Evans blue and ICG dilution curves recorded successively in the same dogs. However, since the appearance times were longer than 5 seconds in these experiments, no conclusions concerning the spectral stabilization of ICG and blood in less than this interval could be drawn from these data.

To place this inference on a firmer footing, Bassingthwaighte and co-workers (1962) devised an experiment in which the dye-blood mixture was sampled simultaneously through two densitometers in tandem at the pulmonary artery (PA) and via a third densitometer at the femoral artery (FA), following injections of ICG into the inferior vena cava. The shortest appearance times of the dilution curves recorded at the proximal PA densitometer were less than 2 seconds (range: 1.5 to 8 seconds); the appearance times of the curves recorded at the distal PA densitometer ranged from 10 to 22 seconds after injection (the appearance times at the distal PA densitometers were arranged to approximate those of the dilution curves recorded simultaneously at the FA in each animal). No significant difference was found between the areas of the curves recorded from the PA by the densitometers in tandem; however, a systematic difference between the areas of the PA and FA curves of 2.5% was found, the PA curves having the smaller areas. These results were considered to rule out delayed stabilization of the dye-blood mixture as cause of the difference in the areas of the PA and FA curves.

In recent years, the entire problem of the spectral characteristics of ICG has been reinvestigated in other laboratories. The conclusion drawn, in some instances, was that large overestimations of flow result when cardiac output in an animal or flow in a model are measured using ICG as indicator. Delayed stabilization of the spectral transmission of the ICG dye-blood mixture has been implicated as the major cause of the overestimation of the true flow in one extensive study by Saunders and co-workers (1970). A dog right heart by-pass preparation with a cannulating flowmeter and three circulation models were used to obtain "actual flows" for comparison with flows calculated from sudden single or constant-rate injection indicator-dilution curves using ICG or Evans blue. In one experiment (Fig. 3A of Saunders and co-workers, 1970), dilution curves recorded simultaneously at the aortic arch and at the femoral artery overestimated the "true flow" by 31.7 and 10%, respectively, and "the error due to slow optical change of indocyanine green" was felt to account "for at least some of the observed discrepancy." In another series of 46 determinations, the ICG dilution curves overestimated the "true flow" by an average of 28%. Experiments using model flow systems and ICG to which a "standard amount" of protein had been added (5 ml of plasma to 20 ml of diluent) resulted in a 32% flow overestimate, whereas dye to which extra protein had been added [15 ml of 25% human serum albumin (HSA) + 10 ml of diluent] resulted in a 46% overestimate of the true flow.

The overestimation of the flow using sudden-single injection ICG dilution curves in a flow model was found by Saunders and co-workers to be temperature-dependent; a 32% overestimate at room temperature was reduced to a 14% flow overestimate when the same experiment was repeated at 37.5°C. Since similar injections of Evans blue into blood and plasma caused only a 2.5 and 2% flow overestimate, respectively, the model was exonerated as cause of the overestimate in the ICG flow measurement. The introduction of a delay loop into the flow model upstream to the densitometer reduced the flow overestimate after a sudden single injection of ICG from 25 to 7%. In the one experiment cited by these authors, in which the flow overestimate using ICG was abolished, a constant-rate injection of the dye was made into a model; a 22% overestimation of the flow resulted, which was reduced to a 6% overestimate by addition of a delay loop. The subsequent use of an integrated sample technique for flow measurement, which would have the effect of allowing a considerable time period for spectral stabilization of the dye-blood mixture, resulted in an accurate flow measurement.

A number of possible sources of error were investigated by these workers. They concluded that "the error was due to delay in changing from the absorption spectrum of concentrated indocyanine green (1000 mg/liter) to the different spectrum of dilute dye (1–40 mg/liter)"; that rapid (< 2 second) stabilization of the spectral transmission of ICG was "not proven"; and finally, that studies using this dye at a "temperature below 37°C or with appearance times under 6 seconds are likely to produce marked overestimates of blood flow." These conditions, of course, include many circumstances under which ICG is currently employed. The major objection to this work is that, in most instances, the ICG dye was not prepared in the recommended manner, i.e. albumin or plasma were added to the concentrated ICG dye.

Other studies have also reported errors in flow rates determined from dilution curves using ICG. The following statement appeared in the abstract of a recent paper by Spangler and co-workers (1971), ". . . pairs of dye-dilution curves [were] recorded from the pulmonary artery (PA) and left atrium (LA) after right atrial injections of indocyanine green. In 345 pairs, the areas of the extrapolated LA curves exceeded those of the PA curves by 8% (S.D., 14%) indicating that delayed color development of the dye-blood mixture is not a serious source of error in standard dye-dilution techniques." Contrariwise, in the section on "color development," these authors state that their data do not support the conclusions of Saunders and co-workers, i.e. "that at least 6 sec are required for the indocyanine green-protein complex to develop." They suggest "a combination of poor mixing, extrapolation errors, time-averaged sampling" to explain this 8% difference with "perhaps some small effect of delayed color development." They also performed experiments using three densitometers in series at the pulmonary artery and reported that the areas of the curves recorded at the distal densitometer exceeded those recorded at the proximal densitometer by 8.6% of the mean of the two areas, which differs from the results of the earlier study by Bassingthwaighte et al. (1962). Jacobs and co-workers (1969) have also reported overestimations of flow as determined by ICG, and their finding of larger errors with shorter intervals to detection of the dye is suggestive of slow stabilization as cause of the errors. Since multiple potential sources of error complicate these studies, and since the good agreement reported between direct flow measurements and values determined by ICG (*vide supra*) involved long stabilization times, we studied the optical stabilization of ICG directly to define the circumstances in which this could cause errors in flow measurement (Tripp *et al.*, 1973).

Concentrated solutions of ICG[1] were prepared in media varying in protein concentration and type and also in ionic composition and strength. This simulated the dye preparations used by other investigators (Saunders *et al.*, 1970; Spangler *et al.*, 1971) and permitted evaluation of the effect of the individual constituents on the rate of optical stabilization of the ICG-blood mixture after dilution.

[1]Cardio-Green® dye, lots 222 and 658, Hynson, Westcott and Dunning, Inc., Baltimore, Maryland.

A standard technique for studying rapid reactions (Estabrook et al., 1972) with a commercially available cuvette-plunger assembly[2] permitted the continuous measurement, on a split-beam, dual-wave length recording spectrophotometer,[3] of the changes in optical density after mixing and diluting ICG in plasma. Mixing-dilution was accomplished with the plunger assembly consisting of a cup on a spring-loaded shaft, which on rapid depression created turbulence, mixing the ICG in the cup with the plasma in the cuvette before re-emerging. Nonspecific optical density changes, for example, due to bubble formation during mixing-dilution, necessitated use of the dual-wave length mode. In this mode, light beams at two wave lengths are alternately passed through the cuvette, and the difference in the optical density of the cuvette contents at these wave lengths is recorded. It was thus possible to minimize nonspecific optical density changes during mixing-dilution by "electronically subtracting" the signal at a wave length at which only nonspecific optical density changes are detected from the signal at a second wave length at which both nonspecific changes and optical density changes due to the presence of ICG are detected (Estabrook et al., 1972). One monochromator was set at 805 nm, which is an isosbestic point for reduced hemoglobin and oxyhemoglobin, the sensitivity peak of densitometers used to detect ICG, and a wave length which is near the absorption peak of a dilute plasma solution of ICG, 800 nm. The second monochromator was set at 890 nm, a wave length at which freshly prepared dilute and concentrated solutions of ICG have no significant absorbance (Fox et al., 1957b; Sutterer et al., 1966; Baker, 1966; Tripp and Fox, 1973). Consequently, when the concentrated dye is diluted, mixed, and bound to protein, the multiple dye states with different absorption spectra, which are potentially but transiently present, should not influence the optical density changes measured by monochromator 2. That ICG as used in the experiment had no significant absorbance at 890 nm was confirmed on the same spectrophotometer operated in the split-beam mode. A slit width of 10 nm was used so that final concentrations of ICG similar to those reached in vivo could be attained without exceeding the high voltage limit of the photomultiplier tube.[4]

[2]Aminco models B2–65085 and B3–65085, American Instrument Co., Bethesda, Maryland.
[3]Aminco model DW-2 Dual-Beam, Dual-Wavelength Spectrophotometer, adapted for RCA 7102 photomultiplier tube, American Instrument Co., Bethesda, Maryland.
[4]RCA 7102 photomultiplier tube.

The degree of compensation achieved for the nonspecific optical density changes associated with the mixing-dilution was assessed by repeating the plunger depressions after the optical density had stabilized, i.e. after a plateau had been reached. Although compensation was incomplete, the return to the optical density plateau was always completed within 1.6 seconds (recorder response: full scale − 100 milliseconds). Without compensation, i.e. in the split-beam mode, the return to the plateau optical density required up to 6 seconds after similar plunger depressions. The dual-wave length mode thus allowed measurement of optical density changes due only to ICG 2 seconds after dilution, a time interval which is shorter than the interval between dilution and detection of the dye in most hemodynamic applications of ICG (Fig. 18.4).

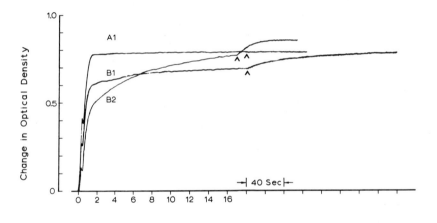

Time - Seconds

Figure 18.4 Original, successive recordings of changes with time in optical density (805 − 890 nm) at 37°C after mixing-dilution with plasma of Indocyanine green (1 mg/ml) prepared in three different solutions. Note the rapid (<2 seconds) spectral stabilization of ICG in distilled water (A1), compared to the slowed stabilization of dye in a solution containing a moderate concentration of protein (B1) or of dye in a low protein-high ionic strength medium (B2). Time base was changed at the *arrows* from 2 to 20 seconds/inch.

Results are reported in per cent of stabilized optical density, because it was found that a variable amount of dye was emptied from the cup during the mixing, resulting in variations of up to 15% in the stabilized optical density attained. Intentional variations in the amount of dye diluted did not affect the time course of stabilization. For each dilution, 9, 10, or 12 μl of the 1 mg/ml dye solution or 4 μl of the 2.5 mg/ml dye solution were diluted in 3 ml of plasma to give a final concentration of 3 to 4 mg/liter, which was within the linear region of the ICG concentration versus optical density plot determined in each experiment. After the plasma had been equilibrated to 25 or 37°C for 5 minutes, the mixing-dilution was accomplished by two rapid plunger depressions.

Our study was directed primarily at comparing the stabilization rate of concentrated ICG prepared in distilled water with that prepared in mixtures similar to those used by others (Saunders et al., 1970; Spangler et al., 1971). Dilutions of these three solutions were performed throughout each experiment to rule out changes in the dye solutions or plasma, both used within 8 hours, and to study the effect of varying anticoagulant, dye lot, and plasma pH.

A total of 131 optical density recordings was made and 22 of these were discarded either because stabilization of the optical density failed to occur (no plateau was established) or because of the appearance of irregular fluctuations in optical density. In 19 of the 22 cases, the aberrations occurred more than 16 seconds after the mixing-dilution, the curves having been typical for the particular dye solution used before that time. Cursory examination revealed that spillage and subsequent drying of plasma on the optical surfaces of the cuvette by the sudden plunger depressions accounted for the majority of the abnormal recordings. Abnormal recordings were fairly uniformly distributed among the different dye solutions. The levels of the optical density plateaus for the various dye solutions were not significantly different. Finally, in approximately 90% of the curves, the maximum spontaneous fluctuation from the mean level during each plateau was less than 1% of this mean value, whereas for the remaining 10% the maximum fluctuation averaged 1.5 (range: 1.1 to 2.2%).

Original recordings of the changes in optical density at 37°C after successive mixing-dilution with plasma of ICG dissolved in each of the three solutions studied most extensively are shown in Fig. 18.4. Note the rapid (< 2 seconds) stabilization of the optical density when concentrated ICG (1 mg/ml) in distilled water (A1) was mixed and diluted with plasma. Compare this to the slowed stabilization when dye (1 mg/ml) in a solution containing a high protein concentration, similar to that of Saunders and co-workers (B1), or in a low protein-high ionic strength medium, similar to that of Spangler and co-workers (B2), was similarly mixed and diluted. Also seen are the typical variations in plateau levels between recordings, the definitive nature of the plateaus, and the optical densities attained after stabilization.

The results from all of the experiments using concentrated ICG dissolved in the same solutions as in the recordings in Fig. 18.4 as well as in other solutions are summarized in Fig. 18.5 and Table 18.1.

EFFECT OF VARYING THE TEMPERATURE AND THE CONCENTRATION OF THE CONCENTRATED DYE IN DISTILLED WATER

For all practical purposes, the optical density of ICG dissolved in distilled water (A1) stabilized within 2 seconds after mixing-dilution with plasma, except when the temperature was lowered to 25°C (A1*) when 3 seconds were required (Fig. 18.5, *left panel*). At 25°C the optical density was 4% below the plateau value at 2 seconds ($p < 0.005$). As seen in Table 18.1, when ICG (1 mg/ml) in distilled water (A1) was diluted in plasma at 37°C, the optical density was below the plateau value by approximately 1% over the initial 16 seconds. Such a difference was not evident with dye at an initial concentration of 2.5 mg/ml (A1†). Variation of plasma pH from pH 6.70 to 7.98, of the dye lot, or of the method of anticoagulation caused no apparent change in the optical density at 2 seconds.

EFFECT OF VARYING THE PROTEIN CONCENTRATION OF THE CONCENTRATED DYE SOLUTION

The addition of protein to the solution [plasma-distilled H_2O, 1:4 (B1)] markedly slowed the rate of stabilization of the ICG-plasma mixture; the values at 2 and 16 seconds were 85 and 93%, respectively, of the plateau value (Table 18.1). Lowering the protein concentration and increasing the ionic strength of the solution [plasma-distilled H_2O-isotonic saline, 1:10:14 (B2)] also caused a marked slowing of the stabilization rate; the values at 2 and 16 seconds were only 76 and 96%, respectively, of the plateau value. For solutions B1 and B2, variations of plasma pH, dye lot, or the plasma donor resulted in small but statistically insignificant ($p > 0.05$) changes in the transient optical densities. This is reflected in the large variability in the values for B1 and B2. Reducing the ionic strength while retain-

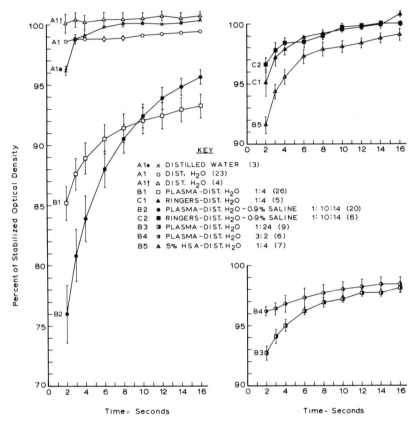

Figure 18.5 Comparison of mean changes with time in optical density (805 − 890 nm) as per cent of stabilized optical density, after mixing-dilution with plasma of all of the different concentrated Indocyanine green solutions listed in the *key*. *Vertical lines* are ± SEM. In the *key: numbers in parentheses* denote the number of dilutions performed with the particular solution; *A1** refers to dilutions performed at 25°C, all others were performed at 37°C; *A1†* denotes ICG solution at 2.5 mg/ml, all others were at 1 mg/ml. *Left panel:* Mean changes in optical density with time of ICG solutions, original recordings of which are shown in Fig. 18.4. Also shown are effects of temperature and of concentration of the dye in distilled water on optical stabilization rate. *Right panel, upper:* Note the small delay in stabilization in an ionic medium, *C1, C2.* Also note the more rapid stabilization of ICG in HSA (*B5*) as compared to ICG in plasma (*B1*). *Right panel, lower:* Note the more rapid stabilization as the protein concentration of the medium was increased, *B3* versus *B4.* See text for details.

Table 18.1 Change with time of optical density, expressed as per cent of plateau optical density, after mixing-dilution of the particular Indocyanine green dye preparation with plasma

No.	Dye preparation[a] Composition	n	Optical density, as per cent of plateau optical density, at the specified time after onset of mixing-dilution						
			2 sec	3 sec	4 sec	6 sec	8 sec	12 sec	16 sec
A1	DW	23	98.6 ± 0.1[b]	98.8 ± 0.2	98.8 ± 0.2	98.8 ± 0.2	98.9 ± 0.2	99.2 ± 0.1	99.4 ± 0.1
A1*[c]	DW	3	96.2 ± 0.4	98.8 ± 0.4	99.1 ± 0.4	99.8 ± 0	100.1 ± 0	100.0 ± 0	100.3 ± 0
A1†[d]	DW	4	100.1 ± 0.8	100.4 ± 0.6	100.2 ± 0.6	100.4 ± 0.6	100.4 ± 0.6	100.7 ± 0.4	100.7 ± 0.4
B1	Pl-DW, 1:4	26	85.2 ± 1.4	87.6 ± 1.3	88.9 ± 1.3	90.5 ± 1.2	91.4 ± 1.2	92.4 ± 1.1	93.2 ± 1.0
C1	Ri-DW, 1:4	5	95.1 ± 1.1	97.2 ± 0.7	97.9 ± 0.5	98.9 ± 0	99.2 ± 0.3	99.7 ± 0.2	100.8 ± 0.2
B2	Pl-DW-Iso, 1:10:14	20	76.0 ± 2.4	80.8 ± 2.1	83.9 ± 1.9	88.0 ± 1.5	90.5 ± 1.2	93.8 ± 1.2	95.6 ± 0.6
C2	Ri-DW-Iso, 1:10:14	6	96.6 ± 0.6	97.8 ± 0.6	98.4 ± 0.5	98.5 ± 0.4	99.0 ± 0.3	99.8 ± 0.2	100.0 ± 0.2
B3	Pl-DW, 1:24	9	92.7 ± 0.6	94.1 ± 0.6	95.0 ± 0.5	96.2 ± 0.3	96.9 ± 0.4	97.7 ± 0.3	98.1 ± 0.4
B4	Pl-DW, 3:2	6	96.2 ± 0.4	96.4 ± 0.5	96.8 ± 0.7	97.3 ± 0.8	97.7 ± 0.7	98.2 ± 0.6	98.4 ± 0.6
B5	HSA-DW, 1:4	7	91.6 ± 0.7	94.4 ± 0.6	95.6 ± 0.6	97.0 ± 0.6	97.9 ± 0.6	98.4 ± 0.4	99.1 ± 0.5

[a] Abbreviations: DW, distilled water; Pl, plasma; Ri, Ringer's solution; Iso, isotonic saline; HSA, 5% human serum albumin.
[b] ±SEM.
[c] Curves were performed at 25°C; all others were at 37°C.
[d] Dye was at a concentration of 2.5 mg/ml; all other preparations were at 1.0 mg/ml.

ing the same protein concentration as in B2 [plasma-distilled H_2O, 1:24 (B3)] caused a significantly more rapid stabilization of the optical density than was found for either B1 or B2, with a 2-second value of 93% of the plateau value. The solution with the highest plasma protein concentration [plasma-distilled H_2O, 3:2 (B4)] was found to have 2- and 4-second values of 96%, both of which were significantly higher than those for B3 but which, at the same time, were below those for the dye in water at 37°C (A1) ($p < 0.05$). This increased rate of stabilization of the solution with the highest protein concentration is seen in Fig. 18.5 (*lower right panel*). Substituting human serum albumin[5] for the plasma protein of solution B1 [5% HSA-distilled H_2O, 1:4 (B5)] also increased the rate of stabilization; the value at 2 seconds was 92% as compared to the value of 85% for B1 at this time ($p < 0.05$).

EFFECT OF VARYING THE IONIC COMPOSITION OF THE CONCENTRATED DYE SOLUTION

When the ionic composition of the concentrated dye solution was similar to that of B1 and B5 but without protein [Ringer's solution-distilled H_2O, 1:4 (C1)], a small effect on the rate of stabilization was found; the 2-second value was 95%. Similarly, when the solution's ionic composition [Ringer's solution-distilled H_2O-isotonic saline, 1:10:14 (C2)] was approximately that of B2, a lesser effect on stabilization rate was found; the 2-second value for C2 was 97%. Optical densities for C1 and C2 were significantly lower than the values for A1 until 6 and 4 seconds after dilution, respectively ($p < 0.05$) (see Table 18.1). The interaction between the effect of the protein concentration and of the ionic composition of the concentrated ICG solutions on the rate of spectral stabilization may be seen in Table 18.2.

These results show that, if ICG is prepared in distilled water, delayed optical stabilization will probably not be a significant source of error in the determination of flow by dye-dilution curves with appearance times of at least 2 seconds, which is less than the time between dilution and detection in most hemodynamic applications. Representative methods of anticoagulation, plasma pH values, and concentration changes upon dilution were shown to have no effect 2 seconds after dilution. The use of plasma instead of whole blood should not alter

[5]Albumisol®, Merck, Sharp & Dohme, West Point, Pennsylvania.

Table 18.2 Comparison of rate of stabilization of optical
density in solutions of varying protein concentration and ionic composition

PLASMA PROTEIN CONCENTRATION

"IONIC STRENGTH"	None	Lo (1:24)[a]	Mod (1:4)	Hi (3:2)
None	A1 O			
Lo (1:24)[b]		B3 + +		
Mod (1:4)	C1 +		B1 +++	
Hi (3:2)	C2 +	B2 ++++		B4 +

Optical density at 2 sec
as % of plateau

O	≥ 98.6 %
+	≥ 95.0 %
+ +	≥ 90.0 %
+ + +	≥ 85.0 %
+ + + +	≥ 75.0 %

[a] Variation in plasma protein concentration by specified plasma:distilled water ratio.
[b] Variation in "ionic strength" by specified isotonic ionic medium:distilled water ratio.

this conclusion (Saunders *et al.,* 1970; Anderson and Sekelj, 1968).
Changes in optical density after the initial stabilization were not
investigated, since earlier studies had established the stability of the
dye over long periods after dilution in plasma and blood (Fox and
Wood, 1960*b;* Simmons and Shephard, 1971). Our results indicate
that the marked temperature effect for dye prepared in protein solutions (Saunders and co-workers, 1970) is much reduced when the dye
is prepared in water.

Most calibration procedures involve the preparation of standards
several minutes before their optical densities are determined, allowing
time for stabilization. When dilution curves are measured on the basis
of such a calibration, the true concentration is underestimated whenever the curves are recorded before the dye has stabilized. In such cases
flow is overestimated since dye concentration enters the denominator
of the equations for flow.

Our results indicate that the erroneous flows reported by Saunders
et al. (1970) were a direct result of the method of dye preparation.
Since a dye curve is not describable by a simple function (Stow and
Hetzel, 1954) only an approximate comparison of results is possible.
They report a 14% error when flow in a circulation model was mea-
sured using a dye solution similar to B1 at 37°C. Our data would pre-

dict an error of between 16 and 12% for such a curve, using the reciprocals of the values for B1 at 2 and 4 seconds, respectively. Our results
also appear to explain the results of Spangler *et al.* (1971). As mentioned earlier, they found that areas of left atrial (LA) dye curves
were 8% larger than the mean of the areas of curves recorded simultaneously from the LA and pulmonary artery (PA). The difference
in the values for B2 at 4.6 and 9.1 seconds, the mean transit times for
their PA and LA curves, respectively, is 7% of the mean of the
two values. Similarly, using the values for the range in mean transit
times to the distal densitometer given for their three-densitometer-in-
series experiments, an approximate range of differences in dye curve
areas between curves recorded at the proximal and distal densitometers
varying between 7 and 11% of the mean of the two areas would be
predicted from our data (B2), which compares favorably with their
difference of 8.6% of the mean of the two areas. These approximate
comparisons suggest that delayed stabilization was the primary cause
of their results, contrary to their conclusion (vide supra). The other
study suggestive of slow stabilization unfortunately did not report the
method of dye preparation (Jacobs *et al.*, 1969).

The mechanism by which proteins and electrolytes in the concentrated dye solution cause slow optical stabilization after dilution in
plasma may be considered a "stabilization" of dye aggregates existing
in the concentrated solution. The colloidal behavior of ICG, which
has the structure of a detergent, and the associated spectral changes
have been determined for aqueous solutions (Baker, 1966; Tripp and
Fox, 1973) (see Fig. 18.3). Similar spectral changes have been shown
for concentrated dye solutions containing protein (Saunders *et al.,*
1970; Tripp and Fox, 1973). The spectral changes that are associated
with dilution of the order seen in this study and in the hemodynamic
applications of the dye have been used as evidence of dissociation into
the monomer form for acridine orange and thiacyanine dyes (West and
Pearce, 1965; Zanker, 1952). It is also known that ICG's absorption
peak shifts from 775 to 800 nm with protein binding at these low concentrations (Fox and Wood, 1960*b*). Therefore, our results for concentrated dye prepared in water can be considered evidence that for
this case, disaggregation and binding to protein are essentially completed in less than 2 seconds, as evidenced by the OD at 805 nm.
After dilution in plasma, dye which was prepared in protein solutions
has an absorption curve closely similar to that for dye prepared in

water (Saunders *et al.,* 1970; Tripp and Fox, 1973). Therefore, the final binding step and dye state are probably the same in both cases. This leaves disaggregation as the rate-limiting step causing the slow increase in optical density at 805 nm after dilution in plasma for all of our solutions except dye in water. Moreover, as shown in our study, protein type, protein concentration, and ionic composition of the concentrated dye solutions significantly influenced the rate of disaggregation. These results suggest that different, slowly reversible dye-protein associations occur at high dye concentrations, as compared to low dye concentrations, when the amount of protein relative to the amount of dye available for binding, is limited, thus "stabilizing" the aggregates, and that electrolytes alone exert a lesser "stabilizing" effect on dye aggregates.

In summary, our experiments demonstrate that the presence of protein and/or electrolyte in the solution used for the preparation of the concentrated ICG dye before dilution slows the rate of optical stabilization of the dye-plasma mixture after dilution. This finding provides a satisfactory explanation for some of the reported discrepancies in flow measurement comparing the ICG dilution technique with other techniques. When the concentrated dye was prepared in distilled water, the rate of spectral stabilization of ICG was rapid (<2 seconds) and would not be expected to be a source of error in flow measurement.

TOXICITY

A single paper has reported the occurrence of allergic reactions after intravascular injection of ICG for determination of cardiac output in three uremic patients (Michie *et al.,* 1971). Since uremic patients on a chronic hemodialysis program, such as the three patients in this report, customarily receive a number of different medications and since further premedication was probably given before the cardiac catheterization, it is very difficult to conclude which medication caused the toxic reaction.

Histamine release by ICG and by other chemical agents such as 48/80 was suggested by Michie and co-workers as the mechanism for the toxic reaction. It is of interest, therefore, to cite the three criteria, listed by Paton (1957), for establishing histamine release by a new therapeutic agent: (1) injection of the substance into the whole animal of a species with a sensitivity to histamine and to histamine liberators comparable to that of man, (2) intradermal testing in man, (3) demonstration of released histamine. With regard to the first criterion, ICG has been injected in massive doses into animals, not only with a sensitivity to histamine and histamine releasers comparable to that of man, but also in those much more exquisitely sensitive to them, e.g. guinea pigs, without any untoward symptoms or physical signs. Intradermal testing in man was also performed with negative results, as in the study of Michie and co-workers. With negative results in these two criteria, it was unnecessary to look for released histamine. The wide-spread use of ICG in animals and man without other such reports of toxicity casts serious doubts on the analogy drawn by the authors between ICG and 48/80, a well known histamine releaser. Paton states that "it has been noted that the toxicity of a compound such as 48/80 is roughly the same in all species," so that it would be very unlikely that some of the animals which received massive doses of ICG, both intravenously and by other routes in our original tests, would not have shown some toxic manifestations.

This leaves the possibility that the uremic state in some way enhances the histamine-releasing property of ICG, a property of the dye which has never been established but only suggested by Michie and co-workers. The fact that ICG is completely excreted by the hepatic route does not favor such enhancement. Furthermore, the failure of many others using ICG in uremic patients to find similar toxic manifestations, Hampers and co-workers (1967) and Goss and co-workers (1967) to cite only two early reports, is strong evidence against this possibility.

COMPLICATIONS

ICG, as available commercially, contains approximately 5% sodium iodide as a contaminant. For this reason, interference with studies of thyroid function in a patient who had recently undergone cardiac catheterization was suspected (Fox and Wood, 1960b). Direct evidence of suppression of thyroidal iodine uptake in normal subjects and in patients with hyperthyroidism has been reported after the use of ICG (Dumlao et al., 1970). It has also been reported that lipemia may result from the use of ICG (Tsubokura and associates, 1970). Recently, Hansing and co-workers (1972) noted a significant, transient increase in coronary sinus blood flow from intravascular injections of ICG for up to 3 hours after administration of hexobendine, a recently synthetized vasodilating agent.

INTERFERING SUBSTANCES

Erratic results in cardiac output measurements using ICG were traced to a particular preparation of heparin (Lipo-Hepin®) which contained sodium bisulfite, a reducing agent, as a preservative (Cobb and Barnes, 1965; Sutterer *et al.,* 1966). Other reducing agents were also shown to reduce markedly the absorbance of ICG at 800 nm (Cobb and Barnes, 1965). The spectral characteristics of cyanine dyes are known to be susceptible to the action of reducing agents (Hamer, 1964).

Contrariwise, cyanine dyes are also susceptible to "oxidative fading" which increases with exposure to light (Hamer, 1964). It is quite likely that oxidation explains the disappearance with time of the absorbance of dilute aqueous solutions of ICG (Fox and Wood, 1960*b;* Sutterer *et al.,* 1966; Gathje *et al.,* 1970) and also explains the similar effect of ferric ion (Usui and Kawamoto, 1971).

Large doses of mixed gamma neutron radiation had no effect on the spectral absorption of concentrated (5 mg/ml) solutions of the dye (Davis *et al.,* 1969).

ADDITIONAL USES

The spectral characteristics of ICG have been utilized for infrared angiography of the dog brain (Kogure and Choromokos, 1969) and the ocular fundus of man (Kulvin *et al.,* 1970; Hochheimer, 1971). No further discussion of this interesting application will be made since it has not been frequently employed. As noted briefly above, ICG can be used to measure hepatic blood flow, to test liver function, and to measure the intravascular volume, either regionally by using transit times of sudden single injections or for the total intravascular volume by extrapolation methods.

ACKNOWLEDGMENTS

This study was supported in part by Public Health Service Grant HL 08068 and grants from the Graduate School of the University of Minnesota and the Minnesota Heart Association.

REFERENCES

Anderson, N. M., and P. Sekelj. 1968. Studies on the determination of dye concentration in nonhemolyzed blood. J. Lab. Clin. Med. 72:705–713.

Andres, R., K. L. Zierler, H. M. Anderson, W. N. Stainsby, G. Cader, A. S. Grayyib, and J. L. Lilienthal, Jr. 1954. Measurement of blood flow and volume in the forearm of man: with notes on the theory of indicator-dilution and on production of turbulence hemolysis, and vasodilation by intra-vascular injection. J. Clin. Invest. 33:482–504.

Baker, C. H., and L. J. O'Brien. 1964. Vascular volume changes in the dog forelimb. Amer. J. Physiol. 206:1291–1298.

Baker, K. J. 1966. Binding of sulfobromophthalein (BSP) sodium and Indocyanine green (ICG) by plasma α_1 lipoproteins. Proc. Soc. Exp. Biol. Med. 122:957–963.

Bassingthwaighte, J. B., A. W. T. Edwards, and E. H. Wood. 1962. Areas of dye-dilution curves sampled simultaneously from central and peripheral sites. J. Appl. Physiol. 17:91–98.

Bradley, E. C., and J. W. Barr. 1968. Determination of blood volume using Indocyanine green dye. Life Sci. 7:1001–1007.

Brooker, L. G. S. 1966. Sensitizing and desensitizing dyes, pp. 198–232. In T. H. James (ed.), The Theory of the Photographic Process. Macmillan Co., New York.

Cherrick, G. R., S. W. Stein, C. M. Leevy, and C. S. Davidson. 1960. Indocyanine green: observations on its physical properties, plasma decay, and hepatic extraction. J. Clin. Invest. 39:592–600.

Cobb, L. A., and S. C. Barnes. 1965. Effects of reducing agents on Indocyanine green dye. Amer. Heart J. 70:145–146.

Davis, L. W., J. A. Brown, and T. A. Strike. 1969. Is the spectral absorption of Indocyanine green altered by radiation? Amer. Heart J. 78: 140–141.

Dumlao, J. S., M. H. Brooks, M. T. Krauss, D. Bronsky, and S. S. Waldstein. 1970. Suppression of thyroidal iodine uptake after Indocyanine green. J. Clin. Endocrinol. 30:531–532.

Estabrook, R. W., J. Peterson, J. Baron, and A. Hildebrandt. 1972. The spectrophotometric measurement of turbid suspensions of cytochromes associated with drug metabolism. In C. F. Chignell (ed.), Methods in Pharmacology, Vol. 2, p. 303. Appleton-Century-Crofts, New York.

Fox, I. J. 1962. Indicators and detectors for circulatory dilution studies and their application to organ or regional blood-flow determination. Circ. Res. 10:447–472.

Fox, I. J. 1966. Indicators and detectors for dilution studies in the circulation and their applications to organ or regional blood-flow measurement. *In* H. A. Zimmerman (ed.), Intravascular Catheterization, 2nd Ed. Charles C Thomas, Springfield, Ill., pp. 371–421.

Fox, I. J., L. G. S. Brooker, D. W. Heseltine, and H. E. Essex, 1957a. A tricarbocyanine dye for continuous recording of dilution curves in whole blood independent of variations in blood oxygen saturation. Fed. Proc. 16:39.

Fox, I. J., L. G. S. Brooker, D. W. Heseltine, H. E. Essex, and E. H. Wood. 1957b. A tricarbocyanine dye for continuous recording of dilution curves in whole blood independent of variations in blood oxygen saturation. Proc. Staff Meet. Mayo Clin. 32:478–484.

Fox, I. J., L. G. S. Brooker, D. W. Heseltine, and E. H. Wood. 1956. A new dye for continuous recording of dilution curves in whole blood independent of variations in blood oxygen saturation. Circulation 14:937–938.

Fox, I. J., A. R. Castaneda, and K. C. Weber. 1964. Abnormal indicator dilution patterns in dogs produced by flowmeter probes on the pulmonary artery. Circ. Res. 15:301–310.

Fox, I. J., W. F. Sutterer, and E. H. Wood. 1957c. Dynamic response characteristics of systems for continuous recording of concentration changes in a flowing liquid (for example, indicator-dilution curves). J. Appl. Physiol. 11:390–404.

Fox, I. J., H. J. C. Swan, and E. H. Wood. 1959. Clinical and physiologic applications of a new dye for continuous recording of dilution curves in whole blood independent of variations in oxygen saturation, pp. 609–636. *In* H. A. Zimmerman (ed.), Intravascular Catheterization, 1st Ed. Charles C Thomas, Springfield, Ill.

Fox, I. J., and E. H. Wood. 1957a. Applications of dilution curves recorded from the right side of the heart or venous circulation with the aid of a new indicator dye. Proc. Staff Meet. Mayo Clin. 32:541–550.

Fox, I. J., and E. H. Wood. 1957b. Use of methylene blue as an indicator for arterial dilution curves in the study of heart disease. J. Lab. Clin. Med. 50:598–612.

Fox, I. J., and E. H. Wood. 1960a. Circulatory system: methods; blood flow measurements by dye-dilution techniques. *In* O. Glasser (ed.), Medical Physics, Vol. 3. Year Book Publishers, Inc., Chicago, Ill., pp. 155–163.

Fox, I. J., and E. H. Wood. 1960b. Indocyanine green: physical and physiologic properties. Proc. Staff Meet. Mayo Clin. 35:732–744.

Fritts, H. W., Jr., P. Harris, C. A. Chidsey III, R. H. Clauss, and A.

Cournand. 1957. Validation of a method for measuring the output of the right ventricle in man by inscription of dye-dilution curves from the pulmonary artery. J. Appl. Physiol. 11:362–364.

Gathje, J., R. R. Steuer, and K. R. K. Nicholes. 1970. Stability studies on Indocyanine green dye. J. Appl. Physiol. 29:181–185.

Goss, J. E., A. C. Alfrey, J. H. K. Vogel, and J. H. Holmes. 1967. Hemodynamic changes during hemodialysis. Trans. Amer. Soc. Artif. Intern. Organs 13:68–74.

Grace, J. B., I. J. Fox, W. P. Crowley, Jr., and E. H. Wood. 1957. Thoracic-aorta flow in man. J. Appl. Physiol. 11:405–418.

Grenvik, A. 1966. Errors of the dye dilution method compared to the direct Fick method in determination of cardiac output in man. Scand. J. Clin. Lab. Invest. 18:486–492.

Groszmann, R. J., B. Kotelanski, J. Kendler, and H. J. Zimmerman. 1969. Effect of sulfobromophthalein and Indocyanine green on bile excretion. Proc. Soc. Exp. Biol. Med. 132:712–714.

Hamer, F. M. 1964. The Cyanine Dyes and Related Compounds. Interscience Publishers, New York, p. 732.

Hamilton, W. F., R. L. Riley, A. M. Attyah, A. Cournand, D. M. Fowler, A. Himmelstein, R. P. Noble, J. W. Remington, D. W. Richards, Jr., N. C. Wheeler, and A. C. Witham. 1948. Comparison of the Fick and dye injection methods of measuring the cardiac output in man. Amer. J. Physiol. 153:309–321.

Hampers, C. L., J. J. Skillman, J. H. Lyons, J. E. Olsen, and J. P. Merrill. 1967. A hemodynamic evaluation of bilateral nephrectomy and hemodialysis in hypertensive man. Circulation 35:272–288.

Hansing, C. E., J. D. Folts, S. Afonso, and C. G. Rowe. 1972. Systemic and coronary hemodynamic effects of hexobendine and its interaction with adenosine and aminophylline. J. Pharm. Exp Ther. 181:498–511.

Hargreaves, T. 1966. Bilirubin, bromsulphthalein and Indocyanine green excretion in bile. Q. J. Exp. Physiol. 51:184–195.

Hochheimer, B. F. 1971. Angiography of the retina with Indocyanine green. Arch. Ophthal. 86:564–565.

Hood, W. B., B. McCarthy, B. Letac, and B. Lown. 1968. Plasma clearance of Indocyanine green following experimental myocardial infarction in dogs. Proc. Soc. Exp. Biol. Med. 129:4–6.

Hunton, D. B., J. L. Bollman, and H. N. Hoffman. 1960. Studies of hepatic function with Indocyanine green. Gastroenterology 39:713–724.

Jacobs, R. R., U. Schmitz, W. C. Heyden, B. Roding, and W. G. Schenk,

Jr. 1969. Determination of the accuracies of the dye-dilution and electro-magnetic flowmeter methods of measuring blood flow. J. Thor. Cardiovasc. Surg. 58:601–608.

Janecki, J., and J. Krawcynski. 1970. Labeling with Indocyanine green of serum protein from normal persons and patients with acute viral hepatitis. Clin. Chem. 16:1008–1011.

Jaross, W., and K. U. Schentke. 1969. Untersuchungen zur Kinetik von Wofaverdin und Bromsulfan am Modell der isolierten perfundierten Rattenleber. Acta Biol. med. Germ. 22:287–294.

Kinsman, J. M., J. W. Moore, and W. F. Hamilton. 1929. Studies on the circulation. I. Injection method: physical and mathematical considerations. Amer. J. Physiol. 89:322–330.

Klocke, F. J., D. G. Greene, and R. C. Koberstein. 1968. Indicator-dilution measurement of cardiac output with dissolved hydrogen. Circ. Res. 22:841–853.

Kogure, K., and E. Choromokos. 1969. Infrared absorption angiography. J. Appl. Physiol. 26:154–157.

Kotelanski, B., R. J. Groszmann, J. Kendler, and H. J. Zimmerman. 1969. Effect of ethanol on sulfobromophthalein and Indocyanine green metabolism in isolated perfused rat liver. Proc. Soc. Exp. Biol. Med. 132:715–721.

Kulvin, S., L. Stauffer, K. Kogure, and N. J. David. 1970. Fundus angiography in man by intracarotid administration of dye. Southern Med. J. 63:998–1000.

Lacy, W. W., C. Ugaz, and E. V. Newman. 1955. The use of indigo carmine for dye dilution curves. Circ. Res. 3:570–574.

Michie, D. D., D. G. Wombolt, R. F. Carretta, A. E. Zencka, and J. D. Egan. 1971. Adverse reactions associated with the administration of a tricarbocyanine dye (Cardio-green) to uremic patients. J. Allergy Clin. Immunol. 48:235–239.

Miller, D. E., W. L. Gleason, and H. D. McIntosh. 1962. A comparison of the cardiac output determination by the direct Fick method and the dye-dilution method using Indocyanine green dye and a cuvette densi-tometer. J. Lab. Clin. Med. 59:345–350.

Moore, J. W., J. M. Kinsman, W. F. Hamilton, and R. G. Spurling. 1929. Studies on the circulation. II. Cardiac output determinations: comparison of the injection method with the direct Fick procedure. Amer. J. Physiol. 89:331–339.

Nicholson, J. W., III, H. B. Burchell, and E. H. Wood. 1951. A method for the continuous recording of Evans blue dye curves in arterial blood, and its application to the diagnosis of cardiovascular abnormalities. J. Lab. Clin. Med. 37:353–364.

Paton, W. D. M. 1957. Histamine release by compounds of simple chemical structure. Pharmacol. Rev. 9:269–328.

Polanyi, M. L., and R. M. Hehir. 1960. New reflection oximeter. Rev. Scient. Instr. 31:401–403.

Rapaport, E., S. G. Ketterer, and B. D. Wiegand. 1959. Hepatic clearance of Indocyanine green. Clin. Res. 7:289–290.

Reemtsma, K., G. C. Hottinger, A. C. DeGraff, and O. Creech. 1960. The estimation of hepatic blood flow using Indocyanine green. Surg. Gyn. Obst. 110:353–356.

Richardson, D. W., E. M. Wyso, A. M. Hecht, and D. P. Fitzpatrick. 1959. Value of continuous photoelectric recording of dye curves in the estimation of cardiac output. Circulation 20:1111–1117.

Saunders, K. B., J. I. E. Hoffman, M. I. M. Noble, and R. J. Domenech. 1970. A source of error in measuring flow with Indocyanine green. J. Appl. Physiol. 28:190–198.

Sekelj, P., A. Oriol, N. M. Anderson, J. Morch, and M. McGregor. 1967. Measurement of Indocyanine green dye with a cuvette oximeter. J. Appl. Physiol. 23:114–120.

Simmons, R., and R. J. Shephard. 1971. Does Indocyanine green obey Beer's law? J. Appl. Physiol. 30:502–507.

Spangler, R. D., T. Yipintsoi, T. J. Knopp, R. L. Frye, and J. B. Bassingthwaighte. 1971. Pulmonary mean-transit-time blood volumes in anethetized dogs. J. Appl. Physiol. 30:56–63.

Stow, R. W., and P. S. Hetzel. 1954. An empirical formula for indicator-dilution curves as obtained in human beings. J. Appl. Physiol. 7:161–167.

Sutterer, W. F., S. E. Hardin, R. W. Benson, L. J. Krovetz, and G. L. Schiebler. 1966. Optical behavior of Indocyanine green dye in blood and in aqueous solution. Amer. Heart J. 72:345–350.

Tripp, M. R., G. M. Cohen, D. A. Gerasch, and I. J. Fox. 1973. Effect of protein and electrolyte on the spectral stabilization of concentrated solutions of Indocyanine green. Proc. Soc. Exp. Biol. Med. 143:879–883.

Tripp, M. R., and I. J. Fox. 1973. Unpublished data.

Tsubokura, A., T. Hirata, and Y. Hosokawa. 1970. Jap. J. Clin. Pathol. 18: Suppl. 26. Cited by Usui and Kawamoto.

Usui, T., and H. Kawamoto. 1971. Decolorization of Indocyanine green in serum by ferric ion. Clin. Chim. Acta 31:477–478.

West, W., and S. Pearce. 1965. The dimeric state of cyanine dyes. J. Phys. Chem. 69:1894–1903.

Wheeler, H. O., W. I. Cranston, and J. I. Meltzer. 1958. Hepatic uptake and biliary excretion of Indocyanine green in the dog. Proc. Soc. Exp. Biol. Med. 99:11–14.

Wolthuis, R. A., H. W. Overbeck, and W. D. Collings. 1969. Measurement of blood flow in the limb of man by cuvette densitometry. J. Appl. Physiol. 26:215–220.

Wood, E. H., D. Bowers, J. T. Shepherd, and I. J. Fox. 1955. O_2 content of "mixed" venous blood in man during various phases of the respiratory and cardiac cycles in relation to possible errors in measurement of cardiac output by conventional application of the Fick method. J. Appl. Physiol. 7:621–628.

Zanker, V. 1952. Über den Nachweis definierter reversibler Assoziate (reversibler polymerisate) des Acridinorange durch Absorptions- und Fluoreszenzmessungen in wässriger Lösung. Z. Phys. Chem. 199:225–258.

Zijlstra, W. G., and G. A. Mook. 1962. Medical Reflection Photometry. Koninklÿke Van Gorcum & Comp. N.V., Assen, The Netherlands, pp. 182–190.

chapter 19/ Dye Densitometers

William F. Sutterer/Dennis A. Bloomfield
Department of Physiology/Mayo Clinic/Rochester, Minnesota/
and Division of Cardiology/Maimonides Medical Center/
Brooklyn, New York

THE DEVELOPMENT OF DYE DENSITOMETERS

Early dye-dilution measurements were accomplished by the withdrawal of dyed blood samples from a systemic artery at short time intervals for subsequent spectroscopic analysis (Hamilton *et al.,* 1948). Continuous recording of dilution curves was reported as early as 1936 by Matthes, but Millikan (1933) had developed a two-color densitometer for the determination of oxygen saturation 3 years previously. Early indicators, such as Evans blue (T-1824) and Coomassie blue, absorbed light in the red spectrum, which is the wave length band also used to measure oxygen saturation by many oximeters. Use of these dyes had the undesirable characteristic that unwanted signals generated by variations in optical density in the red spectrum, due to changes in oxygen saturation, could not be differentiated from changes in dye concentration and frequently were of sufficient magnitude to interfere with the interpretation of diagnostic dilution curves (Fox and Wood, 1960).

It was recognized that a dye bound to protein and with a peak spectral absorption at the isosbestic point of oxy- and reduced hemoglobin (805 nm wave length) could be sensed and recorded by a simple monochromatic densitometer, independent of oxygen saturation; such an indicator, tricarbocyanine green, was developed in 1957 (Fox *et al., a*). Initially, the infrared cell of the Wood oximeter was used to sense the concentration of the new dye, but other monochromatic densitometers were manufactured. These instruments were characterized by their utilization of only one photocell filter assembly, although their name derives from the theoretical intent to respond to only one wave length of light. However, for this application, "monochromatic" implies a wave length band encompassing a number of wave lengths and characterized by the "half band width" measured at the 50% transmission points on the system's spectral response curve. The monochromatic densitometers, although independent of oxygen saturation, are sensitive to changes in transmitted light resulting from variation in the shape and size of red cells caused, in turn, by alteration in the tonicity, temperature, and pH of the plasma. These changes manifest in unwanted signals (optical noise) superimposed on the recorded dilution curve and, under some conditions, are of a magnitude comparable to that of the indicator signal. This stimulated efforts to build a densitometer which would compensate for these nonspecific effects. By the use of a dichroic mirror and associated optical filtering, a two photocell filter system for the detection of Indocyanine green dye was developed (Sutterer and Wood, 1962). The dye-detecting cell was sensitive to light at 800 nm, whereas the compensating cell was sensitive to light at longer and shorter wave lengths, in effect "creating" an isosbestic point. The compensating cell was sensitive, therefore, to nonspecific effects but relatively insensitive to oxygen saturation and dye concentration. Subsequently, the optical filter of the second photocell filter assembly was simplified to transmit light in a band at 900 nm only. The utilization of this wave length band for the compensating cell has the advantage that it is almost totally insensitive to the dye. However, it does exhibit a slight sensitivity to oxygen saturation, and, under conditions of very large changes in this parameter, another source of noise is realized. These conditions can occur, for example, during sudden injections of dye into the coronary artery, causing an increase of blood flow through the coronary bed and, subsequently, a substantial increase in oxygen saturation of approximately 40 to 50% at a sampling site in the coronary sinus.

Nonspecific Optical Density Changes in Whole Blood

Studies have been undertaken to identify chemical and physical phenomena which alter light transmission through whole blood by affecting the shape and size of the red cells (Sutterer, 1967). Sinclair and co-workers (1961) reported that the tonicity of the fluid surrounding the cell has a most profound effect on the red cell morphology and influences the light scattering and, consequently, the light transmission. Earlier, Zylstra (1953) found that the light reflection of blood was also affected. Castaneda and colleagues (1966) and Nevo and associates (1955) reported that polybasic polymers, when added to blood, produce cell agglutination which, because of cell spaces created, cause an increase in light transmission, whereas cell crenation causes the opposite optical effect. Jacobs and colleagues (1936) and Brown (1956) studied the effect that temperature and pH have on the size of the erythrocyte and, hence, the light transmitting properties of whole blood. Variations in optical density due to cessation of whole blood flow also have been studied spectrophotometrically and observed over wave lengths from 600 nm to 2 μm. The effect of these nonspecific changes in producing artifacts in recorded dilution curves has been minimized, but not eliminated by the development of the dichromatic densitometer.

Physical Principles of Densitometry

Blood densitometry is based on the relationship of incident and absorbed light to the length of the light path. The relationship states that the intensity of a parallel beam of monochromatic (light) radiation is reduced logarithmically as it passes rectilinearly through a homogeneous absorbing medium. This relationship, known as Beer's law, can be defined by the equation

$$I = I_0 \cdot 10^{\epsilon c d}, \tag{19.1}$$

where I = intensity of transmitted light, I_0 = intensity of incident light, d = depth of homogeneous substance in centimeters through which light travels before detection, c = concentration in millimoles/liter of substance in solution, ϵ = extinction coefficient, which is the optical density at a specific wave length of a millimole/liter concentration of the substance using an optical pathway (depth of sample) of one centimeter (Drabkin, 1950). Thus, the optical density of a homogeneous substance at wave length λ_1 in solution (D^{λ_1}) is the product of the extinction coefficient, concentration of the substance, and the optical depth; it is written:

$$D^{\lambda_1} = \epsilon_1^{\lambda_1} c_1 d. \tag{19.2}$$

The measured optical density at each wave length consists of the sum of the optical densities at that wave length of all of the substances in the solution.

$$D^{\lambda_1} = \epsilon_1^{\lambda_1} c_1 d + \epsilon_2^{\lambda_1} c_2 d + \ldots + \epsilon_n^{\lambda_1} c_n d. \tag{19.3}$$

The discrimination and measurement of the concentration of one substance (the dye) from a homogeneous solution consisting of three substances (oxyhemoglobin, reduced hemoglobin, and the dye) theoretically requires the utilization of three wave lengths and, consequently, the simultaneous solution of three equations with three unknowns, as follows:

$$D^{\lambda_1} = \epsilon_1^{\lambda_1} c_1 d + \epsilon_2^{\lambda_1} c_2 d + \epsilon_3^{\lambda_1} c_3 d$$
$$D^{\lambda_2} = \epsilon_1^{\lambda_2} c_1 d + \epsilon_2^{\lambda_2} c_2 d + \epsilon_3^{\lambda_2} c_3 d$$
$$D^{\lambda_3} = \epsilon_1^{\lambda_3} c_1 d + \epsilon_2^{\lambda_3} c_2 d + \epsilon_3^{\lambda_3} c_3 d \tag{19.4}$$

Where c_1 = concentration of the dye in millimoles/liter; c_2 = concentration of the oxyhemoglobin in millimoles/liter; c_3 = concentration of reduced hemoglobin in millimoles/liter; ϵ_1^{λ} = extinction coefficient of dye at each wave length λ_1, λ_2, λ_3; ϵ_2^{λ} = extinction coefficient of oxyhemoglobin at each wave length; ϵ_3^{λ} = extinction coefficient of reduced hemoglobin at each wave length; d = depth of cuvette in centimeters (Zylstra, 1958). If the above equations are solved simultaneously, the concentration of the dye in millimoles/liter is

$$c_1 = \frac{K_1 D^{\lambda_1} + K_2 D^{\lambda_2} + K_3 D^{\lambda_3}}{d K_4}, \tag{19.5}$$

where the constants K_1, K_2, K_3, and K_4 are various combinations of the sums and/or differences of the products of different extinction coefficients as shown in the appendix to this chapter D^{λ_1}, D^{λ_2}, and D^{λ_3} are the optical densities of the media at wave lengths λ_1, λ_2, and λ_3.

Note that it does not make any difference theoretically which wave lengths are chosen. However, for practical purposes, the signal-to-noise ratio is maximized and sometimes the mathematics are simplified if the wave lengths are chosen with care. At the wave length of the isosbestic point of oxy- and reduced hemoglobin, reciprocal alterations in concentration of these substances will not affect the over-all light transmission through blood. However, the measured optical density from equation (19.3) will reflect the concentration of total hemoglobin as well as the dye.

Although hemolyzed blood approaches the requirements of an "ideal" fluid from a densitometric point of view, whole blood, with its varied and particulate composition, does not. Several independent investigators have demonstrated that light transmitted through whole blood does not obey Beer's law (Edwards *et al.*, 1963; Anderson and Sekelj, 1967*a* and *b*). This has stimulated the search for another theory on which to base design considerations for future oxygen and dye detection systems for whole blood.

In his doctoral thesis, Cohen (1969) relates that a theory was developed in 1931 which describes the absorption and scatter of light in a heterogeneous medium (Kubelka and Munk). However, it was not until 1966 that this theory was applied to oximetry by Reichert. This "new" concept treats the photons as if they were particles diffusing through the whole blood. These scattered photons are then assumed to obey the diffusion and continuity formulae. (Equations used to calculate the density of the scattered photons for the one dimensional *in vitro* case (cuvette) are given in the appendix to this chapter.)

This theory was transformed to describe electron transport by McKelvey and associates (1961), then adapted to the three dimensional semiconductor case by Shockley (1962), and finally applied to the optics of *in vivo* oximetry (Longini and Zdrojkowski, 1968; Cohen and Longini, 1971; Zdrojkowski and Pisharoty, 1970; Johnson, 1970).

The Kubelka-Munk equation for the description of the transmission of light through an absorbing and scattering material is being adapted to facilitate the calculation of the optical density of dyed whole blood (Cohen, personal communication). The calculation of the optical coefficients is slightly more complicated than that for oxygen saturation. The determination of the numerical values of the optical coefficients at various wave lengths are currently being conducted.

A Simple Blood Densitometer

The basic components of a simple monochromatic blood densitometer are shown in Fig. 19.1. A light source (*A*) emits light which passes through the cuvette (*B*), which is a small transparent or translucent chamber, inside of which the blood to be analyzed for indicator content passes at right angles to the beam in steady flow (*C*).

The light which penetrates the blood-filled cuvette emerges and is collected and collimated by a lens system (*D*) and is then filtered (*E*) to allow passage of only the desired wave lengths. Finally, the light falls on the sensor or detector (*F*), and its intensity is measured. The output from the detector (*G*) is an electrical analogue of the concentration of the dye passing through the cuvette. The electrical signal is amplified and then displayed. At this stage it can be stored, fed into a computer for analysis, or printed out.

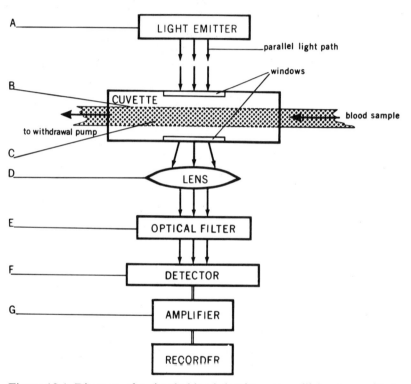

Figure 19.1 Diagram of a simple blood densitometer utilizing transmitted light. See text for detailed description.

The various methods available for the performance of each step are now detailed.

Light Emitter or Photon Source

The requirements of the light source are that it emit light energy at the wave lengths at which the measurements are desired with a maximum light intensity and a minimum amount of current and heat. The light source for instruments designed to detect Indocyanine green have traditionally utilized a tungsten light bulb. However, for transducers designed to detect blood oxygen saturation, light-emitting diodes are currently replacing the tungsten source. A light-emitting diode is a solid state diode which has been specially doped to insure that, when forward biased, electrons which have been free to conduct electricity recombine with atoms of the crystaline structure and, in so doing, must fall through an energy gap of sufficient magnitude, thus emitting light energy of the desired wave length. Such a device capable of emitting light in the range of 800 nm is not commercially available at this time. However, it is technically possible to grow these crystals, and prototype densitometers for the detection of Indocyanine green dye in whole blood, utilizing these new light emitters, are in the process of development at some institutions. Thus, it is expected that these devices will be available for clinical and investigative use in the near future. Ideally, each light emitter should possess identical physical characteristics, so that the exact luminosity is realized from a given amount of electrical current. Also, each device of a particular type should emit light at exactly the same wave length and with the same band widths. The realization of these new physical characteristics, along with envisioned new developments of other densitometer components discussed elsewhere in this communication, will aid in the development of densitometers with identical dye sensitivities.

Blood Cuvette

The blood cuvette is a component frequently made of a plastic material carefully milled to a specific depth, with windows cemented in place. Other units use a piece of very stiff tubing that is optically translucent or transparent. The ideal cuvette should be impervious to any type of saline solution, tolerating both hot and cold sterilization techniques. The interior should be absolutely smooth, to inhibit intraluminal clot formation or trapping of air in the region of the optical pathway. Such artifacts produce erroneous optical density readings. The cuvette is preferably constructed of a material with a high volume elasticity coefficient to minimize any variations in optical depth due to pressure changes. Such changes in optical depth would be manifested by changes in transmitted light, unrelated to the desired signal. Every lumen should possess identical optical characteristics, realized in part by rigid control of external and internal diameters, to facilitate lumens which can be interchanged without necessitating recalibration. Thus, several lumens could be in the process of sterilization while another is being used, without affecting the calibration of the instrument. The dimension of the cuvette blood chamber should be sufficiently small to facilitate maximum flow velocity, thus enhancing the dynamic response of the instrument (Fox, Sutterer, and Wood, 1957, *b*).

Sampling Site to Withdrawal Pump Connection

The tubing leading to and from the cuvette, like the cuvette itself, must be constructed with high volume elasticity coefficient material. The internal volume of the segment from the sampling site to the cuvette must be minimal, and the other dimensions (cross-sectional area and length) are controlled by those factors discussed in Chapter 4.

Lens

A lens system is sometimes placed in a position distal to the blood cuvette to "collect" the maximum amount of the light which has passed through the blood and has been scattered by the blood cells. A lens can also be placed between the light emitter and the cuvette to collimate the incident light, if so desired.

Filter

This device insures that undesired wave lengths of light are blocked, and desired wave lengths are permitted to pass to the blood sample and/or photodetector. Three general types of filters (or their combination) are used: (1) gelatin or Wratten filters which usually have a rather wide pass band; (2) interference filters which can be constructed with an extremely narrow pass band and are made of alternate layers of glass and dielectric material, the distance between the layers of glass determining the wave length passed; and (3) dichroic mirrors used at an angle of 45° to the incident light. These mirrors pass light of one wave length band and reflect the light of another. Use of light emitting diodes for the light source will obviate, at least in part, if not entirely, the need for optical filters.

As discussed previously, very wide spectral responses of optical systems contribute to large deviations from Beer's law (Fig. 19.2). Thus, for maximum accuracy, some care should be taken in the design of multichromatic devices to insure similar spectral half band widths of the photocell filter assemblies.

Detector

This device converts light energy incident on its surface to an electrical analogue. The following is a list of some of the most common photodetectors that have been used in whole blood densitometers and some advantages and disadvantages of each. Figs. 19.3 to 19.6 depict simplified diagrams of four different types of photodetectors and their general associated circuitry to aid signal conditioning.

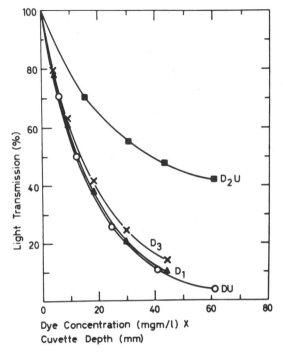

Figure 19.2 Calibration curves of four monochromatic densitometers designed for the detection of Indocyanine green in whole blood. These units consisted of an incandescent light source, an optical interference filter, a milled black lucite lumen, and a vacuum photo diode. Densitometers DU, D_1, and D_3 had a narrow wave length band of incident light (narrow band interference filter). The densitometers have similar calibration curves and exhibit little difference from Beer's law. However, densitometer D_2U had a light leak around the interference filter, hence, a wide wave band of light was incident on the blood, and a marked deviation from Beer's law resulted. The deviation from Beer's law for dye in plasma for these four instruments was: $DU = 1\%$, $D_1 = 2.6\%$, $D_3 = 5.9\%$, and $D_2U = 37\%$. When these same densitometers were studied for their sensitivities to dye in whole blood, a greater deviation from Beer's law was noted: $DU = 6\%$, $D_1 = 6\%$, $D_3 = 8\%$, and $D_2U = 38\%$. (Modified from Edwards, A. W. T., J. Isaacson, W. F. Sutterer, J. B. Bassingthwaighte, and E. H. Wood. 1963. Indocyanine green densitometry in flowing blood compensated for background dye. J. Appl. Physiol. 18:1294.)

PHOTOMULTIPLIER   This device consists of a cathode and 6 or 10 dynodes, each of which is more positively charged than the last by means of voltage dividers (Fig. 19.3). The photons strike the cathode and, if they possess the proper wave length (energy level),[1] they release electrons (photoelectrons) which, due to the relative positive charge on the next dynode, drift in the electric field and strike the next dynode. As the photoelectron strikes the dynode, it releases more electrons, and the process is repeated at successive dynodes. Each electron releases several more, thus, for each electron initially released

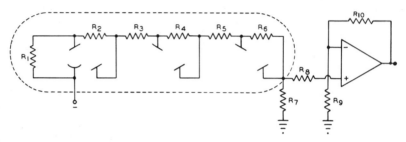

Figure 19.3 The photomultiplier.

at the cathode, many more are collected by the anode—hence, the name "photomultiplier." If the ratio of the number of electrons arriving at the collector to the number released at the cathode (i_0) is denoted by r and there are n dynodes, then the current at the collector i is

$$i = i_0 r^n. \tag{19.6}$$

This device has the advantage of extreme sensitivity, (current amplification of over 3 million has been reported), and the output is linear with light intensity. Disadvantages are its relatively large size and voltage supply requirements.

VACUUM PHOTO DIODE   The number of electrons emitted by the cathode of the vacuum photo diode depends on the energy level and number of incident photons per unit time striking the cathode. These released electrons then drift to the positively charged anode (Fig. 19.4). This detector is stable, relatively insensitive to temperature variation, and its output is linearly related to the intensity of the light incident on the cathode. However, due to the relatively small photoelectron current, a large resistor is necessary, resulting in a large source impedance. The voltage across this resistor is by Ohm's law

---

[1]The wave length (λ) of a photon is related to its energy (E) by $E = hc/\lambda$, where $h =$ Planck's constant and $c =$ the speed of light.

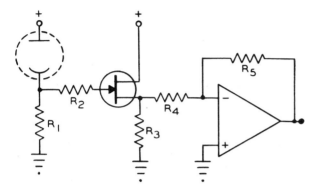

Figure 19.4 The vacuum photo diode.

directly related to the current through the resistance and, hence, the number of electrons liberated at the cathode by the photons. The measurement of this voltage requires a field effect transistor or other device with a high input impedance (several orders of magnitude higher than the source impedance, if possible), so as not to shunt this load resistor with the measuring device and thus reduce measurement accuracy. The vacuum photo diode also has the disadvantage of relatively large size.

PHOTOCONDUCTIVE CELL   An increase of light intensity of the proper wave length incident on a semiconductor causes an increase in covalent bonds broken and "hole-electron" pairs created (above that level generated thermally) to increase the electrical conductivity of the device, hence the name photoconductive cell. The increased number of current carriers decreases the resistance of the device so that it is also often called a "photoresistive cell." The physical principles governing the generation of these hole-electron pairs are well known, but complicated (Bube, 1960). These cells have the advantages that certain mathematical manipulations can sometimes be accomplished with relatively simple analogue circuitry and that they are physically small, enabling the designer to decrease the transducer size. The disadvantages are that a change in resistance is not linearly related to a change in light intensity and that the semiconductor characteristics may drift with age in some devices. However, it is possible under certain conditions to match the nonlinearity of the photocell against that of the optical system to achieve a system which has an output that is a linear function of dye concentration. There are many different ways to convert the resistance of the photoconductor to an analogue voltage or current. A very simple method is shown in Fig. 19.5. The type of circuit used depends on the mathematical solutions and signal conditioning desired by the designer.

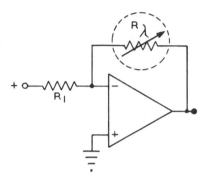

Figure 19.5 The photoconductive cell.

PHOTOVOLTAIC CELL   There are in general two types of photovoltaic cells which generate a current, the magnitude of which is dependent on the number of incident photons per unit time and on the photon energy, and, hence, do not require a power source. One of these detectors, the iron selenium cell, is an extremely capricious device often exhibiting random variations in output due to internal intermittent shorting. Another more dependable type of photovoltaic device is known as the solar cell used in many space probes. This detector has a wide spectral response, and the devices used by one of us (W.F.S.) have proven to be much more dependable than the iron selenium cell. The solar cell is temperature-sensitive and must be sealed from effects of water vapor. The photovoltaic cell generally is treated as a current source, and a current-to-voltage converter is used as shown in Fig. 19.6. This allows the device to "look into" zero

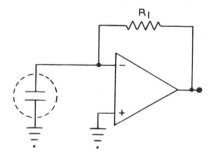

Figure 19.6 The photovoltaic cell.

resistance, thus facilitating maximum linearity with light intensity for subsequent signal conditioning and mathematical manipulation.

Log Linear Conversion and Amplifiers

Indocyanine green dye has a half-life in the body of approximately 10 minutes, so that repeated injections of dye produce a build up of dye concentration in the subject's blood. If the sensitivity of the detector is a linear function of light intensity, then the output of the densitometer will be a nonlinear function of dye concentration (see under "Physical Principles of Densitometry") and, as the background dye concentration in the blood increases, the dye sensitivity of the device will progressively decrease (Fig. 19.2). However, the undesirable effects of this nonlinear response to increasing background dye concentration can be obviated (Edwards et al., 1963). The output of modern densitometers are forced to be linear with dye concentration to at least 30 mg/liter(Fig. 19.7). Consequently, these devices have a constant sensitivity to dye as long as the sum of the background dye and the peak concentration does not exceed 30 mg/liter.

As discussed above, the type of initial amplification and signal conditioning required depends on the type of photo detectors used and the desired subsequent mathematical and electronic manipulations. It is not always necessary to use a logarithmic electronic device to insure that the output of the densitometer utilizing photoconductive cells is linear with dye concentration. For instance, the calibration curves of Fig. 19.7 are from a device which employs a passive method for log-linear conversions. If one of the other photo transducers is used and electronic log conversion is employed, one of the following methods may be utilized. The relationship between the current through a solid state diode (I) and the applied voltage (V) is given by:

$$I = I_s(e^{qV/KT} - 1), \tag{19.7}$$

where $q =$ electronic charge $= 1.602 \times 10^{-19}$ Coulombs, $V =$ applied voltage, $K =$ Boltzmann's constant $= 1.38 \times 10^{-23}$ joules/$^\circ$K, $T =$ temperature in $^\circ$K, $e =$ base of Naperian logarithms (2.7183), $I_s =$ the saturation current reached when the voltage across the diode becomes large and negative and the diode current becomes asymptotic.

Equation (19.7) illustrates the logarithmic nature of the relationship between the current through a diode and the voltage applied across it. This physical phenomenon is used to create systems to linearize the output and may be accomplished by one diode or a system

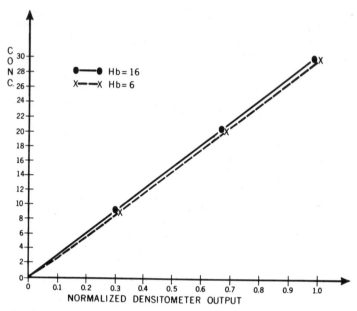

Figure 19.7 Calibration graph of a dichromatic densitometer designed to detect Indocyanine green dye in whole blood and showing some hematocrit sensitivity.

of many. It is also often done by placing a nonlinear element such as a transistor in the negative feedback loop of an operational amplifier, making use of the nonlinear current-voltage relationships to make the output of the amplifier a nonlinear function of the input voltage. In general these "log amplifiers" can be purchased with temperature compensation, which is important, as inspection of equation (19.7) shows the temperature dependency of the diode current-voltage relationship.

The linearity and sensitivity to hematocrit of a dichromatic densitometer is shown in Fig. 19.7. The instrument characteristics shown are the result of a series of tests by the manufacturer to insure that the output is linear within specifications. The spectral characteristics and linearity of the photocell filter assemblies are also adjusted so that sensitivity of the device to variations in hematocrit is minimized. If the spectral characteristics and linearity of the compensating cell are not matched properly to those of the dye-detecting cell, low hematocrit can exhibit an error as much as 40% from that shown by normal hematocrit (Sedivy and Sutterer). The section in this chapter regarding the theory of measuring the dye concentration in blood states that at least three wave lengths of light are required. However, due to the above mentioned physical phenomenon and empirically developed optical and electronic techniques, reasonable compensation is achieved for such nonspecific effects as hematocrit and variation in flow, as well as nonlinearity to the dye concentration.

The demands made on the photocells of a dichromatic dye densitometer for automatic compensation for hematocrit are much less than those made on the cells of the oximeter, because the oximeter must respond to both bloodless and blood conditions, whereas the dye densitometer as used in some models need respond to the blood conditions only. Thus, the light amplitude experienced by the dye densitometer is much less than that experienced by the oximeter. Furthermore, the densitometer's output is balanced to zero on blood before the injection of dye. This can be compared to adjusting an oximeter to read 100% on every subject's arterialized blood. Data from many studies of dichromatic densitometer sensitivity versus hemoglobin indicate that (perhaps due to the above described differences experienced by the two instruments) a much simpler mathematical expression need be solved for hemoglobin-compensated linear dye-dilution curves than for oxygen saturation.

A system utilizing reflection techniques has been developed by using reflection channels which are relatively near the light source (called the near channels) and a "distant" channel which is located at a further distance from the light source (Janssen, 1972). This instrument is designed to measure oxygen saturation independent of hematocrit, hemolysis, and nonspecific effects, such as blood flow velocity, and also hemoglobin independent of hemolysis, oxygen saturation, and flow. For dye measurements, the light output of the distant channel is independent of hemolysis, light scattering, and flow, and, if the isosbestic point is used for the filter, it is also independent of oxygen saturation. Flow insensitivity of this device is realized by a specially designed cuvette which insures turbulent flow of whole blood even at low flows. A linear measurement of the dye concentration in the distant channel is obtained if the logarithm of the measured light output is calculated using the isosbestic point wave length (805 nm). This channel's output is distorted in part at least by the hemoglobin concentration.

A measurement of the dye concentration not influenced by the cell-bound hemoglobin can be obtained from this device by taking the ratio of the light output of two different near channels each at arbitrary wave lengths. This ratio is exponentially related to dye concentration but has the disadvantage of possessing some sensitivity to oxygen saturation and is less sensitive to the dye than to the distant channel. Unfortunately, the optical system is not completely independent of all undesired optical signals.

Voltage amplification is usually accomplished by the use of integrated circuit (I.C.) operational amplifiers. These devices are, in general, inexpensive, and a large selection of amplifier parameters such as temperature stability, input impedance, and amplifier gain (which may be controlled by the designer by proper use of the associated circuit parameters) can be obtained. However, some applications may require that the output of the densitometer control circuit be capable of providing more current than the I.C. can deliver without impairing the amplifier. In this case, a power transistor or amplifier, designed to withstand a large current amplitude, is utilized between the I.C. device and the recorder or load.

Recorders

The output of the modern densitometer control units can drive almost any type of recorder. Photokymographic recorders utilizing ultraviolet light and galvanometers have been popular for many years because light beams can cross, identical zeros may be used, and many different physiological parameters can be recorded simultaneously. Recently, a new ultraviolet photokymographic recorder has been introduced by Honeywell which does not use galvanometers but utilizes a combination of the principles of the cathode ray tube (oscilloscope) and fiber optics. It has many of the advantages of the recorders using galvanometers although its cost is somewhat more. Low pass filtering accomplished by choice of the proper natural frequency and damping coefficient of the galvanometer in the "old" recorder is done by solid state active filters in the "new" unit. Tracings are identified by a dotted line, and the channel identifying number is printed on the "top" of the record.

Single-channel pen writers using both heat and ink styluses have also been used. Analogue tape recorders allow the storage of "live" data and can be edited and/or digitized for subsequent data reduction by a computer. Some laboratories now feed the output of the densitometers directly into analogue and/or digital computers for on-line data analysis (see Chapter 21).

In Vivo Dye Densitometry via an Earpiece

In vivo dye densitometry by an earpiece dichromatic densitometer was reported as early as 1950 (Knutson *et al.*) and several times since (Beard and Wood, 1951; McGregor, Sekelj, and Adam, 1961; Gilmore *et al.*, 1954), but it has never been a widely used technique. A dichromatic earpiece densitometer sensitive to Indocyanine green dye has been used for hepatic function tests (Leevy *et al.*, 1967).

Reed and Wood (1967) compared the cardiac outputs in man, using Indocyanine green as measured by monochromatic and dichromatic earpiece densitometers. The units were calibrated by a spectrophotometric determination of the concentration of dye in a sample withdrawn 1.5 minutes after the injection of the dye. They reported a standard deviation of 31% for the monochromatic earpiece compared to the cardiac output as measured by the cuvette, whereas the dichromatic device exhibited a standard deviation of only 14%.

Cohen and Longini (1971) call attention to the greater complexity for *in vivo* densitometry compared to the *in vitro* situation. Indeed, as discussed previously, considering the photon diffusion theory, the diffusion equation must be solved for only one dimension in the *in vitro* case, but for three dimensions in the case of the earpiece.

The realization of a completely compensated earpiece densitometer for the detection of a dye independent of ear thickness, pigmentation, hematocrit, variations in oxygen saturation, and blood volume would have obvious clinical and investigative implications. The development of such a device, possibly by the insight gained by understanding of the photon diffusion theory and/or by a combination of transmission and reflection techniques, offers some challenge.

DENSITOMETER SPECIFICATIONS

This section contains details of three densitometers in common use in this country. The data was provided by the manufacturers.

Waters D-400 Densitometer (Fig. 19.8)

Manufacturer: Waters Instruments, Inc., Rochester, Minnesota
Class: Dichromatic
Cuvette: flow rate   5 to 30 ml/minute (20 cc recommended)
        lumen volume   0.03 ml

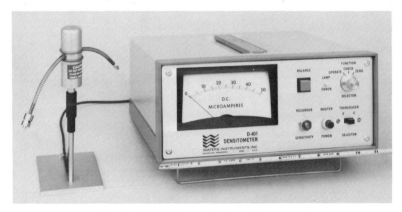

Figure 19.8 Waters D-400 densitometer.

sample volume with 8-inch intake tubing   0.50 ml
construction   plastic tubing
light source   tungsten bulb
sterilization   chemical
size   1 × 3¾ inches
linearity ± 1% to 25 mg/liter dye
background   dye compensation   20 mg/liter
stability (after 30 minutes)   1% full scale
output voltage
   high impedance output   1.2 v.D.C./12mg/liter
   low impedance output   50 mv/12 mg/liter
Earpiece: size   1¼ × ½ × 1½ inches
    weight   3.9 oz
Filter: composite construction
Detectors: photo conductors
        peak sensitivity   (a) 800 nm (b) 910 nm
        half band width   (a) 125 nm (b) 80 nm

Gilford 103–1R Densitometer (Fig. 19.9)

Manufacturer: Gilford Instrument Laboratories, Inc., Oberlin, Ohio
Class: Monochromatic

Figure 19.9 Gilford 103-1R densitometer.

Cuvette: flow rate   20 to 40 ml/minute
    lumen volume   0.3 ml
    sample thickness   0.75 mm
    sample length   0.5 inch
    sample width   0.12 inch
    sterilization   autoclavable
    emitter source   stabilized incandescent (tungsten lamp)

Filter: interference filter with harmonic suppressor includes 630- and 805-nm filters
    half band width   10 nm
    position in light path   before cuvette (with collimator lens)

Detector: Dumont K 1430 photomultiplier tube

Output: type   balanced or unbalanced
    output impedance   3,000 Ohms (without ripple filters)
    maximal output ripple   2.0 mv
    drift (warmed)   0.01 optical density unit/hour

General: sensitivity   4 v/1.0 O.D. unit
    linearity   2% from +1.1 O.D. units to −1.1 O.D. units
    variable density wedge   0 to 3.2 O.D. to 1.0 O.D.
    response time   to 1.0 O.D.
    (to step change)   1.0 second for 95%

Gilson D.T.L. Dye Tracer (Fig. 19.10)

Manufacturer: Gilson Medical Electronics, Inc., Middleton,
     Wisconsin

Class: Monochromatic

Cuvette: material   Teflon tube—elipsoidal cross-section
    flow rate 10 ml/minute minimum
    sample volume 0.02 ml
    sample depth 0.74 mm
    volume with tubing 0.7 ml
    sterilization chemical optical chamber—cold solution—
          complete detector head—gas
    dye tracer head size 2¾ × 2½ × 1 inch with 6-inch
            handle
    emitter source tungsten lamp

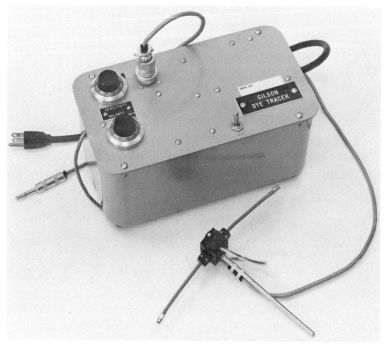

Figure 19.10 Gilson D. T. L. dye tracer.

Filter: Bausch & Lomb interference filter with peak response of 810 nm superimposed with one or two layers of 87 or 88A Wratten filter to suppress wave lengths below 800 nm.

half band width   13.6 nm

Detector: resistor photocell

Output: type   balanced

output voltage   250 mv

output impedance   5,000 ohms

drift $< 1\%$ full scale per 10 minutes

General: sensitivity   8 mg/liter of dye

linearity   within 1% to 8 mg/liter

APPENDIX

The following are the constants for the extinction coefficient equations as determined from the solution for C_1 (concentration of dye) using hemolyzed blood:

$$K_1 = (\epsilon_2^{\lambda_2} \epsilon_3^{\lambda_3} - \epsilon_3^{\lambda_2} \epsilon_2^{\lambda_3})$$

$$K_2 = (\epsilon_3^{\lambda_1} \epsilon_2^{\lambda_3} - \epsilon_2^{\lambda_1} \epsilon_3^{\lambda_3})$$

$$K_3 = (\epsilon_2^{\lambda_1} \epsilon_3^{\lambda_2} - \epsilon_3^{\lambda_1} \epsilon_2^{\lambda_2})$$

$$K_4 = (\epsilon_1^{\lambda_1} \epsilon_2^{\lambda_2} \epsilon_3^{\lambda_3} + \epsilon_2^{\lambda_1} \epsilon_3^{\lambda_2} \epsilon_1^{\lambda_3} + \epsilon_3^{\lambda_1} \epsilon_1^{\lambda_2} \epsilon_2^{\lambda_3}$$
$$- \epsilon_2^{\lambda_1} \epsilon_1^{\lambda_2} \epsilon_3^{\lambda_3} - \epsilon_1^{\lambda_1} \epsilon_3^{\lambda_2} \epsilon_2^{\lambda_3} - \epsilon_3^{\lambda_1} \epsilon_2^{\lambda_2} \epsilon_1^{\lambda_3})$$

The following are calculations of whole blank blood by the means of the photon diffusion theory. These equations use, for the most part, the notation given by Cohen and Longini. The density of the scattered photons (P) can be calculated by:

$$-D\nabla^2 P = g - \frac{P}{\gamma} \tag{19.8}$$

and the flux (F)

$$F = -D\nabla P, \tag{19.9}$$

where

$D = c/[n(a + 2k)]$, the diffusion constant for randomly distributed photons

P = density of scattered photons

γ = photon mean life time

$\gamma = n/ac$

n = refractive index

c = velocity of light in a vacuum

a = absorption coefficient which is similar to the extinction coefficient described previously with the exception that Naperian or natural logarithms (to the base $e = 2.7183$) are used. Its value is dependent on the light wave length, oxygen saturation, and hematocrit of the blood. Thus, at any given wave length λ the absorption coefficient is calculated by $a = (A_1 + B_1 H)V_0 + (A_2 + B_2 H)V_r$, where A_1, B_1, A_2, and B_2 are constants at wave length λ_1 for oxygenated and reduced hemoglobin, respectively. $V_0 + V_r$ are the fractions of oxygenated and reduced hemoglobin (i.e. $V_0 + V_r = 1$).

H = hematocrit

k = scattering coefficient into one hemisphere for randomly distributed photons—dependent on hematocrit as follows: $k = 4\ km\ H(1 - H)$, where H = hematocrit and km = scattering constant.

g = generation function, describing the rate at which collimated photons are converted to diffusing photons. ($g = 0$ in this case as the light source is considered to be a perfectly diffused Lambertion source.)

∇ = the "del operator." This is a mathematical operator which when applied to a function defined on spatial coordinates gives a measure or number that describes how the function is changing at each point in three dimensional space. $\nabla = i(\partial/\partial x) + j(\partial/\partial Y) + k(\partial/\partial z)$, where i, j, and k are unit vectors parallel to the X, Y, and Z axis, respectively.

Only one dimension is considered for the case of *in vitro* blood densitometers. Also, assuming that the sample is uniformly illuminated, the solutions to the diffusion equation giving the description of the transmitted (T) and reflected (R) light, respectively, are as follows:

$$T(d) = \left[\cosh qd + \left(\frac{k + a}{q} \right) \sinh qd \right]^{-1} \tag{19.11}$$

$$R(d) = \frac{k}{q} T(d) \sinh(qd) \tag{19.12}$$

where $q = [a(a + 2k)]^{1/2}$. a and k have previously been defined.

REFERENCES

Anderson, N. M., and P. Sekelj. 1967a. Light-absorbing and scattering properties of nonhemolyzed blood. Phys. Med. Biol. 12:173.

Anderson, N. M., and P. Sekelj. 1967b. Reflection and transmission of light by thin films of nonhemolyzed blood. Phys. Med. Biol. 12:185.

Beard, E., and E. H. Wood. 1951. Estimation of cardiac output by the dye dilution method with an ear oximeter. J. Appl. Physiol. 4:177–187.

Brown, E. A. 1956. The absorption of serum albumin by human erythrocytes. J. Cell Physiol. 47:167.

Bube, R. H. 1960. The photoconductivity of solids. John Wiley & Sons, New York.

Castaneda, A. R., K. C. Weber, G. W. Lyons, and I. J. Fox. 1966. Effect of agglutinating substances on optical density or blood in vitro and in vivo. J. Appl. Physiol. 21:928–932.

Cohen, A. 1969. Photoelectric determination of the relative oxygenation of blood. Ph.D. dissertation, Carnegie-Mellon University, Pittsburgh, Pennsylvania.

Cohen, A., and R. L. Longini. 1971. Theoretical determination of the blood's relative oxygen saturation in vivo. Med. Biol. Eng. 9:61.

Cohen, A. Personal communication to author (W.F.S.)

Drabkin, D. L. 1950. Spectroscopy: photometry and spectrophotometry. In O. Glasser (ed.), Medical Physics, Vol. 2, pp. 1039–1089. Year Book Publishers, Chicago.

Edwards, A. W. T., J. Isaacson, W. F. Sutterer, J. B. Bassingthwaighte, and E. H. Wood. 1963. Indocyanine green densitometry in flowing blood compensated for background dye. J. Appl. Physiol. 18:1294–1304.

Fox, I. J., L. G. S. Brooker, D. W. Heseltine, H. R. Essex, and E. H. Wood. 1957a. A tricarboncyanine dye for continuous recording of dilution curves in whole blood independent of variations in blood oxygen saturation. Mayo Clin. Proc. 32:478–484.

Fox, I. J., W. F. Sutterer, and E. H. Wood. 1957b. Dynamic response characteristics of systems for continuous recording of concentration changes in a flowing liquid (for example, indicator-dilution curves). J. Appl. Physiol. 11:390–404.

Fox, I. J., and E. H. Wood. 1960. Circulatory system: methods—indicator-dilution technics in study of normal and abnormal circulation. In O. Glasser (ed.), Medical Physics, Vol. 3, pp. 163–178. Year Book Publishers, Chicago.

Gilmore, H. R., M. Hamilton, H. Kopelman, and L. S. Sommer. 1954. The ear oximeter: its use clinically and in the determination of cardiac output. Brit. Heart J. 16:301–310.

Hamilton, W. F., R. L. Riley, A. M. Attyah, A. Cournand, D. M. Fowell, A. Himmelstein, R. P. Nobel, J. W. Remington, D. W. Richards, N. C. Wheeler, and A. C. Witham. 1948. Comparison of the Fick and dye injection methods of measuring the cardiac output in man. Amer. J. Physiol. 153:309–321.

Jacobs, M. H., H. N. Glassman, and A. K. Parpart. 1936. Osmotic properties of the erythrocyte. VIII. On the nature of the influence of temperature on osmotic hemolysis. J. Cell Physiol. 8:403.

Janssen, F. J. 1972. The principle design and features of a new Hb-oximeter. Med. Biol. Eng. 10:9–22.

Johnson, C. C. 1970. Optical diffusion in blood. IEEE Trans. Bio-med. Eng. BME-17:129–133.

Knutson, J. R. B., B. E. Taylor, E. J. Ellis, and E. H. Wood. 1950. Studies on circulation time with the aid of the oximeter. Mayo Clin. Proc. 25: 405–412.

Kubelka, P., and F. Munk. 1931. Ein Beitrag zur Optik der Farbanstriche. Z. Tech. Physik. 12:593–601.

Leevy, C. M., F. Smith, J. Longueville, G. Paumgartner, and M. M. Howard. 1967. Indocyanine green clearance as a test for hepatic function: evaluation by dichromatic ear densitometry. J.A.M.A. 200:236–240.

Longini, R. L., and R. J. Zdrojkowski. 1968. A note on the theory of back-scattering of light by the living tissue. IEEE Trans. Bio-med. Eng. BME 15:4–10.

Matthes, K. 1936. Untersuchungen über den Gasaustausch in der menschlichen Lunge. I. Mitteilung: Sauerstoffgehalt und Sauerstoffspannung im Arterienblut und im venosen Mischblut des Menschen. Naunyn Schmiedebergs Arch. Pharmakol. 181:630.

McGregor, M., P. Sekelj, and W. Adam. 1961. Measurement of cardiac output in man by dye dilution curves using simultaneous ear oximeter and whole blood cuvette techniques. Circ. Res. 9:1083–1088.

McKelvey, J. P., R. L. Longini, and T. P. Brody. 1961. Alternative approach to the solution of added carrier transport problems in semiconductors. Phys. Rev. 123:51–57.

Millikan, G. A. 1933. A simple photoelectric colorimeter. J. Physiol. (London) 79:152.

Nevo, A., A. DeVries, and A. Katchalsky. 1955. Interaction of basic polyamino acids with the red blood cell. I. Combination of polylysine with single cells. Biochim. Biophys. Acta 17:536.

Reed, J. H., Jr., and E. H. Wood. 1967. Use of dichromatic earpiece densitometry for determination of cardiac output. J. Appl. Physiol. 23: 373–380.

Reichert, W. J. 1966. The Theory and Construction of Oximeters. International Anesthesiology Clinics, Vol. 4, No. 1. Little, Brown and Co., Boston.

Sedivy, G., and W. F. Sutterer. Unpublished data.

Shockley, W. 1962. Diffusion and drift of minority carriers in semiconductors for comparable capture and scattering mean free paths. Phys. Rev. 125:1570–1576.

Sinclair, J. D., W. F. Sutterer, I. J. Fox, and E. H. Wood. 1961. Apparent dye-dilution curves produced by injection of transparent solutions. J. Appl. Physiol. 16:669–673.

Sutterer, W. F. 1967. Optical artifacts in whole blood densitometry. Pediat. Res. 1:66–75.

Sutterer, W. F., and E. H. Wood. 1962. A compensated dichromatic densitometer for indocyanine green. IRE Trans. Biomed. Electronics BME–9:133–137.

Zdrojkowski, R. J., and N. R. Pisharoty. 1970. Optical transmission and reflection by blood. IEEE Trans. Bio-med. Eng. BME-17:122–128.

Zylstra, W. G. 1953. Fundamentals and Applications of Clinical Oximeters. Assen. Royal Van Gorcum, The Netherlands.

Zylstra, W. G. 1958. A Manual on Reflection Oximetry. Assen. Royal Van Gorcum, The Netherlands.

chapter 20/ Pump Systems for Indicator Dilution

Dennis A. Bloomfield
Division of Cardiology/Maimonides Medical Center/
Brooklyn, New York

The design requirements and considerations for pump systems used with indicator-dilution methods are well defined and widely agreed upon. Such systems for drawing blood samples to obtain and calibrate dilution curves need to possess (1) constant flow rates over a range of selected values, (2) construction from sterilizable and noncorrosive material, (3) ease of setting up and operating, and (4) ease of cleaning. It is clear that such pumps should also be reliable, efficient, small in size, and low in cost.

Displacement pump systems, although reliable and easy to operate, have an overriding disadvantage in that the components are in contact with the blood and are difficult to sterilize and/or assemble. They are also load-dependent and do not compensate for changes in the pressure head.

All of the commercially available pumps therefore utilize syringe-reservoir systems, which are filled by the action of constant speed, load-independent motors. These pumps provide a steady and predictable flow rate to a maximum of 60 cc/minute, a maximum withdrawal or infusion time ranging from 45 to 72 seconds, and a maximum reservoir volume of 50 cc. With the exception of constant infusion methods of indicator injection, these pumps operate exclusively as withdrawal systems, pulling blood from a cardiac catheter or needle through a connecting tube to the densitometer and from there through further connecting tubing to the reservoir. It is these connections which impose the most important limitations on the achievement of optimal steady withdrawal rates (Sherman *et al.*, 1959).

THE MAINTENANCE OF CONSTANT FLOW RATES

In the discussion of distortion of dye curves (Chapter 7), it was recommended that withdrawal systems have the smallest "connecting tube" volume and the fastest withdrawal rates. This was best achieved by providing the shortest, smallest diameter tubing connecting the sampling site and the densitometer. Due to the very high velocities and drag resistance in these connecting tubes and catheters, there is a maximum flow rate at which blood can move without causing cavitation. At flow rates producing high vacuums in the withdrawal syringes, blood becomes "degassed." The flow reaches maximum velocity and then the pressure is equalized by expanding gases. The liquid then has time to fill the space occupied by the gas, which in turn, contracts. Eventually the pressure stabilizes but only at the cost of considerable flow rate fluctuation. This problem was investigated by White (1968), who determined the points of degassing in a series of experiments with various catheter dimensions. An equation to express the relationship was determined to be

$$Q = (5.94 \times 10^9)P^{1.96}D^{5.53}L^{-1.265}, \tag{20.1}$$

where Q = the flow rate of blood at 98°F (cc/minute), P = the vacuum pressure (psig), L = the catheter length (inches), D = the catheter diameter (inches). Significant evolution of gases began at approximately $P = 13$ psig. The flow rate was further limited at this value of vacuum pressure by air leaks into the syringe between the plunger and barrel and at the connections. One atmosphere was considered the ultimate pressure that could be reached before degassing and leakage rendered impossible the maintenance of constant flow rates under normal operating conditions. This equation permits the calculation for any withdrawal system, of the maximum allowable pump flow rate compatible with constant withdrawal. A minor modifying factor is the patient's arterial pressure of 1.7 to 2.9 psi (90 mm

Hg diastolic to 150 mm Hg systolic) which would proportionately reduce the vacuum pressure necessary for a given flow rate if a systemic artery is chosen as the sampling site. The friction between the barrel and plunger of the syringe is usually negligible, but at times during withdrawal, blood may collect in this space and significantly and unpredictably increase drag resistance. A further cause of a nonconstant withdrawal rate is small variations in cross-sectional area in the syringe.

OPERATIONAL ASPECTS

The requirements of sterility and ease of setting up and cleaning are readily met in the syringe-reservoir pumps. They will accept any syringe from 20- to 50-cc capacity, and only the inside glass surface is in contact with the blood. However, all pumps suffer in the compromise of designing a system which, on one hand, holds the syringe (and particularly, the plunger) firmly without longitudinal play, and on the other hand, allows quick release of the withdrawal mechanism for emptying, cleaning, and resetting of the syringe for subsequent operation.

The Harvard dye dilution pumps (Harvard Apparatus Co., Inc., Millis, Massachusetts) effect this compromise by rigidly securing the syringe barrel while containing the plunger flange in a forklike bar adapter (Fig. 20.1). The driving mechanism is engaged by pressure from a large weight attached to the adapter and can be readily disengaged by lifting this weight (and the adapter) away from the syringe. The flange can be wedged or screwed into the adapter to eliminate longitudinal movement synchronous with the arterial pulse (and resultant pulsatile rather than constant flow), but this removes the ability to easily release the plunger for emptying or cleaning.

The Gilford model 105-S constant flow pump (Gilford Instrument Laboratories, Inc., Oberlin, Ohio) secures the plunger flange firmly in a clamp. However, the plunger-clamp assembly can be freed from the drive mechanism at any point by a clutch screw arrangement. This facilitates rapid emptying and recycling. The barrel of the syringe, however, is only supported at the back (Fig. 20.2), and this design is less satisfactory when used with a forward mounted waste syringe as depicted in Fig. 3.1.

The Sage-Waters model SW-367 dye dilution pump (Waters Instruments, Inc., Rochester, Minnesota) has a rigidly held plunger,

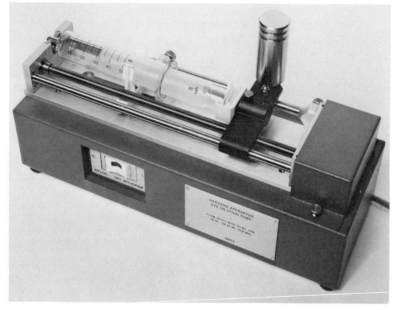

Figure 20.1 The Harvard dye dilution pump.

and the syringe is best emptied by the slower method of reversing the motor. The barrel support is lacking at the front.

The various units have individual feature differences such as limit switches, remote control operation, and presetable infusion volumes, but the practical distinction between these models is made on the basis of the plunger adapter mechanisms discussed above. The pumps are rated, by their manufacturers, to provide reproducibility of flow rates to ±0.5% and stability of flow under load to 5%.

SPECIAL PURPOSE SYSTEMS

Before the time of continuous indicator concentration measurement with flow-through densitometers, dye curves were constructed from concentration analysis of aliquots of blood drawn intermittently from the sampling site after injection of indicator elsewhere in the circulation. A multiple syringe sampler is still required for indicators such as ^{131}I-albumin and tritiated water, for which continuous concentration transducers are not available. A machine which collects 15 fixed-volume blood samples at 1-second intervals without exposing the samples to air was described by Ramsey and co-workers (1964). Five-milliliter syringes are mounted on a disc which intermittently rotates through a single aperture located at the sampling position. A brass adapter on the flange of each plunger interlocks with a sampling lever as the syringe moves into the sampling position. Withdrawal of the lever moves blood into the syringe which then rotates away from the sampling position, disengaging the lever. The syringes remain sealed until removal for analysis of their contents.

A pump system for performing indicator-dilution curves without blood loss was described by Cohn (1969). Although all syringe-reservoir systems can be used with reinfusion of collected blood, the potential dangers of stasis clot formation in the infusate must be con-

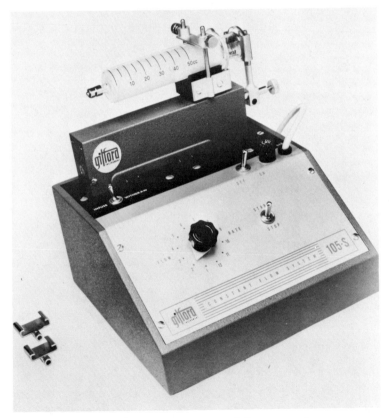

Figure 20.2 The Gilford model 105-S constant flow system.

sidered. Cohn's system used an occlusive roller pump receiving blood from the densitometer and returning it through a venous catheter. The pump is reported as nonpulsatile and unaffected by physiologic blood pressures. It may be indicated in small subjects when analyzing recirculation phenomena or other studies with long withdrawal durations, but has the disadvantage of requiring three intravascular catheters, one each for sampling and return and one for indicator injection.

Withdrawal systems devised for special purposes have also been described. The pump designed by White (1968) for an integrated-sample calibration technique was described in Chapter 6.

REFERENCES

Cohn, J. D. 1969. A pump system for performing indicator dilution curves without blood loss. J. Appl. Physiol. 26:841–843.

Ramsey, L. H., W. Puckett, A. Jose, and W. W. Lacy. 1964. Pericapillary gas and water dilution volumes of the lung calculated from multiple indicator dilution curves. Circ. Res. 15:275–286.

Sherman, H., R. C. Schlant, W. L. Kraus, and C. B. Moore. 1959. A figure of merit for catheter sampling systems. Circ. Res. 7:303–313.

White, R. E. 1968. The design and fabrication of a blood withdrawal system to measure cardiac output. Master of Science thesis, Vanderbilt University, Nashville.

chapter 21/ Computers for Indicator Dilution

Ephraim Glassman
Department of Medicine/
New York University Medical Center/New York, New York

In the past, the ability to measure cardiac output by indicator-dilution methods has been limited by the time required to analyze the curves and perform the necessary computations. Approximately 20 minutes might be spent in measuring deflections, logarithmic replotting, and calculation of results for each curve. Obviously, on line determination of the cardiac output during an experimental procedure or clinical catheterization is impossible under these circumstances. The introduction of small, special purpose, analogue computers designed to determine cardiac output, and in some instances mean transit time as well, has greatly extended the range of applicability of the indicator method. The operating principles of several types of such computers will be the subject of this chapter.

The fundamental problem of obtaining the cardiac output from an indicator-dilution curve consists of determining the area contained between the base line and the recirculation-free concentration curve. Under most conditions, the recorded curve does contain a degree of recirculated indicator, and some method must be used to subtract this value from the total curve area, before calculation of the output. In the Stewart-Hamilton technique (Stewart, 1921; Hamilton *et al.*,

1932), the curve is manually replotted on a semilogarithmic scale, and the onset of recirculation is recognized as the point of deviation of the descending limb from a straight line. It is assumed that, if no recirculation were to occur, the indicator concentration would continue to fall in an exponential fashion, thus, continuing along the straight line (see Chapter 5). Three of the commercially available cardiac output computers utilize this principle (in the form of the time constant of the falling concentration) as the basis of their operation.

THE TIME-CONSTANT COMPUTERS

The Gilford Instrument Laboratories model 104-A dye curve computer (Gilford Instrument Laboratories, Inc., Oberlin, Ohio) consists of two basic units. One of these reconstructs the primary dilution curve, without recirculation, from the curve recorded by the densitometer. The second unit is an integrator which computes the area beneath the primary curve (Fig. 21.1). In practice, the output of the densitometer is fed into the computer which then divides the signal into both of its elements. To reconstruct the primary curve, the basic curve is first converted into a logarithmic form. The rate of decline of the portion of the downslope of the curve between 75 to 45% of peak amplitude is measured, and its time constant is computed. This time constant is then used to generate an exponential curve which decays at the same rate as the selected portion of the downslope, before the appearance of recirculation. The integrator determines the area under the original curve up to the point at which the new curve is generated. Thereafter, the area beneath the generated curve is integrated, and the sum of the two areas is displayed in a digital readout. This sum, expressed in arbitrary units, is proportional to the total area which would exist beneath the dye curve, without recirculation. A calibration factor obtained by passing dyed blood, of known concentration, through the densitometer is used to convert the arbitrary units into milligram-seconds/liter, which is equivalent to $\int_0^\infty c(t)dt$ in the Stewart-Hamilton formulation. With knowledge of the amount of dye injected, calculation of the actual output is simply:

$$\text{cardiac output} = \frac{\text{mg of dye injected}}{\text{calibration factor} \times \text{computed area}} \times 60.$$

A variation of this computation is performed by the Waters Instruments, model CO-4 cardiac output analog computer (Waters Instru-

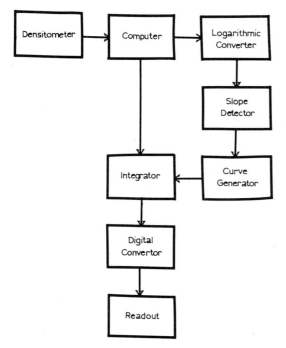

Figure 21.1 A block diagram of the method of operation of the Gilford Instrument Laboratories, model 104-A dye curve computer.

ments, Inc., Rochester, Minnesota). The decaying exponential is continuously measured from 75 to 50% of peak amplitude, and this section of the curve is integrated three times as fast as normal. It is assumed that the slope between 75 and 50% is exponential, and the computation merely completes the line to the base intercept. This method of computation is basically an automated version of the manual Stewart-Hamilton technique. However, when using the manual method, it is essential that the downslope of the replotted curve be examined to determine whether the replotted points do, in fact, fall on a straight line. In the event that a straight line does not exist, the curve must be discarded. The computer, on the other hand, is not designed for continuous checking of the downslope for linear decay. However, that portion of the logarithmic curve used to determine the time constant, which is then used to generate the exponential curve, is checked to ascertain that it is linear. In the event that no linear portion exists, a "Curve Unacceptable" light is displayed by the Gilford computer. In addition, provision is made for recording the logarithmic curve during the period of time in which the exponential decay is being generated. This should then be inspected to be certain of its linearity. Nevertheless, it is possible to obtain curves which decay exponentially for short periods of time and then deviate from this pattern. Depending on the magnitude of such deviation, use of this type computer would introduce a variable amount of error. In actual practice, this type of situation does not occur frequently.

A third method of determining the area beneath the primary curve utilized by the cardiac output computer, manufactured by Columbus Instruments, Columbus, Ohio, differs only moderately from the preceding description. This instrument takes advantage of the principle that the area under an exponentially decaying curve, extrapolated to infinity, may be derived by multiplying its time constant (τ) by the height of the curve at the onset of exponential decay. To perform this the computer first measures the peak height of the dye curve (Fig. 21.2); 83.4% of the peak amplitude is then determined (Point A) and is arbitrarily established as the point at which exponential

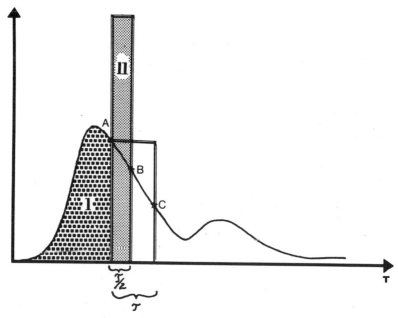

Figure 21.2 Analysis of the dye curve area using the technique employed by the Columbus Instruments cardiac output computer.

decay commences. The time required for the amplitude to decrease from 83.4 to 36.7% (Point C) is equal to one time constant, and the product of this time period τ and the amplitude at Point A is equal to the area under the curve. Because of the possibility that recirculation may have occurred before Point C, Point B, which is equal to 49% of peak amplitude is used to establish $\frac{1}{2}\tau$. The product of 2A and $\frac{1}{2}\tau$ is actually determined by the computer and used in the following step. The integrator portion of the instrument sums the area of the curve from the first appearance of dye until Point A (Area I). It then adds Area I to the area determined as the product $2A \times \frac{1}{2}\tau$ (Area II) to obtain the total area beneath the primary curve. The amount of dye injected into the patient and a calibration factor are then utilized to automatically calculate the cardiac output.

The principles utilized in the design of this computer are mathematically sound. In essence, the computer is performing the steps inherent in the Stewart-Hamilton method. When compared to the Gilford computer, it can be seen that both utilize measurement of the time constant of the logarithmic decay portion of the curve. The Gilford instrument then actually performs an integration of the area beneath a new curve generated by using this time constant, whereas the Columbus instrument simply multiplies τ by the amplitude at the onset of the curve. In fact, closer inspection reveals that these two apparently different methods are identical. The equation of the decaying, first order exponential may be written:

$$f(t) = Ae^{-t/\tau}$$

in which, A = the initial amplitude, t = time, τ = time constant, e = base of natural logarithms. The area beneath the curve is then given as:

$$\begin{aligned}
\text{area} &= \int_0^\infty Ae^{-t/\tau} \\
&= \left| -Ae^{-t/\tau} \right|_0^\infty \\
&= (-Ae^{-\infty/\tau}) - (-Ae^{-0/\tau}) \\
&= 0 - (-A\tau) \\
&= A\tau.
\end{aligned}$$

Thus, it is apparent that the actual integration yields the same result as does the multiplication process.

However, the two computers do differ in one significant regard. Whereas the Gilford computer does require linearity in that portion of the curve used to assess the time constant, the Columbus computer has no such safeguard. An arbitrary Point A is assumed to mark the onset of logarithmic decay and $\frac{1}{2}\tau$ is then simply computed by noting the time required to decay to the appropriate amplitude at Point B. This time is equal to $\frac{1}{2}\tau$ only when the curve has decayed in a truly exponential fashion. If the downslope has been deviated from logarithmic decay by any factor such as early recirculation, excessive valvular regurgitation or extremely low cardiac output then the value for $\frac{1}{2}\tau$ will be spuriously increased resulting in a falsely low value for cardiac output.

THE CONTINUOUS-PREDICTIVE-AREA COMPUTER

The cardiac output computer, manufactured by Lexington Instruments Corporation, Waltham, Massachusetts, differs in its mode of operation from the preceding computers. Again, the basic problem is to sum the area under the curve while excluding the effects of recirculation. To accomplish this the machine sums two functions continuously. The first of these is the integral of the entire dye curve, as it is being recorded. The second of these is a continuous predictive function which determines what the area of the curve would be if exponential decay were to occur at each point in time. When the logarithmic decay portion of the curve is reached, this value becomes constant and is added by the computer to the value obtained for the integral of the curve to that point. This value is displayed in digital readout form by the computer and may also be recorded externally as A_c (Fig. 21.3). So long as logarithmic decay continues, this sum remains constant. With the onset of recirculation, the predictive function changes, and the computer output deviates from its constant value. Should a curve be recorded which never attains exponential decay, the computer signals this fact by not achieving a constant output. The degree of slope of the computer output is a measure of the deviation of the curve from exponential decay. The value of immediate indication that a curve has not demonstrated logarithmic decay, and should therefore be discarded, is apparent. The total area beneath the original curve is also determined by the computer and may be recorded externally. This value, A_u in Fig. 21.3, is utilized for calibration purposes when using the pooled sample method (Theilen et al., 1955; Glassman and associates, 1967).

Several studies have been published comparing the results obtained using these computers with manual calculation by the Stewart-Hamilton technique (Glassman *et al.*, 1967; Sinclair *et al.*, 1965; Leighton and Czekajewski, 1971; Dalby *et al.*, 1967; Benchimol *et al.*, 1965). In general, the results have been closely comparable. Taylor *et al.* (1967) compared the results obtained using Gilford, Lexington, and Sanborn computers to each other, as well as to the manual technique. The variation between manual and automatic values ranged from 1 to 7%. None of the computers had a significantly better correlation with the

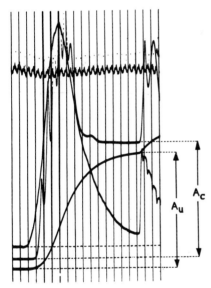

Figure 21.3 A dye curve analyzed using the Lexington Instruments cardiac output computer. See text for explanation.

manual values than the others. In addition, these authors studied the effect of dye curve base line drift, valvular regurgitation, and disturbances in cardiac rhythm on the function of the computers. All three computers were equally sensitive to the errors induced by such changes.

It is now apparent that compact, relatively inexpensive, labor-saving computers are available which greatly simplify and expedite the calculation of cardiac output from indicator-dilution curves. It should be emphasized that these devices do not, in any way, increase the accuracy of the dye method. Errors made in the inscription of the curves will generally not be detected by the computer. Similarly, inaccuracies in calibration methods will lead to incorrect results using manual or automated techniques.

REFERENCES

Benchimol, A., P. R. Akre, and E. G. Dimond. 1965. Clinical experience with the use of computers for calculation of cardiac output. Amer. J. Cardiol. 15:213–219.

Dalby, L., R. McDonald, and G. Sloman. 1967. A comparison of computer and planimetry cardiac output determination. Med. J. Aust. 1: 1255–1257.

Glassman, E., R. Baliff, and C. Herrera. 1967. Comparison of cardiac output calculation by manual and analogue computer methods. Cardiovasc. Res. 1:287–290.

Hamilton, W. F., J. W. Moore, J. M. Kinsman, and R. G. Spurling. 1932. Studies on the circulation. IV. Further analysis of the injection method, and of changes in hemodynamics under physiological and pathological conditions. Amer. J. Physiol. 99:534–551.

Leighton, R. F., and J. Czekajewski. 1971. Use of a new cardiac output computer for human hemodynamic studies. J. Appl. Physiol. 30:914–916.

Sinclair, S., J. H. Duff, and L. D. MacLean. 1965. Use of a computer for calculating cardiac output. Surgery 57:414–418.

Stewart, G. N. 1921. The pulmonary circulation time, the quantity of blood in the lungs and the output of the heart. Amer. J. Physiol. 58:20–44.

Taylor, S. H., H. R. MacDonald, M. C. Robinson, and R. P. Sapru. 1967. Computers in cardiovascular investigation. Brit. Heart J. 29:352–366.

Theilen, E. O., D. E. Gregg, M. H. Paul, and S. R. Gilford. 1955. Determination of cardiac output with the cuvette densitometer in the presence of reduced arterial oxygen saturation. J. Appl. Physiol. 8:330–336.

INDEX

C